Robust Multivariable Feedback Control

Prentice Hall International Series
in Systems and Control Engineering

M.J. Grimble, Series Editor

BANKS, S.P., *Control Systems Engineering: modelling and simulation, control theory and microprocessor implementation*

BANKS, S.P., *Mathematical Theories of Nonlinear Systems*

BENNETT, S., *Real-time Computer Control: an introduction*

CEGRELL, T., *Power Systems Control*

COOK, P. A., *Nonlinear Dynamical Systems*

LUNZE, J., *Robust Multivariable Feedback Control*

PATTON, R., CLARK, R.N., FRANK, P.M., (editors), *Fault Diagnosis in Dynamic Systems*

Robust Multivariable Feedback Control

Jan Lunze

PRENTICE HALL

New York London Toronto Sydney Tokyo

First published 1988 by
Prentice Hall International (UK) Ltd,
66 Wood Lane End, Hemel Hempstead,
Hertfordshire, HP2 4RG
A division of
Simon & Schuster International Group

Printed in the German Democratic Republic

Library of Congress Cataloging-in-Publication Data

Lunze, Jan
Robust multivariable feedback control/Jan Lunze.
p. cm. — (Prentice Hall international series in systems and control engineering)
Bibliography: p.
Includes index.
ISBN 0-13-781949-8
1. Feedback control systems. I. Title. II. Series.
TJ216.L86 1988
629.8'3 – dc19 88-19599

1 2 3 4 5 92 91 90 89 88

ISBN 0-13-781949-8

CONTENTS

PREFACE

Feedback systems are robust, ie relevant properties of the closed-loop system are preserved under perturbations and failures occurring in the plant or the controller. This property is based on the information feedback inherent in each control loop.

The aim of the theory of robust multivariable control is to establish analysis and design methods that take advantage of the robustness properties. The feedback system should be modelled and analysed and the controller should be designed on the basis of an incomplete description of the process to be controlled. This contrasts with the design principles elaborated for linear multivariable control, which presuppose knowledge of a sufficiently precise model of the plant.

The theory of robust control contributes to simplify the process of designing and bringing control loops into operation. It aims to extend the applications of linear feedback control into the areas of non linear and time varying processes. Apart from this primary aim, methods for describing and analysing uncertain systems, which have been developed in the theory of robust control, are being used in other fields of systems analysis and control theory, particularly in the theory of large scale dynamic systems.

To date, the field of robust control is still in a stage of intensive research. To present a complete overview of all the results is impossible in a book of this length. Therefore, I have attempted to describe the fundamental ideas and established theoretical principles that underlie many of these approaches, and to show how these methods can be successfully implemented in a number of practical designs.

One of the goals of this book is to fill the gap between the empirical ways in which practitioners have exploited the robustness of feedback systems for long and the powerful but sometimes rather abstractly formulated techniques that have been developed by theoreticians. This text is directed to a wide range of readers. The broad introduction to and motivation of robust control theory, the general survey of the subject, guidance on the choice of the appropriate design technique as well as many illustrating examples are explained for practising engineers and senior undergraduate and graduate students. For researchers, the book offers an overview of recent developments in the field of robust control, a unified representation of different approaches and results, more sophisticated examples such as the design of robust decentralised controllers as well as a discussion of new directions and subjects where further research work is needed.

This advanced teaching text presupposes some knowledge in modelling and identification of linear control systems and in the design of single-loop feedback systems. The reader should be familiar with basic tools for dealing with multivariable systems and the calculus of matrices. To make the text understandable for a wide range of readers, results from functional analysis are not used although this would sometimes enable us to gain more generality of the treatment.

A part of the material of this book has been collected from lectures given to fourth year undergraduates at the Universities of Technology in Dresden and Ilmenau since 1979.

Many people have helped, by discussion or criticism, to develop my, as I hope, comprehensive view of the rapidly advancing field of robust control. Amongst them I would like to mention my colleagues at the Zentralinstitut für Kybernetik und Informationsprozesse. Two of the application examples have been solved by common efforts. I am especially grateful to Professor Karl Reinisch at the Ilmenau University of Technology for his encouragement and generous support. I have benefited from stimulating and helpful discussions with Professor Ulrich Korn at the University of Technology Magdeburg and from common work, personal contacts or correspondences with Professor J. Ackermann, L. Bakule, B. Kouvaritakis, D. Owens, H. Tolle, H. Schwarz and M. Vallauri, all of whom are active in this field.

Finally, the close cooperation with the publishing houses Akademie-Verlag Berlin, particularly with Mrs G. Lagowitz, and Prentice-Hall Intern. London is gratefully appreciated.

Dresden, October 1986 J. Lunze

INTRODUCTION

This book is devoted to the analysis and design of feedback systems. In the standard configuration of Fig. 0.1 the plant and the controller form a closed loop in which each component exerts an influence upon the other. The plant input **u** depends throughout the controller on the controlled variable **y** which in turn is dependent upon **u** via the plant dynamics.

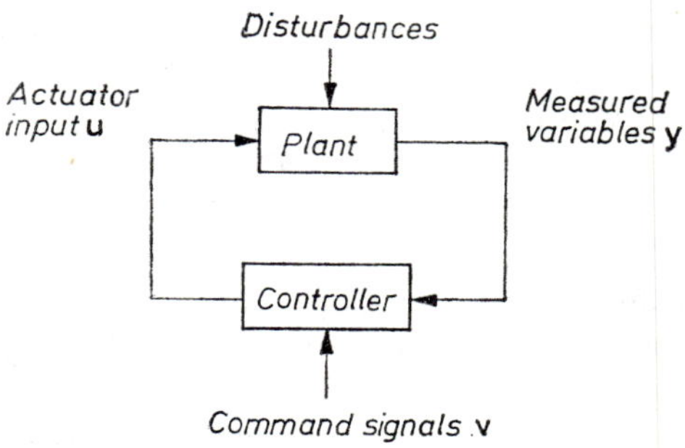

Fig. 0.1 Standard feedback configuration

This phenomenon is called *feedback*. As the plant is a technical entity with dynamical, nonanticipatory properties, the effect **y** of the plant follows the cause **u** in time. The controller feeds its output **u**, which depends on **y**, to a point which has been passed by the signal earlier in time. To feed a signal back means to make it influence an earlier point in the direction of operation.

Feedback has been established as a general concept for controlling technical, ecological or social systems. Chemical plants, for example, have hundreds of control loops in order to make temperatures, pressures or flows remain on given levels despite unmeasurable external influences. More complex multivariable controllers are used for processes with two or more strongly connected signals, such as air conditioning systems, distillation columns, electric power stations, steam boilers or aircrafts.

We want to look at feedback systems from the point of view of the control engineer, who has to find a controller in order to produce or maintain stability, improve robustness, attenuate disturbances, ensure asymptotic regulation etc. For us, a feedback system consists of a given process with fixed properties as well as a controller which can be freely chosen.

The benefits of feedback

Feedback systems exhibit several important properties because the behaviour of the overall system is produced by the properties as well as the interactions of its parts. Feedback makes it possible to stabilise inherently unstable systems, to improve the robustness against variations of the performance of some system part, or to attenuate unmeasurable external disturbances. These are the main reasons for deliberately introducing feedback into already existing processes.

The main benefits of feedback cannot be deduced without considering uncertainties in the behaviour of the plant. There are two reasons why the plant output $\mathbf{y}$ for a given input $\mathbf{u}$ proceeds on a trajectory which is not completely known in advance. At first, the plant dynamics are usually incompletely known so that the plant model can only be considered as an approximate rather than an exact description. Secondly, unknown disturbances may influence the performance of the process. In this case, the output $\mathbf{y}$ is not only an answer to the control input $\mathbf{u}$ but also to the disturbance $\mathbf{d}$, which is generally uncertain in the sense that it can be any one of a set of possible disturbance signals.

Uncertainties of these kinds do occur to a lower or greater extent in nearly all control problems because, for the purpose of modelling and design, the system to be controlled has to be taken out of its environment. Usually it is not obvious which phenomena have to be considered as part of the plant or as major connections between the plant and its environment, and which not. Under these circumstances it is important to know that the main properties of feedback systems depend only weakly on such uncertainties. *Robustness* against unmodelled dynamical elements and disturbances may even be considered as a 'structural' property of feedback systems. As the control input $\mathbf{u}$ is chosen by the controller in dependence upon the actual value of $\mathbf{y}$, it has regard to the current effects of earlier control actions as well as the disturbances.

This fundamental property of feedback systems becomes obvious by the comparison of the simple open-loop and closed-loop control configurations of Fig. 0.2. Open-loop control relies on the accuracy of the plant model since the control input $\mathbf{u}$ has to be chosen by means of the model. The perturbations which result from both kinds of uncertainties appear entirely as deviations of the resulting trajectory from the prespecified response. In contrast to this, closed-loop control compares the actual values of $\mathbf{y}$ with the nominal signal $\mathbf{v}$ and chooses the plant input $\mathbf{u}$ in dependence upon the control error $\mathbf{e}$. In addition to the a-priori information about the plant given in form of a mathematical model, feedback control uses the knowledge about

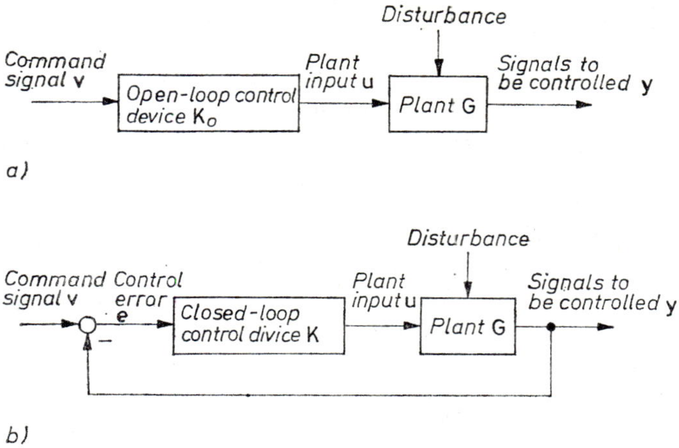

Fig. 0.2 Open-loop vs closed-loop control
(a) Open-loop control configuration
(b) Closed-loop control configuration

the actual behaviour of the plant and the current disturbances, which is implicitly provided by the measured values of y.

The introduction of feedback is mainly motivated by the incompleteness of the knowledge of the system to be controlled and the effect of external disturbances. More strictly, the use of feedback can be regarded as unreasonable if there is no uncertainty in the system because for the unperturbed system open-loop control may produce the same or a probably better output.

Analysis and design in presence of uncertainties

The majority of analysis and design tools for feedback systems presuppose that a sufficiently accurate model of the process to be controlled or the closed-loop system, respectively, has been set up. This model is fixed, once the model structure and parameters are given. Although a few of the methods make some conclusions on the effects of uncertainties possible, the treatment and attenuation of uncertainties is not their primary aim.

However, a more realistic approach to feedback control has to take into account the model inadequacies, which in turn are the main reason for the use of feedback. We want to deal here with principles and methods that enable us to **explicitly** consider the discrepancies between the model and the real process. That is, we make the attempt to analyse and design the controller without having a precise mathematical plant model at hand. We try to get results that are not only valid for the approximate model but do hold with certainty for a given class of plant models and, thus, for the real process.

In more detail, emphasis is laid on the consideration of plant uncertainties as they result from unmodelled dynamics, nonlinearities or time variances.

The second kind of uncertainties which arise from unmodelled excitations of the system by external disturbances or measurement noise is not of primary concern here. It will be shown, however, that good robustness against modelling errors is often accompanied by good disturbance attenuation.

The suitability of theoretical investigations of robustness analysis and design might be called in question by the hint that robustness as a fundamental issue of feedback systems has already been exploited for long. According to the principal statement that an engineer finds a tool and a solution to every problem however rough it may be, many control loops have come into operation without any mathematical modelling. The well known tuning rules by Ziegler and Nichols and others are an obvious example for an engineering tool which was originally based on a simple approximate model but successfully used without confirmation of the appropriatness of this type of model for the process under consideration.

However, without a deep understanding of the benefits of feedback the field of application of time invariant linear controllers in case of incompletely known and time varying or nonlinear processes must remain very much limited. Practical experience as well as academic examples, some of which are included in this book, show that a mere introduction of feedback does not guarantee the required insensitivity and robustness against model uncertainties to be attained. The feedback must be related to the kind and amount of model uncertainties and to the properties that are to be preserved.

From the engineer's point of view, the theoretical investigations of the robustness of feedback systems have to

- determine the fundamental capabilities of linear feedback for controlling uncertain plants, and
- elaborate practical analysis and design procedures.

The first task concerns questions such as: Which model uncertainties can be tolerated by linear feedback controllers? Which structure must the controller have? Under which conditions can the controller be designed so as to ensure stability (integrity) in case of sensor or actuator failures? The second task should clear the way towards a systematic exploitation of robustness.

Aims and structure of the book

It is the aim of this book to give a thorough introduction to the theory of robust feedback control, to demonstrate the theoretical results by practical examples, and to outline the main lines of current research.

The text is divided into 12 chapters. Chapter 1 derives the main questions arising in robustness analysis and design from a detailed characterisation of feedback design under practical circumstances. In Chapter 2 fundamental results from feedback theory, which will be used in the analysis and design of uncertain systems, are briefly summarised. Chapter 3 gives a survey of sensitivity analysis of feedback systems which, compared with robust control, represents an alternative approach but has motivated numerous robustness considerations. Chapter 4 presents a summary and classification of the results of robust control theory. Chapter 5 is devoted to modelling uncertain control systems by sets of possible representatives of the original system. Qualitative robustness considerations are described in Chapter 6, whereas in Chapters 7 and 8 quantitative robustness results are explained. The last three chapters present application examples. Finally, the current trends in robust control are surveyed. The appendices provide short summaries about nonnegative and M-matrices, and of the singular value decomposition, and give a short explanation of the Nichols chart. Relevant literature has been summarised at the end of each chapter. These bibliographical notes are not intended to be a complete review of the literature. The citations concern the most influential papers in which the results presented here have been described for the first time and provide the interested reader with adequate entry points to the rapidly growing literature in this field.

1 THE PROBLEM OF ROBUST CONTROL

1.1 The design problem

In this book we focus our attention to on-line control problems on the lowest hierarchical level. The control aims are determined by the operation conditions of the process and by specifications prescribed by higher level controls. Although they are originally formulated in terms of physical, technological or economical parameters we assume that they have already been transformed into requirements on the static and dynamical behaviour of the closed-loop system. These **design requirements** can be typically arranged in four groups:

(I) *Stability:* The operating point of the closed-loop system has to be asymptotically stable.

(II) *Asymptotic regulation:* For a given class of command inputs and external disturbances asymptotic regulation $\mathbf{y}(t) - \mathbf{v}(t) \underset{t \to \infty}{\to} \mathbf{0}$ has to occur.

(III) *Dynamical requirements:* The performance of the closed-loop system has to satisfy a set of given specifications such as bounds on the step responses, demands of a certain degree of non-interaction between different signals etc.

(IV) *Robustness:* The closed-loop system properties (I)—(III) have to be preserved under a given class of variations of the open-loop dynamics.

The group (II) can be modified in case of stochastic external signals, for which asymptotic regulation is required in the mean.

We will deal here mainly with deterministic signals, because we are primarily concerned with model uncertainties rather than uncertain disturbances. In process control, the design objectives of item (III) are usually formulated as requirements on the closed-loop responses $\mathbf{y}(t)$ and $\mathbf{u}(t)$ to command or disturbance steps (Fig. 1.1). These time domain specifications are often transformed into similar frequency domain specifications (for details see (Krishnam, Cruickshanks, 1977) or (Reinisch, 1979)) or may even be formulated directly as bounds on the transfer functions $l(j\omega)$ and $t(j\omega)$ of the open-loop or the closed-loop system (Fig. 1.2). In the course of the book we will refer to different forms of the design requirements.

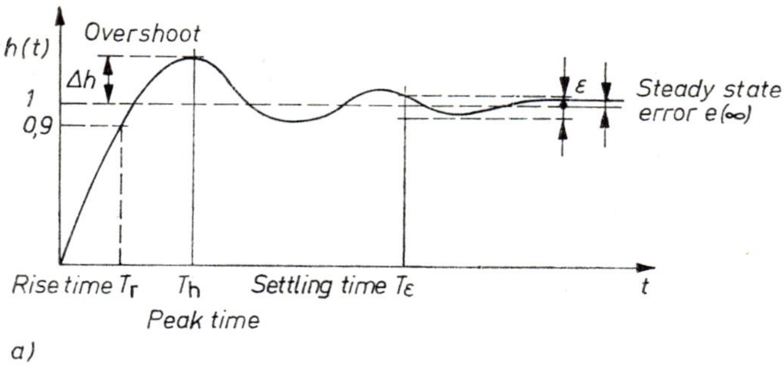

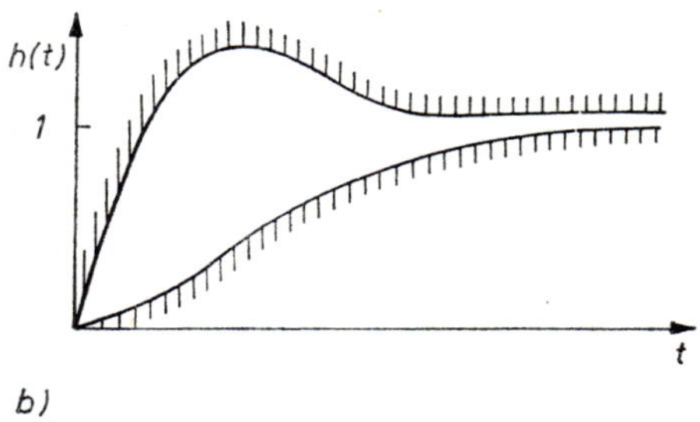

Fig 1.1 Design requirements in the time domain

(a) Characteristic values of the step response that have to satisfy given specifications

(b) Prescribed band for the command step response

The major difficulty of the design problem (I)—(IV) is that both performance and uncertainty aspects have to be weighted to get a suitable compromise. Whereas optimisation theory emphasises the performance aspects (I)—(III) we are primarily concerned with the uncertainty aspect (IV).

In many practical applications, the design specifications should not be considered as definitely fixed. They are often contradictory. The design engineer has to compare the possibilities to satisfy these requirements to a greater or lower extent with the control effort necessary for it and to choose an acceptable compromise. Design is a creative engineering task. Theory should investigate the reachable aims and provide the control engineer with principles and procedures for a quick and satisfactory solution.

The control problem considered so far is to be solved by means of a **feedback-feedforward controller**. In the general configuration of Fig. 1.3 the actuators are considered as part of the plant whereas the sensors are drawn separately in order to mark the measurement noise as an additional input. The entry point of the plant disturbances has been transformed to the plant output. Later, different special forms of this control configuration will be used (Fig. 2.5) and the measurement devices will often be included into the plant.

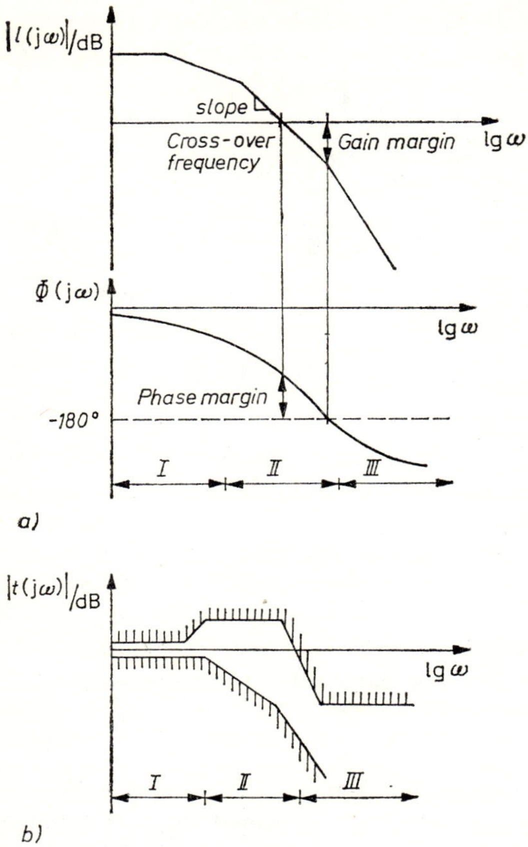

Fig. 1.2 Design requirements in the frequency domain

Region I: Constraints on static and quasistatic behaviour

Region II: Damping constraints on dominant modes

Region III: Constraints on the reinforcement of high frequency disturbances

(a) Characteristic values of the open-loop transfer function $l(j\omega)$

(b) Prescribed band for the closed-loop transfer function $t(j\omega)$

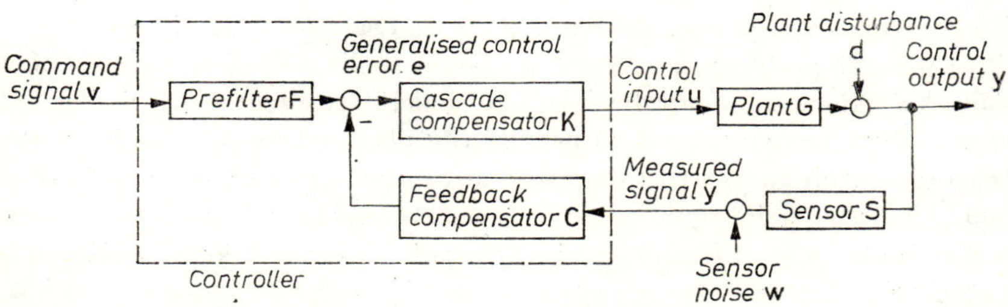

Fig. 1.3 General feedback-feedforward control system

To design the controller means to determine the structure and parameters of the prefilter, the cascade compensator and the feedback compensator such that the closed-loop system will satisfy the design requirements. The main problems are

- the determination of the control configuration (which signals are to be measured, which actuators are to be excited?)
- the determination of the controller structure (which dynamical properties are to be included into the controller?)
- the determination of the controller parameters.

In the following section we very briefly summarise the principles that have been elaborated in linear control theory to solve these problems under the assumption that a precise model of the plant be available.

1.2 The design of feedback controllers for completely known plants

Most of the work in linear multivariable control theory has been done under the assumptions that the process to be controlled has a linear input-output-behaviour and that a nearly precise model of the plant is available. Therefore, only the first three of the aforementioned design specifications are considered.

Guidelines for shaping the control configuration and the controller structure are based on structural investigations. The controllability and the observability of the plant proved to be necessary and sufficient for the existence of stabilising linear feedback controllers and give hints how to choose the location of input and output signals within the plant (Kalman, 1960; Wonham 1967). The Internal Model Principle asserts that the controller cannot ensure asymptotic regulation unless it includes a model of the external signals of the feedback loop, and prescribes a servocompensator as compulsory dynamical part of the controller (Davison, 1976a). The dynamical characteristics that can be produced by feedback controllers are strongly dependent upon the transmission zeros of the plant (Rosenbrock, 1974). The zeros may suggest the introduction of auxiliary control loops and, thus, help to design the control configuration according to the dynamical requirements (III).

Principles and methods for choosing appropriate controller parameters can be classified according to the auxiliary design requirements they use. Accordingly, the time domain approaches aim at the minimisation of some scalar objective function, the assignment of given values to the eigenvalues and eigenvectors of the closed-loop system, dead-beat control, or decoupling of the control signals from the disturbances. The frequency domain design is principally based on the successive closure of the different control loops, where each control loop is designed by means of single-variable control methods. In the Direct and the Inverse Nyquist Array Methods a decoupling compensator is used in order to ensure low interactions between the different control loops. Comprehensive surveys of the design methods can be found in the monographs (Patel, Munro, 1982; Korn, Wilfert, 1982; Tolle, 1983).

Many design methods have been included into computer-aided design packages, which are available now for a great variety of small and personal computers. They are used interactively, where the control engineer has to prescribe the design aims in terms of the auxiliary design requirements such as the poles of the closed loop or a performance criterion, and the computer determines the appropriate controller parameters, simulates the closed-loop system behaviour, and prepares the interpretation of the simulation results in a graphical way.

1.3 Model uncertainties

The design is generally preceded by a modelling step, in which the mathematical descriptions of the plant and the design requirements are extracted from the a-priori information about the process to be controlled and the control aims. In the realisation phase, which succeeds to the design, the control law is implemented by means of appropriate controller devices. In order to make a straightforward application of the resulting control law possible, the design has to take into account the accuracy or the range of validity of the model.

Uncertainties in the system behaviour result from disturbances or from perturbations of the plant dynamics (Fig. 1.4). *Disturbances* are external signals which are neither controllable via the plant inputs nor do they depend on some internal variables of the plant. Their effects can be collected at the plant output and represented as an external input $\mathbf{d}$ as shown in Fig. 1.4(a). As opposed to disturbances, *model uncertainties* describe in-

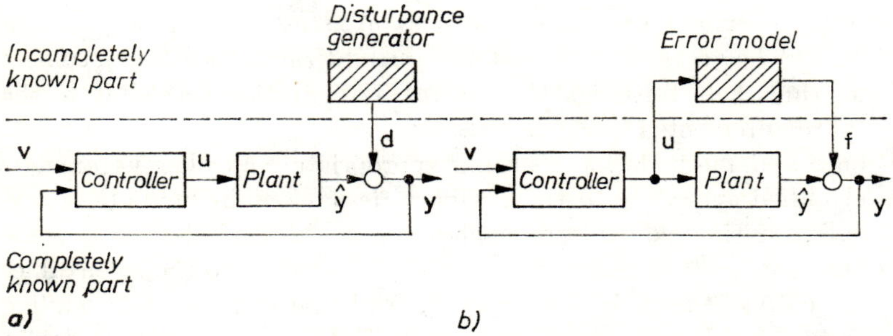

Fig. 1.4 Disturbed vs uncertain closed-loop system

(a) Closed-loop system subject to external disturbances

(b) Closed-loop system with uncertain dynamics

consistencies of the model and the real plant. Their effects depend on the input $\mathbf{u}$. Hence, model uncertainties can be represented by an error model with input $\mathbf{u}$ and output $\mathbf{f} = \mathbf{y} - \hat{\mathbf{y}}$ (Fig. 1.4(b)). We are mainly concerned with this kind of uncertainty.

There are three main **reasons for model uncertainties**.

(a) Model uncertainties occur because of an *incomplete knowledge* of the system. As a typical example, the structure of the model equations can be determined by means of the basic laws of physics and engineering, but the numerical values of the parameters are only known within tolerances. Besides that, if the controller has to be designed when the system to be controlled is still under construction then there might be some estimates of the static or dynamical behaviour but the plant cannot be accurately described. If the model parameters are estimated experimentally, the remaining uncertainties depend on the amount of disturbances that excited the plant during the experiment.

(b) Model uncertainties are brought about by *model aggregation*. Although the original system is exactly known, the model has been reduced in order to simplify calculations or to avoid difficulties that arise from the high complexity of the complete model. For instance, parasitic dynamical elements in actuators, transmitters or measurement devices are often neglected.

(c) Model uncertainties are caused by an *inadequate model structure*. Parameter variations and nonlinearities have to be omitted if the system is to be described by a linear model. Sensor or actuator failures which — from the controller point of view — yield changes or even restrictions of the input-output (I/O)-behaviour of the plant, parameter drifts, or parameter variations caused by moves of the operating point belong to this source of model uncertainties.

In any case, the model at hand yields merely an approximation of the original system performance. This can be reflected by the equation

$$\mathbf{G}(p) = \hat{\mathbf{G}}(p) + \delta\mathbf{G}(p) \tag{1}$$

where the transfer function matrix $\mathbf{G}(p)$ of the real existing process is represented by the transfer function matrices of the approximate model $\hat{\mathbf{G}}(p)$ and the error model $\delta\mathbf{G}(p)$ (cf Fig. 1.4(b)).

Systems with severe model uncertainties $\delta\mathbf{G}$ are called *uncertain systems* regardless of whether $\delta\mathbf{G}$ results from the lack of information about $\mathbf{G}$, nonlinearities or time-variances, or model simplifications. In general, $\delta\mathbf{G}(p)$ is unknown but it can be described by some bounds of its elements or some set of possible representatives. Both these possibilities are included in the set-theoretic concept for modelling uncertain dynamical systems, which will be explained in Chapter 5. Without anticipating a detailed treatment it should be noted here that this deterministic way of modelling uncertain systems leads to a family $\mathcal{S}$ of plants, where each member of the family may represent the original system, but it remains unknown which element does.

'Model uncertainties' are the general notion for deviations of the model dynamics from the original system behaviour. Different specific notions are commonly used in literature to describe special forms of uncertainties. They are explained now for the sake of completeness. Model uncertainties can be caused by *inaccuracies* or *perturbations* of the parameter values or the model structure. While the former means that the parameter values or structure used in the model are not the 'right' ones (as in (a) or (b) above) the latter points to the fact that changes of the parameters or structure have caused the deviation. *Parametric uncertainties* are those that can be compensated for by correcting the model parameter values, whereas *structural uncertainties* are brought about by an imperfect structure of the model, eg by the use of a linear model for an intrinsically nonlinear process or by omitting parasitic dynamics and time lags. With respect to the knowledge available about the causes of the uncertainties, unstructured and structured uncertainties are distinguished. In case of *unstructured uncertainties* it is only known that there are some discrepancies between the model and the real plant and, probably, how large the resulting deviations of the I/O-behaviour may be. For instance, the neglect of some quick dynamical elements or some nonlinear phenomena can be summed up in a global error model $\delta \mathbf{G}$ (Fig. 1.4(b)). Contrary to this, *structured uncertainties* are known insofar as that particular points within the plant are made responsible for the uncertainties. The deviation of the model from the original plant can be described up to some specified tolerances of certain elements. This necessitates a more flexible structure of the model as that depicted in Fig. 1.4(b). Uncertainties of particular parameters, or the model errors that occur because of the linearisation of a specific nonlinearity may serve as examples for structured uncertainties.

1.4 Feedback systems in the presence of model uncertainties

In order to get a first impression of the problems the designer has to face and, simultaneously, to prepare our further considerations, let us combine the plant model $\mathbf{y} = \mathbf{G}\mathbf{u}$ with a feedback controller

$$\mathbf{u} = \mathbf{KFv} - \mathbf{KCy} \tag{1}$$

where $\mathbf{v}$ denotes the command signal. Since $\mathbf{y} = \hat{\mathbf{y}} + \mathbf{f}$ (cf Fig. 1.4(b)) the controller feeds both the approximate output $\hat{\mathbf{y}}$ and the model error $\mathbf{f}$ back to the plant input $\mathbf{u}$. Therefore, the I/O-behaviour of the closed-loop system depends upon the model error $\delta \mathbf{G}$ in a relatively complicated form. From eqns (1.3-1) and (1) we get

$$\mathbf{y} = (\mathbf{I} + \hat{\mathbf{G}}\mathbf{KC})^{-1}\, \hat{\mathbf{G}}\mathbf{KFv} + (\mathbf{I} + \hat{\mathbf{G}}\mathbf{KC})^{-1}\, \mathbf{f} \tag{2}$$

and

$$\mathbf{e} = \left(\mathbf{I} + \mathbf{C}(\hat{\mathbf{G}} + \delta\mathbf{G})\,\mathbf{K}\right)^{-1} \mathbf{F}\mathbf{v} \tag{3}$$

where $\mathbf{e}$ denotes the generalised control error (Fig. 1.3 with $\mathbf{S} = \mathbf{I}$, $\mathbf{d} = \mathbf{0}$, $\mathbf{w} = \mathbf{o}$). Here and in the following the frequency dependence of the variables will be sometimes dropped for notational convenience. We have

$$\mathbf{f} = \delta\mathbf{G}\mathbf{K}\mathbf{e} = \delta\mathbf{G}\mathbf{K}\left(\mathbf{I} + \mathbf{C}(\hat{\mathbf{G}} + \delta\mathbf{G})\,\mathbf{K}\right)^{-1} \mathbf{F}\mathbf{v} \tag{4}$$

and finally

$$\mathbf{y}(p) = (\mathbf{I} + \hat{\mathbf{G}}\mathbf{K}\mathbf{C})^{-1}\hat{\mathbf{G}}\mathbf{K}\mathbf{F}\mathbf{v} + (\mathbf{I} + \hat{\mathbf{G}}\mathbf{K}\mathbf{C})^{-1}\delta\mathbf{G}\mathbf{K}\left(\mathbf{I} + \mathbf{C}(\hat{\mathbf{G}} + \delta\mathbf{G})\,\mathbf{K}\right)^{-1}\mathbf{F}\mathbf{v}(p) \tag{5}$$

provided that the inverses exist. Eqn (5) can be abbreviated by

$$\mathbf{y}(p) = \left(\hat{\mathbf{G}}_{\mathbf{yv}}(p) + \delta\mathbf{G}_{\mathbf{yv}}(p)\right)\mathbf{v}(p)\,. \tag{6}$$

If we neglected the model uncertainty and only combined the approximate model with the controller (1), we would receive the first term of eqn (6)

$$\hat{\mathbf{y}}(p) = \hat{\mathbf{G}}_{\mathbf{yv}}(p)\,\mathbf{v}(p)\,. \tag{7}$$

Although the error model and the approximate plant model lie in parallel the uncertainty $\delta\mathbf{G}_{\mathbf{yv}}$ of the feedback system depends nonlinearly on the uncertainties $\delta\mathbf{G}$ of the plant.

At first sight, the presence of model uncertainties seems to be a similar phenomenon as the presence of unknown disturbances because in both cases the system output $\mathbf{y}$ differs from the model output $\hat{\mathbf{y}}$. In fact, the disturbed closed-loop system in Fig. 1.4(a) has the description

$$\mathbf{y} = (\mathbf{I} + \hat{\mathbf{G}}\mathbf{K}\mathbf{C})^{-1}\,\hat{\mathbf{G}}\mathbf{K}\mathbf{F}\mathbf{v} + (\mathbf{I} + \hat{\mathbf{G}}\mathbf{K}\mathbf{C})^{-1}\,\mathbf{d} \tag{8}$$

if the plant is represented by $\hat{\mathbf{G}}$. This equation coincides with eqn (2) for $\mathbf{d} = \mathbf{f}$. However, whereas disturbances do not depend upon internal signals of the feedback system, the output error $\mathbf{f}$ does. Theorefore, model uncertainties cannot be treated as disturbances, although this is sometimes done, particularly in case of small uncertainties. They raise new questions. For instance, a disturbed linear closed-loop system remains stable unless the disturbance generator produces infinite signals. On the contrary, the stability of the closed-loop system can be affected by the error model so that the second term in eqn (5) becomes unstable.

Example 1.1 Consider the single-loop system (with scalar g rather than matrix $\mathbf{G}$) shown in Fig. 1.5. Assume that the approximate plant model

$$\hat{g}(p) = \frac{1}{1 + p\hat{T}_1} \tag{9}$$

has been set up so that its step response matches the plant step response satisfactorily (for example see Fig. 1.6). Then the closed-loop approximate

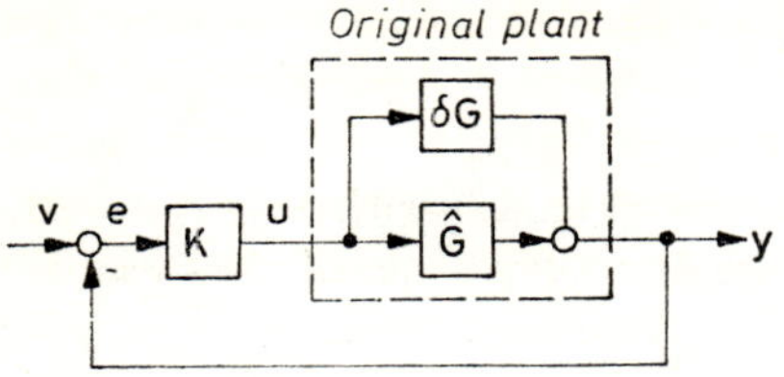

Fig. 1.5 Closed-loop original system

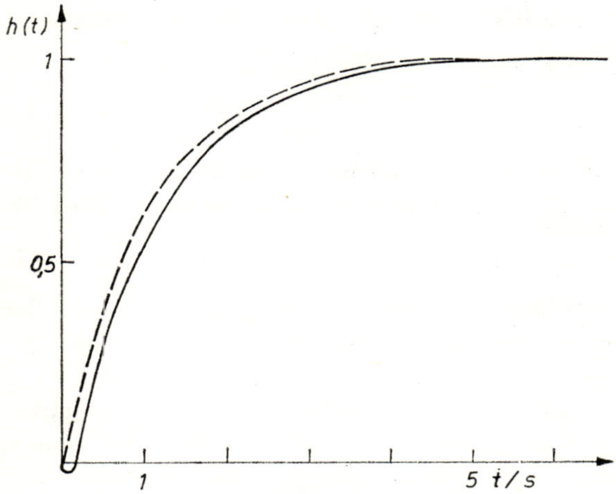

Fig. 1.6 Step response of the original system

$$g(p) = \frac{1 - 0.1p}{(1 + 0.1p)\,(1 + p)}$$

compared with the step response of the approximate model

$$\hat{g}(p) = \frac{1}{(1 + p)}$$

model is stable for any positive controller gain k. However, in connection with the original system

$$g(p) = \frac{1 - pT_0}{(1 + pT_1)\,(1 + pT_2)}$$

the controller can easily cause closed-loop instability. From the characteristic equation

$$T_1T_2p^2 + (T_1 + T_2 - kT_0)\,p + 1 + k = 0$$

we get the upper bound $k < (T_1 + T_2)/T_0$ which must be *necessarily* satisfied in order to ensure closed-loop stability.

This example is important for engineering practice in so far as simplified plant models are customarily used for the controller design. If the plant step response is superimposed by measurement noise, the designer cannot identify the negative part of the step response which indicates the existence of some positive plant zero. Hence, the controller design might end up with an unstable closed loop. □

In which way is feedback control superior to open-loop control in connection with uncertainty attenuation? We want to investigate this problem in more detail now by comparing the general feedback system of Fig. 1.3 with an open-loop control system of Fig. 0.2(a) that exhibits the same nominal performance. For the open-loop control we have

$$\mathbf{y}_o(p) = \mathbf{G}(p)\,\mathbf{K}_o(p)\,\mathbf{v}(p) \tag{10}$$

$$\mathbf{u}_o(p) = \mathbf{K}_o(p)\,\mathbf{v}(p) \tag{11}$$

(with index 'o'). The closed-loop system in Fig. 1.3 is described by

$$\mathbf{y}_c(p) = \mathbf{G}_c(p)\,\mathbf{v}(p) \tag{12}$$

with $\mathbf{S} = \mathbf{I}$,

$$\mathbf{G}_c(p) = (\mathbf{I} + \mathbf{GKC})^{-1}\,\mathbf{GKF} \tag{13}$$

and

$$\mathbf{u}_c(p) = \mathbf{K}(\mathbf{I} + \mathbf{CGK})^{-1}\,\mathbf{Fv} \tag{14}$$

(with index 'c'). We assume that the closed-loop system is stable for all perturbations of the plant dynamics under consideration. Then the inverse matrices exist for all frequencies with $\mathrm{Re}\,[p] > 0$.

Our assumption that both systems have the same behaviour for the approximate model $\hat{\mathbf{G}}(p)$ can be written as

$$\mathbf{u}_o(p) = \hat{\mathbf{u}}_c(p) \tag{15}$$

because this implies

$$\hat{\mathbf{y}}_c(p) = \hat{\mathbf{G}}\hat{\mathbf{u}}_c(p) = \hat{\mathbf{G}}\mathbf{u}_o(p) = \hat{\mathbf{y}}_o(p), \tag{16}$$

where $\hat{\mathbf{y}}_c$, $\hat{\mathbf{u}}_c$ and $\hat{\mathbf{y}}_o$ are given by eqns (12), (14) and (10), respectively, with $\hat{\mathbf{G}}$ instead of $\mathbf{G}$. Due to eqn (15) the feedforward controller $\mathbf{K}_o(p)$ is assumed to be chosen so as to satisfy

$$\mathbf{K}_o(p)\,\mathbf{v}(p) = \mathbf{K}(\mathbf{I} + \mathbf{C}\hat{\mathbf{G}}\mathbf{K})^{-1}\,\mathbf{Fv} \tag{17}$$

for all inputs $\mathbf{v}(p)$ or, for quadratic plants ($\dim \mathbf{u} = \dim \mathbf{y}$),

$$\mathbf{K}_o(p) = \mathbf{K}(\mathbf{I} + \mathbf{C}\hat{\mathbf{G}}\mathbf{K})^{-1}\,\mathbf{F}. \tag{18}$$

If model uncertainties $\delta\mathbf{G}(p) = \mathbf{G}(p) - \hat{\mathbf{G}}(p)$ occur, $\mathbf{y}_o$ and $\mathbf{y}_c$ deviate from their nominals $\hat{\mathbf{y}}_o$ and $\hat{\mathbf{y}}_c$. For the feedback system, eqn (12) yields

$$\begin{aligned}\mathbf{y}_c(p) - \hat{\mathbf{y}}_c(p) &= (\mathbf{I} + \mathbf{GKC})^{-1}\,\mathbf{GKFv} - \hat{\mathbf{G}}_c\mathbf{v}\\ &= (\mathbf{I} + \mathbf{GKC})^{-1}\big(\mathbf{GKF} - (\mathbf{I} + \mathbf{GKC})\,\hat{\mathbf{G}}_c\big)\,\mathbf{v}\\ &= (\mathbf{I} + \mathbf{GKC})^{-1}\big(\mathbf{G}(\mathbf{KF} - \mathbf{KC}\hat{\mathbf{G}}_c) - \hat{\mathbf{G}}_c\big)\,\mathbf{v}\end{aligned} \tag{19}$$

where $\hat{\mathbf{G}}_c$ is given by eqn (13) with $\hat{\mathbf{G}}$ instead of $\mathbf{G}$. The term $(\mathbf{KF} - \mathbf{KC}\hat{\mathbf{G}}_c)$ can be further transformed by the use of eqns (13) and (17)

$$\begin{aligned}(\mathbf{KF} - \mathbf{KC}\hat{\mathbf{G}}_c)\,\mathbf{v} &= \big(\mathbf{KF} - \mathbf{KC}(\mathbf{I} + \hat{\mathbf{G}}\mathbf{KC})^{-1}\,\hat{\mathbf{G}}\mathbf{KF}\big)\,\mathbf{v}\\ &= \mathbf{K}\big(\mathbf{I} - \mathbf{C}(\mathbf{I} + \hat{\mathbf{G}}\mathbf{KC})^{-1}\,\hat{\mathbf{G}}\mathbf{K}\big)\,\mathbf{Fv}\\ &= \mathbf{K}\big(\mathbf{I} - (\mathbf{I} + \mathbf{C}\hat{\mathbf{G}}\mathbf{K})^{-1}\,\mathbf{C}\hat{\mathbf{G}}\mathbf{K}\big)\,\mathbf{Fv}\\ &= \mathbf{K}(\mathbf{I} + \mathbf{C}\hat{\mathbf{G}}\mathbf{K})^{-1}\,(\mathbf{I} + \mathbf{C}\hat{\mathbf{G}}\mathbf{K} - \mathbf{C}\hat{\mathbf{G}}\mathbf{K})\,\mathbf{Fv}\\ &= \mathbf{K}_o\mathbf{v} = \mathbf{u}_o\,.\end{aligned}$$

Hence, eqn (19) leads to

$$\mathbf{y}_c(p) - \hat{\mathbf{y}}_c(p) = (\mathbf{I} + \mathbf{GKC})^{-1}\,(\mathbf{Gu}_o - \hat{\mathbf{G}}_c\mathbf{v})\,.$$

Because of eqns (13) and (16), $\hat{\mathbf{G}}_c\mathbf{v} = \hat{\mathbf{y}}_c = \hat{\mathbf{y}}_o$ holds. Hence,

$$\mathbf{y}_c(p) - \hat{\mathbf{y}}_c(p) = (\mathbf{I} + \mathbf{GKC})^{-1}\,\big(\mathbf{y}_o(p) - \hat{\mathbf{y}}_o(p)\big) \tag{20}$$

is received. This equation relates the output deviations that occur in the closed-loop and the open-loop systems due to model uncertainties, where the nominal behaviour of both configurations are assumed to be identical. The matrix on the right hand side of eqn (20) plays an important role in feedback design. It is often referred to as *sensitivity matrix*

$$\mathbf{S}(p) = \big(\mathbf{I} + \mathbf{G}(p)\,\mathbf{K}(p)\,\mathbf{C}(p)\big)^{-1} \tag{21}$$

for reasons that will be discussed later.

Typically, $\mathbf{S}(p)$ can be made small in the important frequency band, which means $|s(j\omega)| < 1$ for single-input single-output systems or

$$\mathbf{S}'(-j\omega)\,\mathbf{S}(j\omega) - \mathbf{I} = \text{negative semidefinite}$$

for multivariable systems. Then, the open-loop control system undergoes a greater performance deviation than the feedback control system. Eqns (20) and (21) will be often used in our further investigations.

1.5 Different properties describing a tolerance of model uncertainties

Now we want to survey those properties of a system that give to understand that we can expect a small deviation $\delta\mathbf{G}_{yv}$ of the original system performance from the performance $\hat{\mathbf{G}}_{yv}$ of the approximate model. Sensitivity, robustness, reliability, and adaptation will be briefly explained as general engineering terms and then interpreted from the control engineer's point of view.

Sensitivity (or differential sensitivity) is the dependence of the system performance upon the value of the parameter vector **a**. This property is usually investigated in terms of some characteristic property $q(\mathbf{G})$ of the system $\mathbf{G}$. The dominant eigenvalue of the system matrix may serve as an example for such a characteristic q. The *sensitivity function* indicates how strongly this characteristic depends upon the parameter a. It is defined by

$$s = \left.\frac{dq}{da}\right|_{a=\hat{a}}.$$

If $s = 0$, then the property q is insensitive to the parameter a at the nominal point $\hat{a}$. If $s \neq 0$, then q is called sensitive. In this case the value of s is a measure of the degree of dependence of q upon a. Note that sensitivity analysis *deals with (infinitesimally) small parameter* perturbations near a given nominal parameter value $\hat{a}$ and does not refer to the amount of the perturbation.

Robustness describes the ability of the system to satisfactorily perform for all parameter vectors **a** of a given set $\mathcal{A}$. That is, robustness *refers to an acceptable range for the characteristic* $q(\mathbf{G})$ *and to a specified class* $\mathcal{A}$ *of parameter perturbations.* For instance, the system is called robust if the dominant eigenvalue remains in the interval $-2...-1$ when a certain system parameter a varies between -0.5 and $+2.0$.

Both sensitivity and robustness do not refer to some assertion on whether parameter deviations are likely to occur in the system during a certain time interval. Such investigations are made in reliability analysis.

Reliability is the property of the system to satisfactorily perform over a certain period of time. The parameter vector **a** of the system is often considered as a random variable. Then a quantitative measure of reliability of the system is given by the probability for the system to be in operation at a certain instant of time. In contrast to robustness, the system may be reliable even if the characteristic $q(\mathbf{G})$ is out of the acceptable range for some parameter values. We only have to ensure that these values are unlikely to occur.

Adaptation is the ability of a system to (asymptotically) behave in a prescribed way even if the parameter vector is altered. As opposed to sensitivity, robustness and reliability, adaptation assumes that the system is able to actively respond to changes of the parameter by changing its own properties. Some time after the parameter vector has changed its value the system has adjusted its behaviour and again performs in the required way.

In feedback control, sensitivity, robustness, and adaptation play an important role. The controller is chosen for a given plant and given parameter perturbations so as to ensure insensitivity, robustness or adaptation, respectively, of the closed-loop system. In Chapter 3 we will deal with the sensitivity approach in more detail. The further chapters are devoted to robustness considerations.

In adaptive control, the controller plays an active role in that it compensates the effects of a parameter deviation within the plant. This leads to nonlinear control laws which are in contrast to our linearity assumptions on the controller. Interested readers are referred to the monographs (Goodwin, Sin, 1984; Unbehauen, 1985) for an introduction to adaptive control.

Reliability studies for control systems are mainly considered in connection with the realisation of the controller by means of analog or digital controller devices rather than with the suitability of the control law. They depend heavily on the reliability of the control equipment. We will not refer here to this property.

1.6 The main problems of robustness analysis and design

The preceding discussion of the different approaches to uncertain systems analysis and design points out that the robustness of a feedback system must be related to a certain tolerance of system properties and a given set of possible system perturbations. In feedback control, the properties to be preserved are given by the design requirements (I)—(III), whereas the model uncertainties are specified by a set $\mathcal{S}$ of candidate descriptions of the plant.

Definition 1.1 A *robust multivariable controller* is a linear time invariant feedback controller that satisfies the design requirements in connection with all plants of a given set $\mathcal{S}$.

This definition explains the differences between robust control and reliable, insensitive, and adaptive control, respectively. In the robustness approach the design specifications must be fulfilled with certainty for the whole set $\mathcal{S}$ of possible plant dynamics and not only with a certain probability or in the mean as in reliable control. It contrasts also with the sensitivity approach, in which the effect of differentially small parameter deviations on the quality of the closed-loop system is considered. The difference to adaptive control becomes obvious by the restriction to linear time invariant control laws.

In our definition the attribute 'robust' is used in connection with the controller, although the property of robustness concerns the performance of the closed-loop system. This incorrectness should be tolerated since the notion of robust controllers came into use in the literature in the way defined above.

The aim of robustness analysis and design is to cope with the difficulties of model uncertainties when designing feedback controllers. Constant-gain feedback can yield good performances if during the design stage all the critical situations that may be encountered in the working phase are checked.

The explicit consideration of the model uncertainties poses novel problems in the analysis and the design of feedback systems. The main problems can be summarised in four questions:

How can uncertain control systems be modelled?

The first question concerns the description of the uncertain original system $\mathbf{G}$ and, in particular, the error model $\delta\mathbf{G}$. The previous sections have made clear that in modelling uncertain systems we must deeply investigate which declarations can be made concerning the system dynamics and in which way this knowledge can be brought into a mathematically manageable form. On the one hand, the model must be chosen so that it can be really used on the basis of the information available in practical situations. On the other hand, the model should have a form which can be tackled by powerful analysis and synthesis methods. To underline this difficulty, even the choice of the model structure turns out to be a problem of its own, because it is questionable whether the parallel structure of the model in Fig. 1.4(b) is actually an appropriate one or, however reasonable it seems to be, whether a more general structure must be adopted.

In connection with robust control *set-theoretic models* have proved to be a reasonable compromise between these requirements. A detailed development of such models is given in Chapter 5.

How can uncertain control systems be analysed?

If we have a description of the system under consideration the question arises how to determine the properties of the original system. Think of eqn (1.4-5) in the case that $\delta\mathbf{G}$ is described only by bounds of its elements. Under which conditions is the closed-loop original system stable for all these $\delta\mathbf{G}(p)$? Which I/O-specifications does it meet?

Chapters 6, 7 and 8 deal with the analysis of uncertain feedback systems. Further analysis methods are explained in connection with application studies in Chapters 9, 10 and 11.

How can controllers be designed for uncertain plants?

Due to the model uncertainties the controller has to be designed for a set of plants rather than a single plant. Therefore, new design principles have to be elaborated.

Only few results are known concerning the existence of robust controllers and the determination of a reasonable controller structure. Some hints are given in Chapters 6, 7, 10 and 11. Most of the work has been done for the parameter design which is surveyed in Chapter 4 and explained in Chapters 6, 7, 8 and 11.

What is the field of application of robust control?

This question summarises a lot of problems concerning the solution of practical control tasks by means of robust multivariable linear feedback controllers. The existence conditions of robust controllers should provide a boundary between model uncertainties that are tolerable by linear feedback and those that necessitate adaptive control. For example, it is important to know which classes of nonlinear and time varying processes may be tackled by linear approximate models and controlled by linear time invariant regulators. As a second example, the control engineer wants to assess at a first sight which design requirements can be satisfied by appropriately choosing the controller although the process under consideration is incompletely known. This problem has particular emphasis in connection with decoupling procedures whose results are likely to depend very strongly on the accuracy of the plant model. Furthermore, constant-gain feedback may not produce optimal performance under plant perturbation. Which specifications may be satisfied?

These four basic questions make clear that the problem of designing controllers for uncertain systems necessitates *new concepts of modelling, analysing and designing multivariable control systems*. In the following we deal with methods and ideas that are suitable to answer these questions.

2 FUNDAMENTAL RESULTS FROM FEEDBACK THEORY

2.1 Models of control systems

In feedback control, models have to describe the performance of the plant or the closed-loop system around a given operating point as well as possible, the dynamical interactions between the input signals (actuator inputs, commands, external disturbances) and the output signals (measured outputs, signals to be controlled) being the most important issues. Throughout this book we will refer to models from two different points of view. At first, assuming the values of the model parameters are given, the models are used as an exact mathematical representation of a given real process or some part of it. Alternatively, the models are used to define the class of systems under consideration. In this case, the parameters are assumed to belong to a certain parameter space. No reference is made to particular numerical values.

This section surveys standard forms of the system description without considering uncertainties of the model structure or parameters.

Time domain description

Nonlinear dynamical systems are described in the **state space** by a set of n first-order differential equations, which can be written in the compact form

$$\begin{aligned} \dot{\mathbf{x}} &= \mathbf{f}(\mathbf{x}, \mathbf{u}, \mathbf{d}, t), \qquad \mathbf{x}(0) = \mathbf{x}_0 \\ \mathbf{y} &= \mathbf{g}(\mathbf{x}, \mathbf{u}, \mathbf{d}, t) \end{aligned} \tag{1}$$

where $\mathbf{x}$, $\mathbf{u}$, $\mathbf{y}$, and $\mathbf{d}$ are the vectors of system states, inputs, outputs, and disturbances of dimensions n, m, r, and q, respectively (Fig. 2.1). $\mathbf{f}$ and $\mathbf{g}$ are nonlinear vector functions which, for every instant of time t, unambiguously determine vectors $\dot{\mathbf{x}}(t)$ and $\mathbf{y}(t)$ in terms of $\mathbf{x}(t)$, $\mathbf{u}(t)$ and $\mathbf{d}(t)$. We assume that

$$\mathbf{u}(t) = \mathbf{0}, \qquad \mathbf{d}(t) = \mathbf{0} \quad \text{for} \quad t < 0 \tag{2}$$

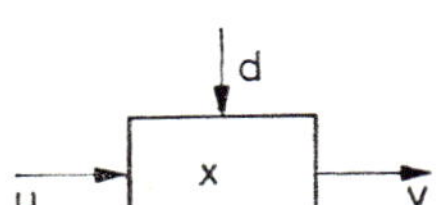

Fig. 2.1 System described by eqn (1)

holds, because the effects of the inputs for $t < 0$ are described by the initial state $\mathbf{x}_0$.

For linear systems, eqn (1) has the form

$$\begin{aligned} \dot{\mathbf{x}}(t) &= \mathbf{A}(t)\,\mathbf{x}(t) + \mathbf{B}(t)\,\mathbf{u}(t) + \mathbf{H}(t)\,\mathbf{d}(t), \qquad \mathbf{x}(0) = \mathbf{x}_0 \\ \mathbf{y}(t) &= \mathbf{C}(t)\,\mathbf{x}(t) + \mathbf{D}(t)\,\mathbf{u}(t) + \mathbf{L}(t)\,\mathbf{d}(t) \end{aligned} \tag{3}$$

where **A, B, C, D, H, L** are matrices of appropriate dimensions. For linear time invariant plants the entries of these matrices do not depend on time t

$$\begin{aligned} \dot{\mathbf{x}}(t) &= \mathbf{A}\mathbf{x}(t) + \mathbf{B}\mathbf{u}(t) + \mathbf{H}\mathbf{d}(t), \qquad \mathbf{x}(0) = \mathbf{x}_0 \\ \mathbf{y}(t) &= \mathbf{C}\mathbf{x}(t) + \mathbf{D}\mathbf{u}(t) + \mathbf{L}\mathbf{d}(t). \end{aligned} \tag{4}$$

In general, we assume that the model (4) has the least possible order n, ie it presents a minimal realisation of the I/O-properties of the system.

The alternative **input-output (I/O)-description** has the form of convolution integrals. For linear time invariant systems such a model is given by

$$\mathbf{y}(t) = \mathbf{G}_0(t)\,\mathbf{x}_0 + \mathbf{G}_{\mathbf{yu}}(t) * \mathbf{u}(t) + \mathbf{G}_{\mathbf{yd}}(t) * \mathbf{d}(t) \tag{5}$$

where $\mathbf{G}_{\mathbf{yu}}$ and $\mathbf{G}_{\mathbf{yd}}$ are the impulse response matrices of the system with respect to input $\mathbf{u}$ and disturbance input $\mathbf{d}$, respectively. The star symbolises the convolution operation

$$\mathbf{G}(t) * \mathbf{u}(t) = \int_0^\infty \mathbf{G}(t-\tau)\,\mathbf{u}(\tau)\,\mathrm{d}\tau .$$

Eqn (5) represents a very general form of description because any matrix function $\mathbf{G}(t)$ with

$$\mathbf{G}(t) = \mathbf{0} \quad \text{for} \quad t < 0 \tag{6}$$

can be interpreted as impulse response matrix. Therefore, the class of systems described by eqn (5) is larger than that modelled by eqn (4) and includes, for instance, distributed parameter systems and some kind of time delay systems. This will be utilised in Section 5.3 in the determination of model uncertainty bounds. Nevertheless, linear time invariant systems (4) are of particular importance in our analysis and design tasks. For them we have

$$\begin{aligned} \mathbf{G}_0(t) &= \mathbf{C}\,\mathrm{e}^{\mathbf{A}t} && t \geqq 0 \\ \mathbf{G}_{\mathbf{yu}}(t) &= \begin{cases} \mathbf{D}\delta(t) + \mathbf{C}\,\mathrm{e}^{\mathbf{A}t}\mathbf{B} & t \geqq 0 \\ \mathbf{0} & t < 0 \end{cases} \\ \mathbf{G}_{\mathbf{yd}}(t) &= \begin{cases} \mathbf{L}\delta(t) + \mathbf{C}\,\mathrm{e}^{\mathbf{A}t}\mathbf{H} & t \geqq 0 \\ \mathbf{0} & t < 0. \end{cases} \end{aligned} \tag{7}$$

The model form (5) will often be used in the following chapters because the determination of the impulse response matrices does not necessitate the knowledge of the internal structure of the system and is, therefore, more appropriate to incompletely known systems. Moreover, it represents the time domain analogue of the well known transfer function model of linear systems. A drawback of this model is that the use of the inverse convolution operation is not customary. The impulse response matrix $\mathbf{G}$ of the closed loop in Fig. 2.2 can be given only implicitly by

$$\bar{\mathbf{G}} = \mathbf{G}_1 - \mathbf{G}_1 * \mathbf{G}_2 * \bar{\mathbf{G}} \tag{8}$$

so that

$$\mathbf{y} = \bar{\mathbf{G}} * \mathbf{v}.$$

For numerical calculations the equation for $\bar{\mathbf{G}}(t)$ must be recursively solved for $\mathbf{G}(kT)$, $k = 0, 1, 2, \ldots$

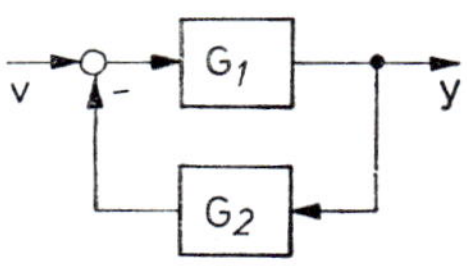

Fig. 2.2 Feedback system

If an initially quiescent system ($\mathbf{x}_0 = \mathbf{0}$) is excited by a step input $\mathbf{u} = \bar{\mathbf{u}}\sigma(t)$, where $\sigma(t)$ denotes the unit step function, the output is given by

$$\mathbf{y}(t) = \int_0^t \mathbf{G}_{\mathbf{yu}}(\tau)\, \mathrm{d}\tau\, \bar{\mathbf{u}}.$$

The matrix

$$\mathbf{H}(t) = \int_0^t \mathbf{G}_{\mathbf{yu}}(\tau)\, \mathrm{d}\tau \tag{9}$$

denotes the step response matrix of the system (5). If the system (4) has no direct throughput ($\mathbf{D} = \mathbf{0}$) then

$$\dot{\mathbf{H}}(0) = \left.\frac{\mathrm{d}\mathbf{H}}{\mathrm{d}t}\right|_{t=0} = \mathbf{CB} \tag{10}$$

holds. If the system is stable, then

$$\mathbf{K}_s = \mathbf{H}(\infty) = \mathbf{D} - \mathbf{CA}^{-1}\mathbf{B} \tag{11}$$

is the static transmission matrix (matrix of static reinforcement). The static I/O-behaviour is described by

$$\mathbf{y} = \mathbf{K}_s\mathbf{u}. \tag{12}$$

The same kind of models as described so far for the system to be controlled can be used for the **controller**. In a rather general form the linear control law is written as

$$\begin{aligned} \dot{\mathbf{x}}_r &= \mathbf{A}_r\mathbf{x}_r + \mathbf{B}_y\mathbf{y} + \mathbf{B}_v\mathbf{v}, \qquad \mathbf{x}_r(0) = \mathbf{x}_{ro} \\ \mathbf{u} &= \mathbf{C}_r\mathbf{x}_r - \mathbf{K}_y\mathbf{y} + \mathbf{K}_v\mathbf{v} \end{aligned} \tag{13}$$

where $\mathbf{x}_r$ represents the n_r-dimensional vector of controller states and $\mathbf{v}$ denotes the r-vector of command signals. Proportional output or state feedback controllers are described by

$$\mathbf{u} = -\mathbf{K}_y\mathbf{y} \tag{14}$$

or

$$\mathbf{u} = -\mathbf{K}\mathbf{x}, \tag{15}$$

respectively.

Frequency domain description

Linear time invariant systems are often described in the frequency domain by means of transfer function matrices $\hat{\mathbf{G}}_{yu}(p)$ and $\hat{\mathbf{G}}_{yd}(p)$. For $\mathbf{x}_o = \mathbf{0}$ the equation

$$\hat{\mathbf{y}}(p) = \hat{\mathbf{G}}_{yu}(p)\,\hat{\mathbf{u}}(p) + \hat{\mathbf{G}}_{yd}(p)\,\hat{\mathbf{d}}(p) \tag{16}$$

is an equivalent to eqn (5), where the transformation of (5) into (16) is given by the Laplace transform $\mathscr{L}$

$$\hat{\mathbf{y}}(p) = \mathscr{L}(\mathbf{y}(t)) = \int_0^\infty \mathbf{y}(t)\, e^{-pt}\, dt$$

$$\hat{\mathbf{G}}_{yu}(p) = \mathscr{L}(\mathbf{G}_{yu}(t)) = \int_0^\infty \mathbf{G}_{yu}(t)\, e^{-pt}\, dt$$

etc for complex p. To avoid confusion the functions in the frequency domain are sometimes indicated by '$\hat{\ }$'. The transformation of eqn (4) for zero initial conditions into the form (16) yields

$$\begin{aligned} \hat{\mathbf{G}}_{yu}(p) &= \mathbf{D} + \mathbf{C}(p\mathbf{I} - \mathbf{A})^{-1}\mathbf{B}. \\ \hat{\mathbf{G}}_{yd}(p) &= \mathbf{L} + \mathbf{C}(p\mathbf{I} - \mathbf{A})^{-1}\mathbf{H}. \end{aligned} \tag{17}$$

Note that $\hat{\mathbf{G}}_{yu}$ and $\hat{\mathbf{G}}_{yd}$ are rational function matrices.

The controller (13) has the form ($\mathbf{x}_{ro} = \mathbf{0}$)

$$\hat{\mathbf{u}}(p) = -\hat{\mathbf{K}}_{ry}(p)\,\hat{\mathbf{y}}(p) + \hat{\mathbf{K}}_{rv}(p)\,\hat{\mathbf{v}}(p) \tag{18}$$

where $\hat{\mathbf{K}}_{ry}(p)$ and $\hat{\mathbf{K}}_{rv}(p)$ will be specified later in dependence upon the control configuration used.

If no cancellations have occured in the product of eqn (17) the *poles* of $\hat{\mathbf{G}}_{yu}(p)$ and $\hat{\mathbf{G}}_{yd}(p)$ are the roots of det $(p\mathbf{I} - \mathbf{A}) = 0$ and, hence, equivalent to the eigenvalues of the system matrix $\mathbf{A}$ of eqn (4). The invariant *zeros*

p_0 are such frequencies for which rank $\hat{\mathbf{G}}(p)$ is lower than its maximum. For the system (4) they correspond to p for which rank $\begin{pmatrix} p\,\mathbf{I} - \mathbf{A} & \mathbf{B} \\ \mathbf{C} & \mathbf{D} \end{pmatrix} < n + \min(r, m)$ holds. If $\hat{\hat{\mathbf{G}}}(p)$ approaches a finite matrix for $|p| \to \infty$, ie it does not have any poles at infinity, then $\mathbf{G}(p)$ is said to be *proper*. If $\hat{\hat{\mathbf{G}}}(p) \underset{|p|\to\infty}{\to} \mathbf{0}$ holds then $\hat{\hat{\mathbf{G}}}(p)$ is *strictly proper*. For scalar rational transfer functions 'proper' means that the degree of the numerator at most equals the degree of the denominator. For strictly proper systems the numerator degree is less than the denominator degree. Technically realisable systems have one of these properties.

A real rational matrix $\hat{\hat{\mathbf{G}}}(p)$ presents a stable system if its elements are analytic in the closed right half plane. The system is minimum-phase if the elements of its inverse are analytic in the open right-half plane.

Functional analytic description

A comparison of eqns (5) and (16) makes evident that the I/O-description of linear dynamical systems has a similar form in both domains. We can consider the system as a transformation $\mathbf{S}$ of the inputs $\mathbf{u}$ and $\mathbf{d}$ into the output $\mathbf{y}$. Ignoring the input $\mathbf{d}$ for the sake of simplicity, the system is represented by

$$\mathbf{y} = \mathbf{S}\mathbf{u} \tag{19}$$

where $\mathbf{S}$ is the system operator. Such a general form enables us to consider different forms of the system description (time and frequency domain models) and different kinds of system (discrete-time and continuous-time) within the same framework and to extend our field of considerations to time varying or nonlinear systems.

Although functional analytic methods are standard in modern texts on large scale control systems or stability theory we avoid such general considerations to make our treatment understandable without knowledge in this particular field of mathematics. As, however, a few results could not be presented without using the operator norm, we have to introduce here a few notions which have been primarily defined in functional analysis but can, in the restricted form as presented here, be understood without knowing methods that are typical in this field.

In the general interpretation of eqns (5) and (16) in the spirit of eqn (19),

$$\mathbf{y}(t) = \mathbf{G}_{\mathbf{yu}}(t) * \mathbf{u}(t) \tag{20}$$

and

$$\hat{\mathbf{y}}(p) = \hat{\hat{\mathbf{G}}}_{\mathbf{yu}}(p)\, \hat{\mathbf{u}}(p) \tag{21}$$

describe systems whose input $\mathbf{u}$ and output $\mathbf{y}$ are members of spaces of functions of time t or frequency p. The system is represented by the operator $\mathbf{G}_{\mathbf{yu}}(t)\, *$ or $\hat{\hat{\mathbf{G}}}_{\mathbf{yu}}(p)$, respectively, whose application on $\mathbf{u}(t)$ or $\hat{\mathbf{u}}(p)$ is defined as a matrix convolution or a matrix-vector multiplication, respectively. The null and unity operators are given by $\mathbf{0}*$ and $\delta(t)\mathbf{I}*$ or $\mathbf{0}$ and $\mathbf{I}$, respectively, with $\mathbf{0}$ and $\mathbf{I}$ denoting the null and unity matrices.

With this interpretation in mind it is obvious that we can define a measure of the system dynamics by introducing a norm of the signals $\mathbf{u}$ and $\mathbf{y}$ and of the system operator $\mathbf{G_{yu}}$. For example, if we consider square-integrable functions

$$\begin{aligned} \|\mathbf{u}\| &= \sqrt{\int_0^\infty \mathbf{u}'\mathbf{u}\,\mathrm{d}t} \\ \|\mathbf{y}\| &= \sqrt{\int_0^\infty \mathbf{y}'\mathbf{y}\,\mathrm{d}t} \end{aligned} \tag{22}$$

then the system operator norm

$$\|\mathbf{G_{yu}}\|_{\mathrm{Op}} = \sup_{\|\mathbf{u}\| \neq 0} \frac{\|\mathbf{y}\|}{\|\mathbf{u}\|} \tag{23}$$

is given by

$$\|\mathbf{G_{yu}}\|_{\mathrm{Op}} = \sup_{p=\mathrm{j}\omega} \sqrt{\lambda_{\max} \hat{\hat{\mathbf{G}}}_{\mathbf{yu}}(p)^* \, \hat{\hat{\mathbf{G}}}_{\mathbf{yu}}(p)} \tag{24}$$

where $\lambda_{\max}[\cdot]$ denotes the maximum eigenvalue of the Hermitean matrix. Alternatively, if the operator $\mathbf{G_{yu}}$ is given as impulse response matrix $\mathbf{G_{yu}}(t)$ then a possible operator norm is

$$\|\mathbf{G_{yu}}\|_{\mathrm{Op}} = \int_0^\infty \|\mathbf{G_{yu}}(t)\|_{\mathrm{M}}\,\mathrm{d}t \tag{25}$$

where $\|\cdot\|_{\mathrm{M}}$ is some matrix norm.

Note that both definitions concern the whole function $\mathbf{G_{yu}}(\cdot)$. To avoid a confusion with the matrix norm $\|\mathbf{G_{yu}}\|_{\mathrm{M}}$, which has regard only to the value of $\mathbf{G_{yu}}(t)$ at the indicated value of the argument t, we use the index 'Op' if necessary. The norms (24) and (25) exist only if the system is stable. The operator norm observes the norm axioms

$$\|\mathbf{G_{yu}}\|_{\mathrm{Op}} > 0 \quad \text{if} \quad \mathbf{G_{yu}} \neq \mathbf{0}$$

$$\|a\mathbf{G_{yu}}\|_{\mathrm{Op}} = |a| \, \|\mathbf{G_{yu}}\|_{\mathrm{Op}} \quad \text{for scalar } a$$

$$\|\mathbf{G}^1_{\mathbf{yu}} + \mathbf{G}^2_{\mathbf{yu}}\|_{\mathrm{Op}} \leqq \|\mathbf{G}^1_{\mathbf{yu}}\|_{\mathrm{Op}} + \|\mathbf{G}^2_{\mathbf{yu}}\|_{\mathrm{Op}}.$$

To make possible an application of the norm definition (23) to unstable systems we consider the set of quadratic rational transfer function matrices $\mathbf{G}(p)$ of order m and denote it by $\mathcal{R}(p)^{m\times m}$. $\mathcal{S}(p)$ be the set of proper rational functions whose poles are all in the left-hand complex plane, $\mathcal{S}(p)^{m\times m}$ the set of $(m \times m)$-matrices whose elements belong to $\mathscr{S}(p)$. Then every proper transfer function matrix $\mathbf{G}(p) \in \mathscr{R}(p)^{m \times m}$ can be represented by its right-coprime or left-coprime factorisation over $\mathscr{S}(p)^{m \times m}$

$$\mathbf{G}(p) = \mathbf{N}(p)\,\mathbf{D}(p)^{-1} = \tilde{\mathbf{D}}(p)^{-1}\,\tilde{\mathbf{N}}(p). \tag{26}$$

Note that both $\mathbf{N}(p)$ and $\mathbf{D}(p)$ represent stable systems whereas $\mathbf{G}(p)$ may be unstable. Coprimeness means that there exist matrices $\mathbf{X}(p)$, $\mathbf{Y}(p)$ or $\tilde{\mathbf{X}}(p)$, $\tilde{\mathbf{Y}}(p)$ in $\mathscr{S}(p)^{m \times m}$ such that

$$\mathbf{X}(p)\,\mathbf{N}(p) + \mathbf{Y}(p)\,\mathbf{D}(p) = \mathbf{I} \tag{27}$$

$$\tilde{\mathbf{N}}(p)\,\tilde{\mathbf{X}}(p) + \tilde{\mathbf{D}}(p)\,\tilde{\mathbf{Y}}(p) = \mathbf{I} \tag{28}$$

hold for all p. With this representation, the norm definition (23) can be applied to an unstable system $\mathbf{G}(p)$ by using it for $\mathbf{N}(p)$ and $\mathbf{D}(p)$ or $\tilde{\mathbf{N}}(p)$ and $\tilde{\mathbf{D}}(p)$. This will be utilised in Section 6.5 to consider structural properties of uncertain systems.

2.2 Stability of feedback systems

We begin this survey of the stability analysis with two definitions.

Definition 2.1 (*Lyapunov stability*). The equilibrium $\mathbf{x}_e = \mathbf{0}$ of the autonomous system (2.1-1)

$$\dot{\mathbf{x}} = \mathbf{f}(\mathbf{x}, \mathbf{0}, \mathbf{0}, t), \qquad \mathbf{x}(0) = \mathbf{x}_0$$

is said to be asymptotically stable if for all $\varepsilon > 0$ there exists some δ such that $\|\mathbf{x}_0\| < \varepsilon$ implies $\|\mathbf{x}(t)\| < \delta$ for all $t \geqq 0$ and

$$\lim_{t \to 0} \|\mathbf{x}(t) - \mathbf{x}_e\| = 0.$$

Definition 2.2 (*Input-output-stability*). The system

$$\hat{\mathbf{y}}(p) = \hat{\hat{\mathbf{G}}}(p)\,\hat{\mathbf{u}}(p) \tag{1}$$

is said to be $\mathscr{L}_2$-stable (bounded-input bounded-output stable) if $\|\mathbf{G}\|_{\mathrm{Op}} < \infty$ holds where the operator norm is used as defined in eqn (2.1-24).

For linear systems the relation between both stability definitions is very simple.

Lemma 2.1 If the linear system

$$\dot{\mathbf{x}} = \mathbf{A}\mathbf{x} + \mathbf{B}\mathbf{u}, \qquad \mathbf{x}(0) = \mathbf{x}_0 \tag{2}$$

is completely controllable and observable then it is asymptotically stable if and only if it is $\mathscr{L}_2$-stable.

The stability of linear systems can be proved by the following equivalent conditions.

Theorem 2.1 A linear system is asymptotically stable if and only if one of the following conditions hold

(1) If the system is described by

$$\dot{\mathbf{x}} = \mathbf{A}\mathbf{x} + \mathbf{B}\mathbf{u}, \qquad \mathbf{x}(0) = \mathbf{x}_0$$

all eigenvalues of $\mathbf{A}$ must have negative real parts.

(2) If the system is described by

$$\mathbf{y} = \mathbf{G}(t) * \mathbf{u}(t)$$

all the uncontrollable and unobservable modes must be stable and there must exist a finite constant M such that

$$\int_0^\infty \|\mathbf{G}(t)\|_M \, dt < M. \tag{3}$$

(3) If the system is described by

$$\hat{\mathbf{y}}(p) = \hat{\mathbf{G}}(p)\, \hat{\mathbf{u}}(p)$$

all the uncontrollable and unobservable modes must be stable and there must exist a finite constant M such that

$$\|\hat{\mathbf{G}}(p)\|_M < M \quad \text{for all } p \text{ with} \quad \mathrm{Re}\,[p] > 0 \tag{4}$$

or, equivalently, all poles of $\hat{\mathbf{G}}(p)$ must have negative real parts.

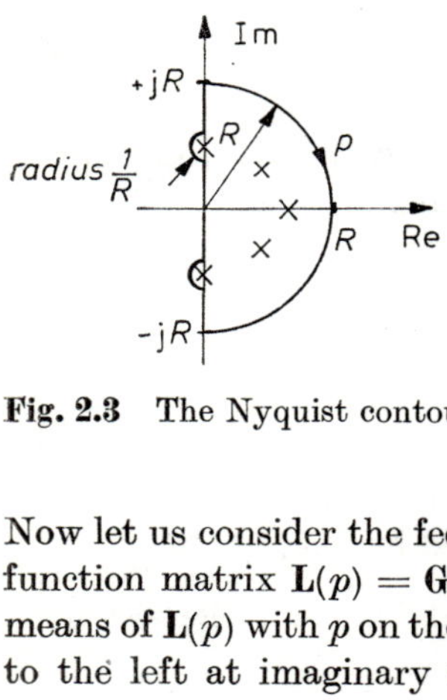

Fig. 2.3 The Nyquist contour $\mathcal{D}$

Now let us consider the feedback system of Fig. 2.2 with open-loop transfer function matrix $\mathbf{L}(p) = \mathbf{G}_1(p)\, \mathbf{G}_2(p)$. Then the stability can be checked by means of $\mathbf{L}(p)$ with p on the Nyquist contour $\mathcal{D}$ (Fig. 2.3). $\mathcal{D}$ has indentations to the left at imaginary axis poles of $\mathbf{L}(p)$ and clockwise orientation. R is choosen large enough so that all finite right half plane poles of $\mathbf{L}(p)$ are encircled by $\mathcal{D}$.

Theorem 2.2 (*Multivariable Nyquist stability criterion.*)
The feedback system of Fig. 2.2 is asymptotically stable if and only if the image of the Nyquist contour $\mathcal{D}$ under $\det\big(\mathbf{I} + \mathbf{L}(p)\big)$ encircles the origin n_+ times anticlockwise, where n_+ is the number of poles of $\mathbf{L}(p)$ in the closed right-half complex plane.

If the open-loop system is stable then $\det\big(\mathbf{I} + \mathbf{L}(p)\big)$ must not encircle the origin. If $\mathbf{L}(p)$ is strictly proper then $\det\big(\mathbf{I} + \mathbf{L}(p)\big) \to 1$ for $|p| = R$ and R large enough. Therefore, the Nyquist criterion must be checked only for the frequency response matrix $\mathbf{L}(\mathrm{j}\omega)$ with $\omega = -\infty \ldots +\infty$ and $\mathbf{L}(p)$ along the indentations at imaginary axis poles. For systems with only one input and one output the Nyquist criterion states the well known condition that the frequency response $l(\mathrm{j}\omega)$ must not encircle the point $-1 + \mathrm{j}0$ in the complex plane.

Assuming that the inverse of $\mathbf{L}(p)$ exists, an alternative stability criterion is given in the following theorem.

Theorem 2.3 (*Inverse Nyquist stability criterion*).
The feedback system of Fig. 2.2 is asymptotically stable if and only if the image of the Nyquist contour $\mathscr{D}$ under $\det(\mathbf{I} + \mathbf{L}(p)^{-1})$ encircles the point $(-1 + \mathrm{j}0)$ n_1 times and the origin n_2 times anticlockwise, where $n_1 - n_2 = n_{0+}$ is the number of zeros of $\mathbf{L}(p)$ in the closed right-half complex plane.

We conclude this survey on stability analysis by considering a linear single-input single-output system with nonlinear static feedback (Fig. 2.4).

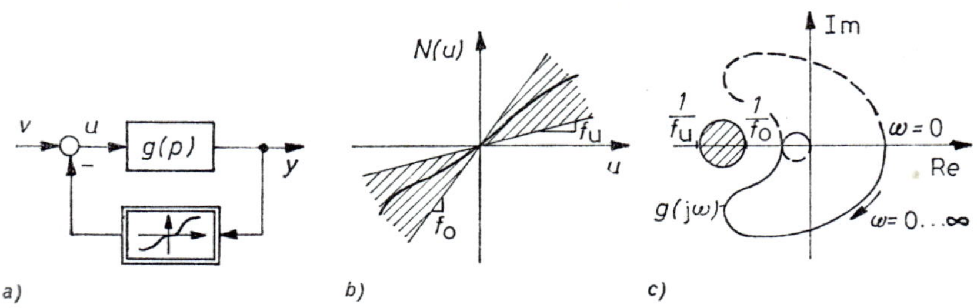

Fig. 2.4 Stability analysis of a nonlinear feedback system
(a) Nonlinear feedback system
(b) Sector bounds for the static nonlinearity
(c) Generalised circle criterion

Theorem 2.4 (*Generalised circle criterion*). Consider the linear system

$$\hat{y}(p) = \hat{q}(p)\,\hat{u}(p)$$

with nonlinear static feedback

$$u = -n(y)\,. \tag{5}$$

Assume that the nonlinearity lies inside the sector

$$f_\mathrm{u} \leq \frac{n(y)}{y} \leq f_\mathrm{o} \tag{6}$$

and that the linear part $\hat{g}(p)$ is strictly proper and asymptotically stable. Then the closed-loop system is asymptotically stable if $g(j\omega)$ for $\omega = -\infty \ldots +\infty$ does not encircle or cross the area defined by the circle through the points $-\frac{1}{f_u}$ and $-\frac{1}{f_o}$ with centre at the real axis.

That is, $g(j\omega)$ has to satisfy the Nyquist criterion with regard to the given circular area (Fig. 2.4(c)). Theorem 2.4 proves the *absolute stability* of the closed-loop system, ie the closed loop is asymptotically stable for all static nonlinearities inside the sector (6).

2.3 Fundamental limitations of the closed-loop properties

The feedback system in the general configuration is shown in Fig. 2.5. As the controller is linear and proper, the properties of the closed loop may not be shaped arbitrarily by appropriately designing the controller parameters. This fundamental restriction will be outlined in this section.

The frequency domain description of the complete configuration of Fig. 2.5(a) is given by

$$\mathbf{y} = (\mathbf{I} + \mathbf{GKC})^{-1}\,\mathbf{GKFv} - (\mathbf{I} + \mathbf{GKC})^{-1}\,\mathbf{GKCw} + (\mathbf{I} + \mathbf{GKC})^{-1}\,\mathbf{d}\,.$$

Using the *sensitivity matrix* as defined in eqn (1.4-21)

$$\mathbf{S} = (\mathbf{I} + \mathbf{GKC})^{-1} \tag{1}$$

and the *complementary sensitivity matrix*

$$\mathbf{T}(p) = \mathbf{I} - \mathbf{S}(p) = (\mathbf{I} + \mathbf{GKC})^{-1}\,\mathbf{GKC} \tag{2}$$

we have

$$\mathbf{y} = \mathbf{SGKFv} - \mathbf{Tw} + \mathbf{Sd}\,. \tag{3}$$

At first, eqn (3) shows that stability, disturbance rejection, and model uncertainty attenuation are '*feedback properties*', that is they can only be influenced by feedback whereas command tracking can be produced by open-loop control. This becomes obvious by considering eqn (3) with $\mathbf{C} = \mathbf{0}$ (no feedback). Then the stability of the controlled system necessitates the stability of the plant. The disturbance $\mathbf{d}$ adds to the output with full magnitude. Command tracking can be reached by appropriately designing the feedforward part $\mathbf{F}$ of the controller, but all model uncertainties are reflected as control error $\mathbf{y} - \mathbf{v}$. For considering robustness issues we have to focus our attention on the feedback properties.

Secondly, in view of eqn (2) plant disturbance attenuation and measurement noise attenuation are contradictory design aims. The product $\mathbf{K}(p)\,\mathbf{C}(p)$ has to be chosen so as to find a compromise between these two objectives.

As opposed to that, command tracking and disturbance rejection can be reached independently. When the feedback part of the controller has been designed, the prefilter $\mathbf{F}(p)$ can be shaped in order to get the required I/O-properties of the closed-loop system.

The independence of command tracking and disturbance rejection is the merit of the complete configuration shown in Fig. 2.5(a). This configuration presents a **two-degrees-of-freedom system** since there are two functions to be chosen independently. Note that there are actually only two rather than three independent parts since $\mathbf{K}(p)$ does only occur in the products $\mathbf{K}(p)\,\mathbf{C}(p)$ and $\mathbf{K}(p)\,\mathbf{F}(p)$. Therefore, the configurations depicted in Fig. 2.5(b)

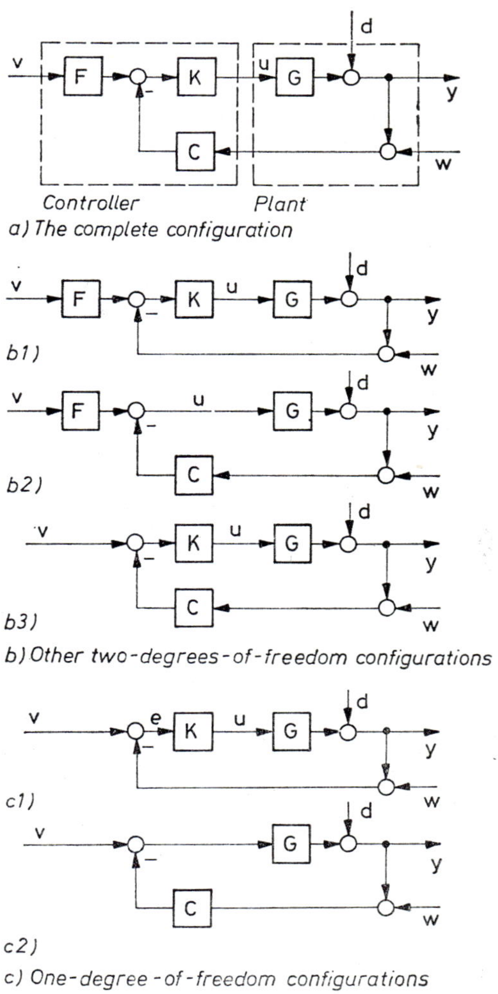

Fig. 2.5 Feedback configurations

have two degrees of freedom as well. It can be generally said that if the plant has only input $\mathbf{u}$ and output $\mathbf{y}$ and there are no further measured signals or actuators available, then all configurations, even more complicated ones than those of Fig. 2.5(a) and (b), have two degrees of freedom.

If the restrictions $\mathbf{C} = \mathbf{I}$ and $\mathbf{F} = \mathbf{I}$ are imposed on the controller then

$$\mathbf{y} = \mathbf{T}\mathbf{v} - \mathbf{T}\mathbf{w} + \mathbf{S}\mathbf{d}. \tag{4}$$

The system has only **one degree of freedom**. The closed-loop system tracks the command signals in the same way as the measurement noise. Disturbance attenuation and command tracking cannot be influenced independently. Decreasing or even minimising the effect of the measurement noise $\mathbf{w}$ inevitably results in bad command tracking. This demonstrates the inadequacy of this simple structure (Fig. 2.5(c)) for command tracking. Nonetheless, the feedback properties of stability, disturbance attenuation and robustness can be achieved by this simple configuration. Therefore, we will often restrict our consideration on such a structure.

A further restriction appears in the choice of $\mathbf{S}(p)$ and $\mathbf{T}(p)$ due to the nonanticipatory properties of the controller. We want to discuss it for single-input single-output systems. Our aim is to make $|s(p)|$ as low as possible in order to attain plant disturbance rejection. Eqn (1) relates $s(p)$ to the open-loop transfer function $l(p) = g(p)\, k(p)\, c(p)$,

$$s(p) = \frac{1}{1 + l(p)}. \tag{5}$$

The value of $s(\mathrm{j}\omega)$ is indirectly proportional to the distance between the point $l(\mathrm{j}\omega)$ and the point $-1 + \mathrm{j}0$ in the complex plane (Fig. 2.6(a)). As long as the frequency response $l(\mathrm{j}\omega)$ lies outside the unit circle centred in the point -1 at the real axis the inequality

$$|s(\mathrm{j}\omega)| < 1 \tag{6}$$

holds, and thus, the feedback system damps the plant disturbances. However, since the frequency response of all strictly proper systems approaches the origin of the complex plane and, at least for pole excess of 2 or more, goes through the left half plane, the Nyquist diagram of stable closed-loop systems crosses the unit circle centred in the origin and remains inside for high frequencies (Fig. 2.6(a)). Then we have

$$|s(\mathrm{j}\omega)| > 1$$

ie the feedback system amplifies the high frequency disturbances and is worse than if no feedback at all has been implemented.

Theorem 2.5 (*Balance Theorem of single-loop systems*).
If the pole excess of the open loop $l(j\omega)$ is at least two and the closed-loop system is stable then the achievable disturbance attenuation is limited by

$$\int_0^\infty \ln |1 + l(j\omega)|\, d\omega = -\int_0^\infty \ln |s(j\omega)|\, d\omega = 0. \tag{7}$$

Fig. 2.6(b) shows the typical form of the graph of $|s(j\omega)|$. Accordingly, the frequency interval $0 \ldots \infty$ can be divided into three regions

I	$\omega = 0 \ldots \omega_o$	$\lvert s(j\omega)\rvert < 1$
II	$\omega = \omega_o \ldots \omega_p$	$\lvert s(j\omega)\rvert > 1$
III	$\omega = \omega_p \ldots \infty$	$\lvert s(j\omega)\rvert \approx 1$

where ω_o is the cross-over frequency. Only in the first interval the feedback system satisfies our requirement (6). It is the region in which the controller represents in fact a negative feedback and decreases the effects of external disturbances. In the second region the feedback system transmits the disturbances "better" than the plant. In the third region the feedback can be considered as ineffective.

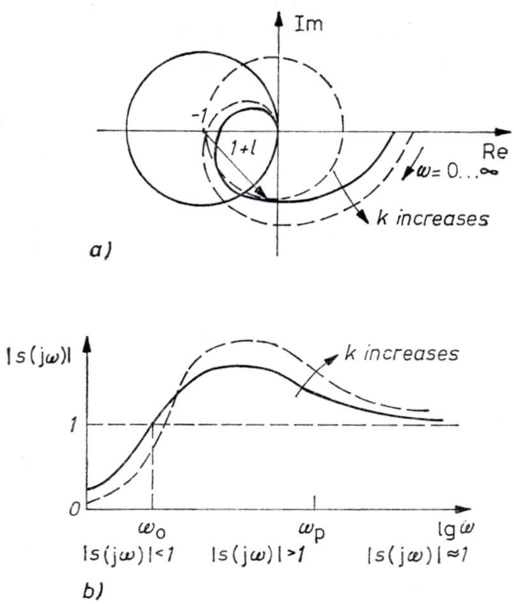

Fig. 2.6 Limitations on the achievable disturbance attenuation and sensitivity improvement

(a) Nyquist diagram of the open-loop transfer function $l(p)$
(b) Absolute value of the sensitivity function $s(j\omega)$

The problem of designing controllers consists, therefore, in assigning the cross-over frequency a suitable value so as to make the feedback effective in the main frequency interval of the disturbances, and in shaping the function $s(j\omega)$ in a suitable way. Note that this attempt is quantitatively restricted by eqn (7) which says that the areas between the graph $|s(j\omega)|$ and the frequency axis in a logarithmic scale are the same in regions I and II—III. This is a fundamental restriction in the design of linear controllers where each improvement in region I has to be paid for in region II.

Fortunately, most of the processes that are to be controlled have low-pass filter characteristics and are excited by signals of limited bandwidth. Then the design aim described in eqn (6) has only to be satisfied over the frequency band of interest (cf Fig. 1.2, but note the different notions of cross-over frequency).

This section shows that even if the plant is completely known the appropriate controller represents a compromise between several partly contradictory design aims. In the presence of uncertainties the robustness requirement (IV) will occur as a new dimension in this compromise.

2.4 LQ-regulation

In this section we briefly summarise the design of feedback controllers by the principle of optimal control. This principle will be used in several ways for designing robust controllers in the succeeding chapters. The plant is described by

$$\begin{aligned}\dot{\mathbf{x}} &= \mathbf{A}\mathbf{x} + \mathbf{B}\mathbf{u}, \qquad \mathbf{x}(0) = \mathbf{x}_0 \\ \mathbf{y} &= \mathbf{C}\mathbf{x}\end{aligned} \tag{1}$$

where the pair $(\mathbf{A}, \mathbf{B})$ is assumed to be controllable. The design requirements are given in form of the performance index

$$I = \frac{1}{2}\int_0^\infty (\mathbf{x}'\mathbf{Q}\mathbf{x} + \mathbf{u}'\mathbf{R}\mathbf{u})\, dt \to \min_{\mathbf{u}(t)} \tag{2}$$

where $\mathbf{Q}$ and $\mathbf{R}$ are symmetric positive semidefinite and symmetric positive definite matrices. The pair $(\mathbf{A}, \mathbf{L})$ is assumed to be observable where $\mathbf{L}'\mathbf{L} = \mathbf{Q}$.

The **solution** of this design problem is given by the state feedback

$$\mathbf{u}(t) = -\mathbf{K}\mathbf{x}(t) \tag{3}$$

with

$$\mathbf{K} = \mathbf{R}^{-1}\mathbf{B}'\mathbf{P} \tag{4}$$

where $\mathbf{P}$ denotes the symmetric positive definite solution of the matrix Ricatti equation

$$\mathbf{P}: \mathbf{A}'\mathbf{P} + \mathbf{P}\mathbf{A} - \mathbf{P}\mathbf{B}\mathbf{R}^{-1}\mathbf{B}'\mathbf{P} + \mathbf{Q} = \mathbf{0}. \tag{5}$$

Under the assumptions made above a unique solution exists and has the following properties: The closed-loop system (1), (3) and (4)

$$\begin{aligned} \dot{\mathbf{x}} &= (\mathbf{A} - \mathbf{B}\mathbf{R}^{-1}\mathbf{B}'\mathbf{P})\,\mathbf{x}, \qquad \mathbf{x}(0) = \mathbf{x}_0 \\ \mathbf{y} &= \mathbf{C}\mathbf{x} \end{aligned} \tag{6}$$

is asymptotically stable. The eigenvalues of $\mathbf{A}$ and $\bar{\mathbf{A}} = \mathbf{A} - \mathbf{B}\mathbf{R}^{-1}\mathbf{B}'\mathbf{P}$ satisfy the relation

$$\prod_{i=1}^{n} |\lambda_i[\mathbf{A}]| \leqq \prod_{i=1}^{n} |\lambda_i[\bar{\mathbf{A}}]|$$

ie the closed-loop system tends to be quicker than the open-loop system. The function

$$w(t) = \mathbf{x}(t)'\,\mathbf{P}\mathbf{x}(t) \tag{7}$$

is a Lyapunov function of the closed-loop system (6), ie the derivative $\dot{w}$ along the trajectory $\mathbf{x}(t)$ of (6) is negative definite.

LQ-regulation (where the feedback (3) is determined for linear plant (1) and quadratic criterion (2)) is only one of several design principles in multivariable feedback control (cf Section 1.2). It is often used in theoretical investigations because it is inherently a multiloop procedure and the solution can be characterised in detail on the basis of eqn (5). An obvious advantage of LQ-regulation is the well defined division of activities in computer-aided design. The designer has to give the model and the weighting matrices $\mathbf{Q}$ and $\mathbf{R}$. Then the computer can solve the design problem and prepare the evaluation of the closed-loop performance by the engineer. The difficulties lie in the choice of the matrices $\mathbf{Q}$ and $\mathbf{R}$ in accordance with the design specifications (I)—(IV) described in Section 1.1 and the practical realisation of the state feedback (3).

Bibliographical notes

The results presented in this chapter are standard in control theory. They are derived and thoroughly explained in monographs and textbooks like (Patel and Munro, 1982; Korn and Wilfert, 1982; Tolle, 1983; Unbehauen, 1985). An easy-to-read introduction to the functional analytic approach to control system analysis is given by Desoer and Vidyasagar (1975).

3 SENSITIVITY OF FEEDBACK SYSTEMS

As in many fields of engineering, sensitivity analysis represents a standard treatment of control systems subject to uncertainties. After the classical work by Bode appeared in 1945, sensitivity reduction by means of feedback attracted the interest of many researchers and practitioners so that nowadays sensitivity analysis can be considered as a special part of control and systems theory.

Section 3.1 gives a short survey of the aim and the principal way of sensitivity analysis. Sections 3.2 and 3.3 consider some aspects of the sensitivity of dynamical systems in general and feedback systems in particular. These sections are not intended to give a thorough introduction to sensitivity analysis of feedback systems but merely to illustrate this approach and, simultaneously, to prepare some results for later use. Section 3.4 is devoted to sensitivity design. It will become obvious later in this book that several design methods, which aim at producing robust controllers, are based on or motivated by sensitivity considerations.

3.1 The principle of sensitivity analysis

The motivation of sensitivity analysis is given by the practical experience that parameters often drift during the operation of the system due to wear, ageing or external influences such as the increase of the surrounding temperature, and that they typically remain in the vicinity of their nominal values. It is interesting to know whether such parameter variations cause severe changes of the system behaviour or have merely a minor influence on the important system properties.

For the analysis, the system behaviour is characterised by some specific measure $q(\mathbf{a})$, eg by the dominant eigenvalue of the system matrix, the overshoot of the step response, the value of a performance index evaluating the dynamics in a global way, or even by some characteristic function such as the frequency response $g(j\omega, \mathbf{a})$. This measure depends upon the parameter vector $\mathbf{a}$, or, for the sake of simplicity of presentation, upon a single parameter a. For the nominal parameter $\hat{a}$ we get the nominal value $\hat{q} = q(\hat{a})$.

Parameter changes δa induce changes δq of the performance measure. Hence, instead of $\hat{q}$ we get the new value $\hat{q} + \delta q = q(\hat{a} + \delta a)$.

In order to make a more detailed treatment of these changes possible we assume that the function $q(a)$ is differentiable. Then the Taylor expansion around the nominal point $\hat{a}$ yields

$$q(\hat{a} + \delta a) = q(\hat{a}) + \left.\frac{\mathrm{d}q}{\mathrm{d}a}\right|_{a=\hat{a}} \delta a + \cdots. \tag{1}$$

If we restrict our considerations to terms of at most first order we get

$$\delta q \approx \left.\frac{\mathrm{d}q}{\mathrm{d}a}\right|_{a=\hat{a}} \delta a\,. \tag{2}$$

The function

$$s(a) = \frac{\mathrm{d}q}{\mathrm{d}a} \tag{3}$$

is called *sensitivity function*, because it provides a measure of how strongly the performance index q is changed by deviations of the parameter from its nominal value. It represents the basic notion of sensitivity analysis.

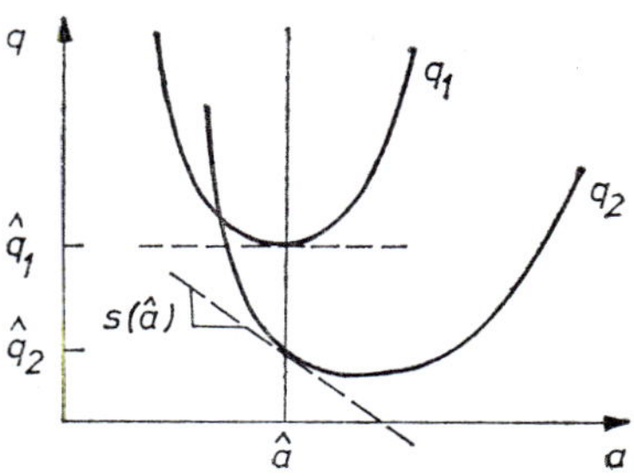

Fig. 3.1 Graphical interpretation of the sensitivity function $s(a)$

In Fig. 3.1 the sensitivity function is depicted in a graphical way. $q_1(a)$ and $q_2(a)$ are two different performance indices. Their sensitivity upon the parameter a at the point $\hat{a}$ is given by the direction of the tangents of the functions $q_1(a)$ and $q_2(a)$ in the nominal point $\hat{a}$. The function $q_1(a)$ is insensitive because the tangent is horizontal. $q_2(a)$ is sensitive because the tangent is not horizontal. The ascent $s(\hat{a})$ of the tangent measures the sensitivity since at least for infinitesimally small parameter perturbations the change δq of the performance index is given by the product of the sensitivity function and the parameter deviation δa (cf eqn (2)).

This short explanation makes clear that in sensitivity analysis the effect of infinitesimally small parameter variations on the system performance are considered. In a qualitative sense, sensitivity is the property that small parameter perturbations do have an effect on a given measure q of the dynamical behaviour. The degree of sensitivity is described by the sensitivity function. For uncertain parameter vectors $\mathbf{a}$ the sensitivity function is generalised as sensitivity vector

$$\frac{dq}{d\mathbf{a}} = \left(\frac{dq}{da_1}, \frac{dq}{da_2}, \ldots, \frac{dq}{da_p}\right)'.$$

That is, sensitivity analysis answers the questions of system analysis in the presence of model uncertainties, which have been formulated in Section 1.6, in the following way: The model of the uncertainties is set up by determining *which* parameters are uncertain and assuming that the parameter perturbations are small. The possible effects of the uncertainties are analysed by determining the sensitivity function $s(a)$. In the design of insensitive control systems a controller is looked for which minimises the sensitivity function for a given nominal parameter value $\hat{a}$.

All such investigations would be useless if the results held actually only for infinitesimally small parameter variations. However, by experience the behaviour of many systems depends on the parameters continuously. Then the sensitivity function is a reasonable characteristic of the effects of small rather than infinitesimally small parameter perturbations. However, each sensitivity analysis must be accompanied by the estimation of the subset $\mathcal{A}$ of the parameter space for which the approximation (2) may be used. We will return to this problem several times.

3.2 Sensitivity of dynamical systems

To begin with let us consider dynamical systems described by the state space model

$$\begin{aligned} \dot{\mathbf{x}} &= \mathbf{A}(a)\,\mathbf{x}(t) + \mathbf{B}\mathbf{u}(t), \qquad \mathbf{x}(0) = \mathbf{x}_0 \\ \mathbf{y} &= \mathbf{C}\mathbf{x}(t). \end{aligned} \tag{3}$$

As known from theoretical process analysis all the matrices $\mathbf{A}$, $\mathbf{B}$ and $\mathbf{C}$ depend on the parameters of the process under consideration. We assume here that the uncertain parameter a may have influence only on the system matrix $\mathbf{A}$. At first, we want to determine the **sensitivity of the state trajectory** $\mathbf{x}(t)$ upon the parameter a. We define the sensitivity vector

$$\mathbf{s}(t, a) = \frac{d\mathbf{x}(t, a)}{da}. \tag{4}$$

From eqn (3) we have

$$\frac{d\dot{\mathbf{x}}}{da} = \frac{d\mathbf{A}(a)}{da}\,\mathbf{x} + \mathbf{A}\,\frac{d\mathbf{x}(t, a)}{da}$$

and

$$\dot{\mathbf{s}} = \mathbf{A}(a)\,\mathbf{s}(t, a) + \left.\frac{\mathrm{d}\mathbf{A}}{\mathrm{d}a}\right|_{a=\hat{a}} \mathbf{x}(t, a), \qquad \mathbf{s}(0, a) = \mathbf{0}. \tag{5}$$

This equation is called *sensitivity equation* because it determines the motion of the sensitivity vector over time.

The sensitivity of the output trajectory

$$\mathbf{s_y}(t, a) = \frac{\mathrm{d}\mathbf{y}(t, a)}{\mathrm{d}a} \tag{6}$$

is given by

$$\mathbf{s_y}(t, a) = \mathbf{C}\mathbf{s}(t, a). \tag{7}$$

Eqns (5) and (7) depend on the whole state vector of the system and form a new dynamical system of order n (Fig. 3.2). Altogether, eqns (3), (5) and (7) form a $2n$-th order system for the determination of $\mathbf{s_y}$. The sensitivity function depends on the input $\mathbf{u}(t)$ and can only be calculated for specified input signals.

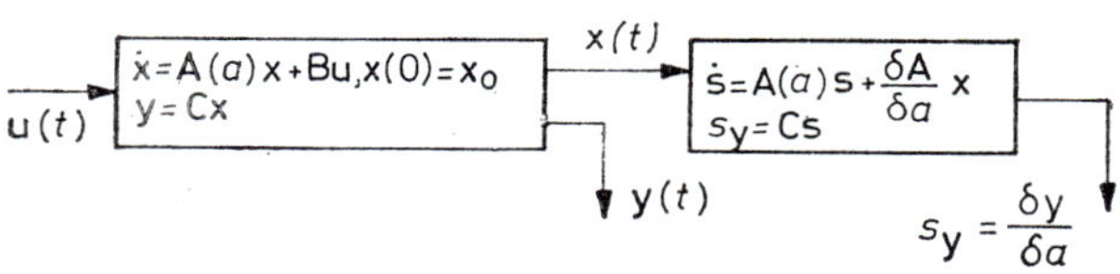

Fig. 3.2 Sensitivity of the output trajectory

As a second example for the sensitivity analysis of dynamical systems we consider the **sensitivity of the eigenvalues** of $\mathbf{A}(a)$. As opposed to the investigations above, the eigenvalues are independent of the external disturbances and, thus, represent properties of the system dynamics alone. The eigenvalues $\lambda[\mathbf{A}]$ of an $(n \times n)$-matrix $\mathbf{A}$ are defined as complex scalars for which complex n-vectors $\tilde{\mathbf{u}}$ exist, such that

$$\mathbf{A}\tilde{\mathbf{u}} = \lambda[\mathbf{A}]\,\tilde{\mathbf{u}} \tag{8}$$

holds. $\tilde{\mathbf{u}}$ is called right eigenvector of $\mathbf{A}$ belonging to $\lambda[\mathbf{A}]$. If the matrix $\mathbf{A}$ depends on the parameter a then both $\lambda[\mathbf{A}]$ and $\tilde{\mathbf{u}}$ are dependent on a. Differentiation according to a yields

$$\frac{\mathrm{d}\mathbf{A}}{\mathrm{d}a}\,\tilde{\mathbf{u}} + \mathbf{A}\,\frac{\mathrm{d}\tilde{\mathbf{u}}}{\mathrm{d}a} = \frac{\mathrm{d}\lambda}{\mathrm{d}a}\,\tilde{\mathbf{u}} + \lambda\,\frac{\mathrm{d}\tilde{\mathbf{u}}}{\mathrm{d}a}. \tag{9}$$

This equation will be multiplied by the left eigenvector $\tilde{\mathbf{v}}$, which satisfies

$$\tilde{\mathbf{v}}'\mathbf{A} = \lambda[\mathbf{A}]\,\tilde{\mathbf{v}}'. \tag{10}$$

Then

$$\tilde{\mathbf{v}}' \frac{\mathrm{d}\mathbf{A}}{\mathrm{d}a} \tilde{\mathbf{u}} + \tilde{\mathbf{v}}'\mathbf{A} \frac{\mathrm{d}\tilde{\mathbf{u}}}{\mathrm{d}a} = \frac{\mathrm{d}\lambda}{\mathrm{d}a} \tilde{\mathbf{v}}'\tilde{\mathbf{u}} + \lambda\tilde{\mathbf{v}}' \frac{\mathrm{d}\tilde{\mathbf{u}}}{\mathrm{d}a} \tag{11}$$

holds. Because of eqn (10)

$$\tilde{\mathbf{v}}'\mathbf{A} \frac{\mathrm{d}\tilde{\mathbf{u}}}{\mathrm{d}a} = \lambda\tilde{\mathbf{v}}' \frac{\mathrm{d}\tilde{\mathbf{u}}}{\mathrm{d}a}.$$

Hence, eqn (11) leads to

$$\frac{\mathrm{d}\lambda}{\mathrm{d}a} = \frac{\tilde{\mathbf{v}}' \dfrac{\mathrm{d}\mathbf{A}}{\mathrm{d}a} \tilde{\mathbf{u}}}{\tilde{\mathbf{v}}'\tilde{\mathbf{u}}}. \tag{12}$$

For multiple eigenvalues the formula for $\frac{\mathrm{d}\lambda}{\mathrm{d}a}$ becomes more involved if the corresponding eigenvectors are not linearly independent since in this case $\tilde{\mathbf{v}}'\tilde{\mathbf{u}}$ vanishes. Eqn (12) presents the sensitivity of the eigenvalue $\lambda[\mathbf{A}(a)]$ in relation to the sensitivity of the matrix $\mathbf{A}(a)$ and the right and left eigenvectors that belong to $\lambda[\mathbf{A}(a)]$.

As an example consider the matrix

$$\mathbf{A} = \begin{pmatrix} 0.2 + a & -0.8 \\ 4.8 & -4.2 \end{pmatrix}$$

where a reflects the perturbation of $\mathbf{A}$. For $\hat{a} = 0$ this matrix has the eigenvalues and corresponding eigenvectors

$$\lambda_1 = -1, \qquad \tilde{\mathbf{u}}_1 = \begin{pmatrix} 2 \\ 3 \end{pmatrix}, \qquad \tilde{\mathbf{v}}_1 = \begin{pmatrix} 0.8 \\ -0.2 \end{pmatrix}$$

$$\lambda_2 = -3, \qquad \tilde{\mathbf{u}}_2 = \begin{pmatrix} 1 \\ 4 \end{pmatrix}, \qquad \tilde{\mathbf{v}}_2 = \begin{pmatrix} -0.6 \\ 0.4 \end{pmatrix}.$$

Obviously,

$$\frac{\mathrm{d}\mathbf{A}}{\mathrm{d}a} = \begin{pmatrix} 1 & 0 \\ 0 & 0 \end{pmatrix}.$$

Hence, the dominant eigenvalue λ_1 is sensitive to a because

$$\frac{\mathrm{d}\lambda_1}{\mathrm{d}a} = \frac{(0.8 \;\; -0.2) \begin{pmatrix} 1 & 0 \\ 0 & 0 \end{pmatrix} \begin{pmatrix} 2 \\ 3 \end{pmatrix}}{(0.8 \;\; -0.2) \begin{pmatrix} 2 \\ 3 \end{pmatrix}} = 1.6.$$

The value of $\frac{\mathrm{d}\lambda_1}{\mathrm{d}a}$ gives the direction in which the eigenvalue λ_1 moves for small positive perturbations δa. It means that the dominant eigenvalue changes its position in direction towards the stability border (imaginary axis)

(Fig. 3.3). For the other eigenvalue we have

$$\frac{d\lambda_2}{da} = -0.6.$$

To conclude this section, the **sensitivity index of Bode** should be mentioned here. It addresses the sensitivity of the transfer function $g(p, a)$, which, as in the Bode plot, is considered in a logarithmic scale. Therefore, the sensitivity index is defined as

$$s = \frac{d \ln g(p, a)}{d \ln a} = \frac{\dfrac{dg(p, a)}{g(p, a)}}{\dfrac{da}{a}}. \tag{13}$$

The term on the right hand side of the definition shows that this index measures the relative change of $g(p, a)$ and a. Therefore, it is sometimes called '*percentage change sensitivity function*'. Its definition is reasonable since various feedback systems are customarily compared in terms of percentage variation of the parameters. In Section 3.3 we will see that this index provides a meaningful analytical measure for the design of insensitive or even robust feedback systems.

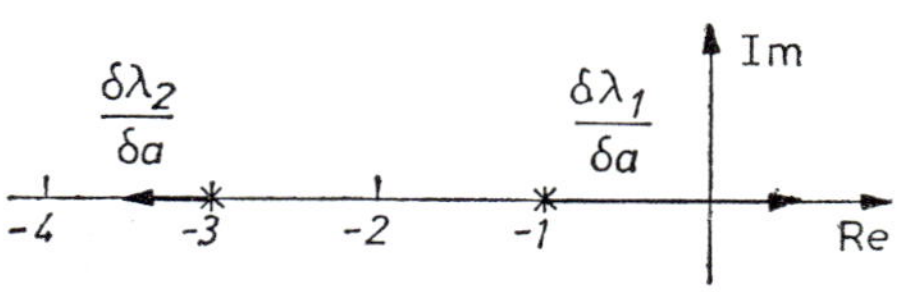

Fig. 3.3 Sensitivity of the eigenvalues

Horowitz has generalised this index in order to include small rather than infinitesimally small parameter variations

$$s_H = \frac{\dfrac{\delta g(p, a)}{g(p, a)}}{\dfrac{\delta a}{a}}.$$

This index, however, depends upon the value of parameter variations δa under consideration.

3.3 Sensitivity analysis of feedback systems

Now we consider the sensitivity of feedback systems in terms of the controller parameters and ask the question to which degree feedback can reduce the output variations that are caused by parameter variations in the plant. The answer to this question should give us guidelines for the selection of controllers that minimise the sensitivity of the closed-loop performance.

We want to emphasise that the sensitivity matrix $\mathbf{S}$ introduced in eqn (1.4-21) plays an important role in the description of the effects of model uncertainties and disturbances on the performance of the closed-loop system. At first, the importance of the matrix $\mathbf{S}$ can be seen by comparing the feedback system with the open-loop control system that exhibits the same nominal performance. Eqn (1.4-20) states

$$\mathbf{y}_c(p, \mathbf{a}) - \mathbf{y}_c(p, \hat{\mathbf{a}}) = \big(\mathbf{I} + \mathbf{G}(p, \mathbf{a})\, \mathbf{K}(p)\, \mathbf{C}(p)\big)^{-1} \big(\mathbf{y}_0(p, \mathbf{a}) - \mathbf{y}_0(p, \hat{\mathbf{a}})\big) \tag{1}$$

where

$$\mathbf{S}(p, \mathbf{a}) = \big(\mathbf{I} + \mathbf{G}(p, \mathbf{a})\, \mathbf{K}(p)\, \mathbf{C}(p)\big)^{-1} \tag{2}$$

is defined to be the *sensitivity matrix*. In both equations the dependence upon the parameter vector $\mathbf{a}$ is explicitly indicated. Eqn (1) compares the closed-loop and the open-loop output variations brought about by parameter variations in the plant. If $\mathbf{S}(p, \mathbf{a})$ is low, then the closed-loop deviations are smaller than the performance deviations in the open-loop system.

Secondly, we want to show that the sensitivity matrix $\mathbf{S}$ is actually a sensitivity indicator according to the definition in eqn (3.1-3). This can be done in two ways. On the one hand, eqns (1) and (2) can be reformulated as

$$s(p, \mathbf{a}) = \frac{\mathrm{d}\mathbf{y}_c(p, \mathbf{a})}{\mathrm{d}\mathbf{y}_0(p, \mathbf{a})}$$

if only single-input single-output systems and infinitesimally small parameter deviations are considered. The only difference between this equation and eqn (3.1-3) is that the differentiation is performed here in relation to the output $\mathbf{y}_0(p, \mathbf{a})$ rather than the parameter $\mathbf{a}$, which caused the output deviation.

On the other hand, it can be easily shown that in the scalar case the sensitivity function

$$s(p, \mathbf{a}) = \frac{1}{1 + g(p, \mathbf{a})\, k(p)\, c(p)} \tag{3}$$

coincides with the Bode index (3.2-13) if this index is applied for assessing the sensitivity of the closed-loop transfer function

$$g_{\mathbf{yv}}(p, \mathbf{a}) = \frac{g(p, \mathbf{a})\, k(p)\, f(p)}{1 + g(p, \mathbf{a})\, k(p)\, c(p)}$$

(cf eqn (1.4-13)) to variations of the plant transfer function $g(p, \mathbf{a})$. Then

$$\frac{\dfrac{\mathrm{d}g_{\mathbf{yv}}(p, \mathbf{a})}{g_{\mathbf{yv}}(p, \mathbf{a})}}{\dfrac{\mathrm{d}g(p, \mathbf{a})}{g(p, \mathbf{a})}} = \frac{1}{1 + g(p, \mathbf{a})\, k(p)\, c(p)} = s(p, \mathbf{a})\,. \tag{4}$$

A similar relation can be derived in the multivariable case if $\mathbf{C} = \mathbf{I}$. From eqn (1.4-5) we have

$$\begin{aligned}\delta G_{\mathbf{yv}}(p, \mathbf{a}) &= \mathbf{S}(p, \mathbf{a})\, \delta\mathbf{G}\mathbf{K}\big(\mathbf{I} + \mathbf{G}(p, \mathbf{a})\, \mathbf{K}(p)\big)^{-1}\, \mathbf{F}(p) \\ &= \mathbf{S}(p, \mathbf{a})\, \delta\mathbf{G}\mathbf{G}(p, \mathbf{a})^{-1}\, \mathbf{G}(p, \mathbf{a})\, \mathbf{K}\big(\mathbf{I} + \mathbf{G}(p, \mathbf{a})\, \mathbf{K}(p)\big)^{-1}\, \mathbf{F}(p) \\ &= \mathbf{S}(p, \mathbf{a})\, \delta\mathbf{G}\mathbf{G}(p, \mathbf{a})^{-1}\, \big(\mathbf{I} + \mathbf{G}(p, \mathbf{a})\, \mathbf{K}(p)\big)^{-1}\, \mathbf{G}(p, \mathbf{a})\, \mathbf{K}(p)\, \mathbf{F}(p) \\ &= \mathbf{S}(p, \mathbf{a})\, \delta\mathbf{G}\mathbf{G}(p, \mathbf{a})^{-1}\, \mathbf{G}_{\mathbf{yv}}(p, \mathbf{a})\end{aligned}$$

for all frequencies for which the inverse of $\mathbf{G}(p, \mathbf{a})$ exists. Hence,

$$\delta\mathbf{G}_{\mathbf{yv}}(p, \mathbf{a})\, \mathbf{G}_{\mathbf{yv}}(p, \mathbf{a})^{-1} = \mathbf{S}(p, \mathbf{a})\, \delta\mathbf{G}(p, \mathbf{a})\, \mathbf{G}(p, \mathbf{a})^{-1} \tag{5}$$

holds.

Thirdly, the role of the sensitivity matrix in disturbance attenuation should be outlined. We consider the two-degree-of-freedom configuration of Fig. 2.5(b1) where

$$\mathbf{y} = (\mathbf{I} + \mathbf{G}\mathbf{K})^{-1}\, \mathbf{G}\mathbf{K}\mathbf{F}\mathbf{v} - (\mathbf{I} + \mathbf{G}\mathbf{K})^{-1}\, \mathbf{G}\mathbf{K}\mathbf{w} + (\mathbf{I} + \mathbf{G}\mathbf{K})^{-1}\, \mathbf{d}.$$

Using the sensitivity matrix of this configuration

$$\mathbf{S}(p, \mathbf{a}) = \big(\mathbf{I} + \mathbf{G}(p, \mathbf{a})\, \mathbf{K}(p)\big)^{-1} \tag{6}$$

and the complementary sensitivity matrix

$$\begin{aligned}\mathbf{T}(p, \mathbf{a}) &= \mathbf{I} - \mathbf{S}(p, \mathbf{a}) \\ &= \big(\mathbf{I} + \mathbf{G}(p, \mathbf{a})\, \mathbf{K}(p)\big)^{-1}\, \mathbf{G}(p, \mathbf{a})\, \mathbf{K}(p)\end{aligned} \tag{7}$$

we have

$$\mathbf{y}(p, \mathbf{a}) = \mathbf{T}(p, \mathbf{a})\, \mathbf{F}(p)\, \mathbf{v}(p) - \mathbf{T}(p, \mathbf{a})\, \mathbf{w}(p) + \mathbf{S}(p, \mathbf{a})\, \mathbf{d}(p). \tag{8}$$

The sensitivity matrix coincides with the transfer function matrix between the disturbance $\mathbf{d}$ and the output $\mathbf{y}$. The complementary sensitivity matrix is identical to the transfer function matrix between the measurement noise $\mathbf{w}$ and the output y. This shows that variations of the plant dynamics propagate towards the output of the feedback system, in principle, in the same way as the disturbances that affect the plant output. Feedback systems which are insensitive to plant parameter deviations are also insensitive to plant disturbances. Both design objectives can be dealt with in the same manner. However, this does not hold for other disturbances. The measurement noise is transmitted to the closed-loop system output through the complementary sensitivity matrix. Systems which are insensitive to plant uncertainties are sensitive to this kind of external disturbances.

On this occasion, a remark should be made concerning the use of the terms 'sensitivity' and 'robustness'. Some authors, particularly in the American literature, limit the use of the term 'sensitivity' to the analysis of the effects of external disturbances and use the term 'robustness' in the assessment of the perturbation of the feedback system performance caused by plant uncertainties. In contrast to these definitions we will continue to use both terms in connection with uncertainties of the plant dynamics. While sensitivity analysis considers very small perturbations of the parameters around a given nominal value $\hat{\mathbf{a}}$, robustness analysis concerns large perturbations. As eqn (8) points out, sensitivity in this sense and disturbance attenuation are related in a complicated way, which depends on the entry point of the external signals. We will come back to this relation in Section 8.3.

All these remarks should explain why the sensitivity matrix $\mathbf{S}(p, \mathbf{a})$ represents a basic notion in the analysis of feedback systems subject to uncertainties although it is usually derived as in Section 1.4 by comparing the open-loop and closed-loop variations rather than by considering the effect of parameter variations on the feedback system alone.

Eqns (1), (5) and (8) provide deep insight into the mechanisms of uncertainty propagation through feedback systems because they hold for arbitrarily large parameter variations. In sensitivity analysis, however, the plant transfer function matrix $\mathbf{G}(p, \mathbf{a})$ is assumed to be known only for the nominal parameter vector $\hat{\mathbf{a}}$. Therefore, instead of (1), (5) and (8) the following approximations are used

$$\mathbf{y}_c(p, \mathbf{a}) - \mathbf{y}_c(p, \hat{\mathbf{a}}) \approx \mathbf{S}(p, \hat{\mathbf{a}}) \big(\mathbf{y}_0(p, \mathbf{a}) - \mathbf{y}_0(p, \hat{\mathbf{a}})\big) \tag{9}$$

$$\delta\mathbf{G}_{\mathbf{yv}}(p, \mathbf{a})\, \mathbf{G}_{\mathbf{yv}}(p, \hat{\mathbf{a}})^{-1} \approx \mathbf{S}(p, \hat{\mathbf{a}})\, \delta\mathbf{G}(p, \mathbf{a})\, \mathbf{G}(p, \hat{\mathbf{a}})^{-1} \tag{10}$$

$$\mathbf{y}(p, \mathbf{a}) \approx \mathbf{T}(p, \hat{\mathbf{a}})\, \mathbf{F}(p)\, \mathbf{v}(p) - \mathbf{T}(p, \hat{\mathbf{a}})\, \mathbf{w}(p) + \mathbf{S}(p, \hat{\mathbf{a}})\, \mathbf{d}(p)\,. \tag{11}$$

These equations provide the background for a lot of design rules and design methods for insensitive or even robust controllers. Therefore, it should be stressed here again that these equations yield reasonable approximations only in a restricted (sufficiently small) surrounding of the nominal parameter $\hat{\mathbf{a}}$.

3.4 Sensitivity design

In this section we take up the problem of designing a feedback controller for the plant with nominal parameters $\hat{\mathbf{a}}$ such that the closed-loop system satisfies the design objectives (I)—(III) given in Section 1.1 and, simultaneously, has a low sensitivity to plant parameter variations. The sensitivity requirement can be specified in terms of the sensitivity matrix $\mathbf{S}(p, \mathbf{a})$ since a small sensitivity matrix ensures that the command tracking properties are weakly affected by parameter variations (cf eqn (3.3-5)). For single-variable systems the least requirement is given by

$$|s(j\omega, \mathbf{a})| < 1 \tag{1}$$

ie we are looking for a controller such that the closed-loop system performance is less affected by plant uncertainties than a nominally equivalent open-loop control system. In other words, eqn (1) guarantees the superiority of the feedback system over feedforward control.

For multivariable systems this requirement has to be formulated in terms of the vector-valued output $\mathbf{y}$ or the matrix-valued sensitivity index $\mathbf{S}(p, \mathbf{a})$. A reasonable form is given by

$$\int_{-\infty}^{+\infty} [\mathbf{y}_c(-j\omega, \mathbf{a}) - \mathbf{y}_c(-j\omega, \hat{\mathbf{a}}]' [\mathbf{y}_c(j\omega, \mathbf{a}) - \mathbf{y}_c(j\omega, \hat{\mathbf{a}})] \, d\omega$$

$$\leq \int_{-\infty}^{+\infty} [\mathbf{y}_o(-j\omega, \mathbf{a}) - \mathbf{y}_o(-j\omega, \hat{\mathbf{a}})]' [\mathbf{y}_o(j\omega, \mathbf{a}) - \mathbf{y}_o(j\omega, \hat{\mathbf{a}})] \, d\omega .$$

Because of eqn (3.3-9), this inequality is satisfied for small parameter variations if

$$\mathbf{S}'(-j\omega, \hat{\mathbf{a}}) \, \mathbf{S}(j\omega, \hat{\mathbf{a}}) - \mathbf{I} = \text{positive semidefinite.} \tag{2}$$

In a more general form in which the output variables are weighted by some constant or frequency-dependent matrix $\mathbf{Q}$, eqn (2) requires

$$\mathbf{S}'(-j\omega, \hat{\mathbf{a}}) \, \mathbf{Q}\mathbf{S}(j\omega, \hat{\mathbf{a}}) - \mathbf{Q} = \text{positive semidefinite.} \tag{3}$$

Now, we want to investigate the design problem for the two-degree-of-freedom configuration shown in Fig. 2.5(b1). There, the command tracking properties can be shaped by means of the prefilter $\mathbf{F}(p)$ once the feedback compensator $\mathbf{K}(p)$ has been chosen. Moreover, the sensitivity of command tracking $\mathbf{G}_{\mathbf{yv}}(p, \mathbf{a})$ is given independently of $\mathbf{F}(p)$ by the sensitivity matrix

$$\mathbf{S}(p, \mathbf{a}) = \big(\mathbf{I} + \mathbf{G}(p, \mathbf{a}) \, \mathbf{K}(p)\big)^{-1} \tag{4}$$

(cf eqn (3.3-10)). Therefore, we can focus our attention here on the choice of the feedback compensator $\mathbf{K}(p)$ in connection with the sensitivity requirements and leave the determination of $\mathbf{F}(p)$ as the second design step.

Besides, by our aim to minimise the sensitivity matrix the choice of $\mathbf{K}(p)$ is influenced by the fundamental restrictions of feedback design discussed in Section 2.3 and by the necessity to arrive at a compromise between sensitivity reduction and measurement noise attenuation (cf eqn (3.3-8)). We proceed with an outline of different methods for solving this problem.

At first, if we consider the sensitivity reduction as the primary aim the controller has to be chosen so as to make the sensitivity matrix (4) as small as possible. This can be done by using feedback compensators $\mathbf{K}(p)$ which produce a **high feedback gain**. For single-variable systems high loop gain means a high value of $|g(p, \mathbf{a}) \, k(p)|$, whereas in the multivariable case the loop gain can be measured, for example, in terms of the matrix norm $\|\mathbf{G}(p, \mathbf{a}) \, \mathbf{K}(p)\|$. This leads to a rule of thumb, which is widely accepted by control engineers: *Increasing the loop gain reduces the closed-loop sensitivity* to plant parameter uncertainties. It makes the loop 'tight'.

Although this rule may apply in many single-loop systems we must keep in mind that there is a fundamental restriction of its use. All considerations are true only as long as the closed-loop system remains stable. This, however, can be ensured merely for a restricted range of $|k(p)|$. The Nyquist diagram in Fig. 2.6(a) shows that the loop gain $|l(p)| = |g(p, \mathbf{a})\, k(p)|$ is limited by the value which gives rise to an encirclement of the point $-1 + \mathrm{j}0$. In Section 6.4 the idea of high-gain feedback is taken up again in connection with robustness considerations and investigations on system properties that ensure stability even for very high loop gains.

Secondly, we consider **LQ-regulation** as introduced in Section 2.4. For the plant

$$\dot{\mathbf{x}} = \mathbf{Ax} + \mathbf{Bu}, \qquad \mathbf{x}(0) = \mathbf{x}_0 \tag{5}$$

the LQ-regulator has the form

$$\mathbf{u}(t) = -\mathbf{Kx}(t) \tag{6}$$

with

$$\mathbf{K} = \mathbf{R}^{-1}\mathbf{B}'\mathbf{P}, \tag{7}$$

where $\mathbf{P}$ is the positive definite solution of the matrix Riccati equation

$$\mathbf{A}'\mathbf{P} + \mathbf{PA} - \mathbf{PBR}^{-1}\mathbf{B}'\mathbf{P} + \mathbf{Q} = 0. \tag{8}$$

For sensitivity considerations this controller has an important property:

Theorem 2.1 For the LQ-regulator (6), (7) the return difference matrix of the open-loop system (5), (6) satisfies the inequality

$$\left|\det\left(\mathbf{I} + \mathbf{L}(\mathrm{j}\omega)\right)\right| \geqq 1 \quad \text{for all } \omega \tag{9}$$

where

$$\mathbf{L}(\mathrm{j}\omega) = \mathbf{K}(\mathrm{j}\omega\mathbf{I} - \mathbf{A})^{-1}\,\mathbf{B}. \tag{10}$$

Proof. Adding $\pm p\mathbf{I}$ and $\pm p^*\mathbf{I}$, the Riccati equation (8) can be transformed into

$$(p^*\mathbf{I} - \mathbf{A}')\,\mathbf{P} + \mathbf{P}(p\mathbf{I} - \mathbf{A}) + \mathbf{PBR}^{-1}\mathbf{B}'\mathbf{P} = \mathbf{Q} + (p^* + p)\,\mathbf{P},$$

where p^* denotes the complex conjugate of p. Premultiplication by $\left((p\mathbf{I} - \mathbf{A})^{-1}\mathbf{B}\right)^*$ and postmultiplication by $\left((p\mathbf{I} - \mathbf{A})^{-1}\mathbf{B}\right)$ yield

$$\begin{aligned}&\left((p\mathbf{I} - \mathbf{A})^{-1}\mathbf{B}\right)^*(p^*\mathbf{I} - \mathbf{A})'\,\mathbf{P}\,(p\mathbf{I} - \mathbf{A})^{-1}\mathbf{B} + \left((p\mathbf{I} - \mathbf{A})^{-1}\mathbf{B}\right)^*\mathbf{P}(p\mathbf{I} - \mathbf{A})\,(p\mathbf{I} - \mathbf{A})^{-1}\mathbf{B}\\ &\quad + \left((p\mathbf{I} - \mathbf{A})^{-1}\mathbf{B}\right)^*\mathbf{PBR}^{-1}\mathbf{RR}^{-1}\mathbf{B}'\mathbf{P}(p\mathbf{I} - \mathbf{A})^{-1}\mathbf{B}\\ &= \left((p\mathbf{I} - \mathbf{A})^{-1}\mathbf{B}\right)^*(\mathbf{Q} + 2\mathrm{Re}[p]\,\mathbf{P})\,(p\mathbf{I} - \mathbf{A})^{-1}\mathbf{B},\end{aligned} \tag{11}$$

where $\mathbf{A}^*$ denotes the complex conjugate transpose of $\mathbf{A}$. The right-hand side of eqn (11) is abbreviated by $\mathbf{H}(p)$. Since $\mathbf{B}'\mathbf{P} = \mathbf{RK}$ holds, we get

$$\begin{aligned}\mathbf{H}(p) &= \mathbf{RK}(p\mathbf{I} - \mathbf{A})^{-1}\mathbf{B} + \mathbf{B}'(p^*\mathbf{I} - \mathbf{A}')^{-1}\mathbf{K}'\mathbf{R} + \mathbf{B}'(p^*\mathbf{I} - \mathbf{A}')^{-1}\mathbf{K}'\mathbf{RK}(p\mathbf{I} - \mathbf{A})^{-1}\mathbf{B}\\ &= \mathbf{RL}(p) + \mathbf{L}(p)^*\,\mathbf{R} + \mathbf{L}(p)^*\,\mathbf{RL}(p).\end{aligned}$$

Addition of $\mathbf{R}$ leads to

$$(\mathbf{I} + \mathbf{L}(p))^* \mathbf{R}(\mathbf{I} + \mathbf{L}(p)) = \mathbf{H}(p) + \mathbf{R}. \tag{12}$$

For $p = j\omega$ on the Nyquist contour (with sufficiently small indentations around imaginary poles of $\mathbf{L}(p)$, see Fig. 2.3) $\mathbf{Q} + \mathrm{Re}\,[p]\,\mathbf{P}$ in eqn (11) and, thus, $\mathbf{H}(p)$ is positive semidefinite. Therefore, eqn (12) yields

$$\det (\mathbf{I} + \mathbf{L}(j\omega))^2 \det \mathbf{R} \geqq \det \mathbf{R}$$

and eqn (9) which completes the proof. □

For single-input systems (5), eqn (9) reads as

$$|1 + l(j\omega)| \geqq 1 \qquad \text{for all } \omega. \tag{13}$$

This condition is necessary and even sufficient for the feedback (6) to be optimal with respect to an appropriately chosen quadratic cost index. It ensures that the modulus of the sensitivity function is less than unity for all frequencies

$$|s(j\omega, \mathbf{a})| \leqq 1. \tag{14}$$

That is, for all weighting factors $r > 0$ and all matrices $\mathbf{Q}$ satisfying the observability assumption of Section 2.4 the LQ-regulator leads to low-sensitive feedback systems. For multi-input plants, this property is described by

$$|\det \mathbf{S}(j\omega)| \leqq 1. \tag{15}$$

Note that our design requirement (1) is, in fact, satisfied for all frequencies ω. In view of our considerations in Section 2.3, low sensitivity over the whole frequency domain is achievable only if the open loop has pole excess of at most two. Hence, the property (9) of the LQ-regulator is reached because state feedback is used. It cannot be directly extended to optimal output feedback systems. We will come back to this problem in connection with robustness considerations in Section 6.1.

As a further method, **sensitivity optimisation** is currently receiving considerable attention. The aim is not only to reduce sensitivity, but to minimise a weighted sensitivity index such as

$$\sup_{\omega} |s(j\omega)\, v_1(j\omega)| \to \min_{k(j\omega)} \tag{16}$$

where $v_1(j\omega)$ represents a suitably chosen weighting function. This optimisation problem requires that $|s(j\omega)|$ is small for almost all frequencies and that the local peaks have sufficiently small areas.

The objective (16) is motivated by several observations. First, the Nyquist plot in Fig. 2.6(a) shows that the minimum distance of the curve from the point $(-1 + j0)$ or, equivalently, the highest value of $|s(j\omega)|$ describes the critical domain for stability. Secondly, LQ-minimisation or its stochastic counterpart, LQG-minimisation, presupposes that the disturbance is known to be an impulse function or to have a Gaussian distribution density with known covariance matrix. If no spectral information on the disturbance is available it is desirable to limit the maximum value of the disturbance frequency response $|s(j\omega)|$.

As the design aim (16) may produce trivial solutions and does not include essential design objectives it is extended to form the combined performance index

$$\sup_{\omega} |s(j\omega)\, v_1(j\omega)|^2 + |t(j\omega)\, v_2(j\omega)|^2 \rightarrow \min_{k(j\omega)} \tag{17}$$

where $s(j\omega)$ and $t(j\omega)$ are the sensitivity and complementary sensitivity functions and $v_1(j\omega)$ and $v_2(j\omega)$ are weighting functions. Now, many design requirements can be transformed into bounds on $s(j\omega)$ and $t(j\omega)$. For instance, for good disturbance attenuation $|s(j\omega)|$ should be made small over the frequency range of the disturbances (cf eqn (3.3-8)). This can be ensured by choosing sufficiently large $v_1(j\omega)$ in this frequency band. Power limitation of the input u can be translated into requirements on $v_2(j\omega)$ since

$$u(j\omega) = t(j\omega)\,\frac{f(j\omega)}{g(j\omega)}\,v(j\omega) - t(j\omega)\,\frac{1}{g(j\omega)}\,\big(d(j\omega) + w(j\omega)\big).$$

For the multivariable case, an extension can be made by considering the set of stable, proper, real-rational matrices with the spectral norm

$$\|\mathbf{M}(j\omega)\| = \max_{i}\, \big(\lambda_i[\mathbf{M}(-j\omega)'\,\mathbf{M}(j\omega)]\big)^{1/2}$$

where λ_i denotes the eigenvalues of the indicated matrix product. Then eqn (17) can be extended to

$$\sup_{\omega} \|\mathbf{S}(j\omega)\,\mathbf{V}_1(j\omega)\|^2 + \|\mathbf{T}(j\omega)\,\mathbf{V}_2(j\omega)\|^2 \rightarrow \min_{\mathbf{K}(j\omega)}. \tag{18}$$

Since

$$\sup_{\omega}\max_{i}\, \big(\lambda_i[\mathbf{M}(-j\omega)'\,\mathbf{M}(j\omega)]\big)^{1/2} = \|\mathbf{M}(p)\|_\infty$$

represents the norm of the $\mathcal{H}^\infty$-space, this approach is often referred to as *H^∞-optimisation*.

The analytical development of the problems (17) and (18) can be carried out very far. For instance, necessary and sufficient conditions can be given in terms of the polynomial matrices that describe the plant and the feedback compensator. However, the numerical solution turns out to be a rather difficult problem. Search procedures as well as the minimisation of upper bounds of the performance index have been proposed. The interested reader is referred to the currently increasing literature on this problem.

3.5 Summary

Sensitivity is a local property that describes how strongly the system performance is affected by very small parameter perturbations around a given nominal point $\hat{\mathbf{a}}$. No information about the amount of perturbations is used. However, as many properties depend continuously on the parameter vector **a** extrapolations can be made from very small to finite deviations. Therefore, sensitivity analysis yields guidelines for the attenuation of severe parameter perturbation and, hence, the achievement of robustness. But sensitivity analysis alone cannot ensure this robustness because the range of validity of its results is not known.

Table 3.1 Sensitivity and robustness

	Sensitivity	**Robustness**
Definition	Dependence of the system performance upon the parameter values	Property of the system to perform satisfactorily for all parameter vectors $\mathbf{a} \in \mathcal{A}$
Necessary information about the model uncertainties for the analysis	Which parameters are uncertain?	Which parameters are uncertain and what are the ranges of their possible values?
Aim of feedback control synthesis	Small plant parameter deviations result in small deviations of the closed-loop system performance from its nominal	The closed-loop system performance lies in the acceptable range for all possible parameter values

The following part of the book is devoted to robustness analysis and design of feedback systems. As opposed to sensitivity considerations we aim at reaching and remaining within given tolerance bounds on the design specifications for all perturbations of a given class. The plant uncertainties may be very large, but they are described explicitly by specimens or upper bounds. This new point of view brings about new problems and refines the answer to the problem which closed-loop system properties indicate uncertainty tolerance. For example, the problem of stability robustness must be investigated explicitly whereas in sensitivity analysis the closed-loop stability could be taken for granted. As a further example, the robustness against model uncertainties will turn out to be no equivalent property of disturbance attenuation as sensitivity was.

Bibliographical notes

The foundation of sensitivity analysis has been laid by Bode (1945), who introduced the sensitivity function in order to describe the benefits of feedback in presence of uncertainties. He presented it in terms of the open-loop transmission function. Although he was primarily concerned with feedback amplifiers rather than control systems his results on the capability and limitations of minimum-phase systems and the gain and phase margin are of great impact in control theory. Cruz and Perkins (1964) introduced the sensitivity index into control systems analysis and launched an intensive research in this field. Valuable summaries of the tools and results of sensitivity analysis of dynamical systems are given in the book edited by Cruz (1973) and the monograph by Frank (1976).

The sensitivity of LQ-regulators has been considered by Kalman (1964), who derived the basic properties from the matrix-Riccati-equation.

The H^∞-approach has been initiated by Zames (1981), who introduced the weighted H^∞-norm for systems without right-half plane zeros and single-variable systems with at most one right-half plane zero (for an introductory course on H^∞-theory see (Francis, 1987)). Zames et al. (1983) and Francis et al. (1984) investigated the disturbance attenuation problem in the framework of sensitivity optimisation, whereas Kwakernaak (1984, 1985) and Foo et al. (1984) considered the sensitivity of the closed loop against plant parameter variations. Freudenberg et al. (1986) tackled the problem of stability of the H^∞-compensator if the plant has right-half plane zeros. Another line of research is devoted to the relationship between H^∞- and the more conventional LQG- or $\mathcal{L}_2$-optimisation, see eg Grimble (1986).

4 A SURVEY OF THE METHODS FOR ANALYSING AND DESIGNING UNCERTAIN FEEDBACK SYSTEMS

In this chapter we want to survey the known principles and methods for analysing uncertain feedback systems and designing robust controllers. This intention turns out to be very intricate, because in the current state of research there is no general theoretical background of all these methods. On the contrary, the known approaches are highly divergent. They presuppose different assumptions concerning the form of the model uncertainties and the information available to describe them, they deal with different design requirements and aspects of robustness, and — last but not least — they are based on different systems theoretical concepts.

In order to describe reasonably accurately and concisely the essential features of the different lines of research, we look at the field of robust control from two points of view. At first, we consider the *results* received and classify them into qualitative and quantitative robustness issues. Then we survey the *methods* used and group them according to their systems theoretical background. The survey should provide the reader with a classification scheme which enables him to understand the relations between the different approaches that we are going to explain in the following chapters according to the modelling, analysis and design concepts they use.

4.1 Classification according to the kind of robustness considered

As opposed to linear control in general, robust control methods consider the robustness requirement (IV) as a primary aim of design. To which degree this new requirement affects the design process and the results depends on the amount of model uncertainties compared with the width of tolerances admitted for the closed-loop system properties. The design process and its solution are mainly influenced by the robustness requirement, if the model uncertainties are large. Then, quantitatively specified relations between the amount of uncertainties and the properties to be attained dominate the design process. If, on the other hand, rather small model uncertainties are encountered, the robustness requirement plays a minor role and the design engineer can centre his attention primarily on the dynamics specifications

(III), while observing some general guidelines concerning the robustness improvement.

These situations motivate two different lines of theoretical investigation. In *qualitative robustness* considerations, the system properties are to be preserved in some neighbourhood of the nominal parameter vector $\hat{\mathbf{a}}$. No effort is made to give quantitative bounds on the permissible deviation $\mathbf{a} - \hat{\mathbf{a}}$. *Quantitative robustness* has regard to the amount of model uncertainties and concerns the question whether the properties of interest are preserved under a completely specified class $\mathcal{A}$ of plant perturbations. In what follows, this classification is used to get some order into the results of robustness considerations although there is no sharp boundary between both classes, and between qualitative robustness and sensitivity.

Qualitative robustness

Qualitative robustness considerations tackle the question which kind of systems may have robustness properties:

Given: Structural information about the plant and the model uncertainties, nominal parameter vector $\hat{\mathbf{a}}$

Find: Conditions on the control configuration for the existence of robust controllers,
General guidelines to the choice of controller parameters so as to ensure robustness.

The results hold for a large class of systems with the given structure but it remains unknown how large this class actually is. In contrast to the sensitivity approach, qualitative robustness refers to substantial model uncertainties and the preservation of the properties within given tolerances.

Existence of robust controllers

The question whether a constant-gain controller may ensure closed-loop stability for all plants of a given class regards mainly the structure of the plant and of the model uncertainties. The known conditions range from rigorous statements that hold for 'almost all' perturbations of a certain class to specific results which concern a special controller structure or a particular class of uncertainties.

An example of the former kind of results is given by Theorem 8.16. Accordingly, if the entry point of the model uncertainties is restricted by some '*matching conditions*' the system is robustly stabilisable for arbitrarily large uncertainties $\delta\mathbf{A}$ of the system matrix. Since very large perturbations are permitted, conditions of this kind refer to uncertainties whose range of influence on the plant might be considered, from a practical point of view, as highly restricted.

More practical results refer to particular plant properties or controller structures. If the transmission zeros of the plant are known to lie in the left-half complex plane, robust stabilisation can be ensured by high-gain state feedback (Section 6.4). Robust PI-controllers can be found if the plant is stable and the static characteristics are sufficiently accurately known (Section 10.1) or if the transmission zeros of the possibly unstable plant lie in the left-half plane (Theorem 6.5).

Most of the results represent conditions of a form, that enables the designer to check the existence of appropriate controller parameters once the control configuration, the controller structure and the class of model uncertainties are given. Only very few results, as eg Theorem 7.2, help to determine a suitable controller structure. Until now, shaping the controller structure remained an open question to be solved by experience.

The reason for this situation can be outlined by considering the stabilisation problem. For fixed plant parameters, the necessary controller dynamics are given by the minimal-order observer which, due to the separation theorem, can be designed independently of the state feedback. By contrast, for incompletely known plants the separation principle does not hold and the complete controller has to be designed in a common step (see examples in Section 7.1).

Guidelines for designing robust controllers

Qualitative results tell the design engineer for which controller parameters the closed-loop system tends to be robust and, consequently, can be expected to tolerate the given uncertainties. For example, as the stability margin represents a measure of the degree of robustness, it is reasonable to select controller matrices that maximise or at least enlarge the stability margin of the approximate closed-loop system (Section 6.1). Sensitivity optimisation yields some robustness because systems with low sensitivity functions are likely to tolerate substantial plant perturbations (Section 3.4). Singular perturbation methods use a low-order approximate model to design a controller that satisfies the design requirements in spite of unmodelled quick dynamics (Section 6.3). High-gain feedback methods rely on the realisation that the controller will be sufficiently robust if the loop gain is made sufficiently high (Section 6.4). The last two methods take advantage of qualitative robustness results as far as they do not refer to bounds of the tolerable additional dynamics or the least necessary loop gain, respectively.

Robust solution of the servomechanism problem

According to the Internal Model Principle the controller must necessarily include a model of the external signals if command tracking and disturbance attenuation are to be ensured by appropriately choosing the remaining controller parameters. Thus, a certain part of the controller (the servocompensator) has to be fixed in dependence upon the command and disturbance signals under consideration. The second part of the controller (stabilising compensator) has to be designed so as to stabilise the whole system.

The importance of this result for control system design is based on the robustness behind it. We use a strong internal model, that is, roughly speaking, we do not reduce the model dynamics due to particular properties of the nominal plant. Then, asymptotic regulation occurs in spite of all plant perturbations that do not destabilise the closed-loop system. This robustness makes it possible to ignore the design requirement (II) in the stabilisation problem once the unstable servocompensator has been chosen. Therefore, we need not deal with servomechanism problems in more detail but can focus our attention on the stabilisation problem which follows the servocompensator design. However, these assertions hold only for linear original systems. Similar results on nonlinear or time-varying systems are rare.

Generic properties of feedback systems

A field of research which is closely related to qualitative robustness analysis concerns structural (generic) properties of control systems, which hold for 'almost all' systems of a given class. That is, systems that do not have these properties are encountered very seldom or, mathematically more exactly, lie on some hypersurface in the parameter space. Well-known examples of such properties are the controllability and the observability of linear time invariant systems.

Until now nearly all of these results concern analysis problems and only a few provide some help for an effective design of robust controllers. Therefore, we will not deal with such investigations in more detail. Basic results can be found in (Davison and Wang, 1973; Tchon, 1983 Reinschke 1988) or papers on graphtheoretic methods of control systems analysis.

Quantitative robustness

Quantitative robustness considerations aim at a direct relationship between quantitatively specified plant uncertainties and the resulting closed-loop tolerances. The following problem is tackled:

Given: Quantitative information about the plant and the model uncertainties

Find: Conditions on the model accuracy for the existence of robust controllers,
Systematic methods for designing robust controllers.

Existence of robust controllers

Starting from a completely specified set $\mathcal{A}$ of model uncertainties, and, in most cases, from a given controller structure, conditions on the uncertainties are derived for a robust feedback to exist. Several results of this kind will be presented in Sections 7.1, 10.1 and 11.2. For example, if the original plant is stable, stability-robust PI-controllers exist whenever the uncertainties of the static reinforcement do not prevent the controller from being definitely chosen so as to ensure negative feedback (Section 10.1). Decentralised controllers for which the stability of the closed-loop system is not jeopardised by disconnecting some control stations from the plant do exist provided that the subsystems of the plant satisfy an autonomy condition (Section 11.2).

Methods for designing robust controllers

Most of the design procedures for finding a suitable compromise between all the given design requirements are based on quantitative robustness investigations. They are reviewed in the following section.

4.2 Survey of the design methods

The organisation of this survey of the methods for designing robust feedback controllers corresponds with the methodical background of the methods and does not consider whether these methods result in qualitative or quantitative statements on the robustness. As a matter of fact, most of the methods yield a mixture of qualitative and quantitative results.

Root locus techniques

A classical method for designing robust controllers is based on root locus techniques. It exploits the fundamental assertion that the poles of the closed-loop system approach the zeros of the open loop if the loop gain increases. Accordingly, uncertain linear systems whose zeros lie in the left half complex plane (minimum-phase systems) can be stabilised by using a sufficiently high controller gain (Section 6.4).

The merits and shortcomings of this high-gain philosophy are the subject of discussions among practitioners. The decoupling effect of different control loops seems to be the most favourable byproduct of the use of high gains, whereas the restriction to minimum-phase systems, the possible appearance of highly undamped closed-loop oscillations, and the technical problems of ensuring the high loop gain in a sufficiently broad frequency band make an application of this philosophy in process control questionable.

Designing controllers with a high stability margin

The stability margin describes how much the gain or phase of the open-loop system may vary without causing instability in the closed loop. Its suitability as a robustness measure of single-variable systems initiated a large amount of research work for generalising the notion of stability margin to multivariable systems. For example, the optimal regulator was proved to have an infinite gain margin, at least 50% gain reduction tolerance and at least $\pm 60°$ phase margin in each control input channel (Section 6.1). However, since the stability margin represents merely an indirect indicator of robustness, there is no general relation between its value and the amount of model uncertainties tolerated. Simple examples demonstrate that relatively small parameter perturbations may cause instability despite a large stability margin.

Feedback design in presence of unknown-but-bounded uncertainties

Most of the recent progress in robust feedback control has been made for systems whose uncertainties are described by upper bounds. Likewise, the design specifications are given by tolerances of the acceptable closed-loop system response. In the *quantitative feedback design* of single-input single-output systems the open-loop transfer function and its tolerances are transformed into the closed-loop counterparts using the Nichols chart (Section 8.1). For multi-input multi-output systems the problem becomes much more involved because the choice of a suitable measure of the plant tolerances turned out to be a crucial step. The more detailed the uncertainties may be described by the form of presentation used the more information about the uncertainties must be available and the more difficult is the analysis of the closed-loop system. The *singular value approach* yields necessary and sufficient stability conditions and bounds of the I/O-behaviour of the closed loop which can be easily derived but refer merely to scalar bounds (Sections 8.3, 8.5). By using *element-by-element bounds* of the uncertainties a more detailed characterisation of the closed-loop system may by obtained. However, the derivation and use of necessary and sufficient conditions for robust stability and the determination of tight bounds of the system response pose severe problems (Sections 8.4, 8.5). Further methods of this kind use *sector bounds* of the uncertainties or even generally bounded sets of the matrices which occur in the state equations (Sections 5.3 and 8.7). All the design procedures have the advantage of producing explicit statements on the behaviour of the closed-loop original system, which are directly related to the design requirements (I)—(IV). However, the results may suffer from a considerable degree of conservatism, ie the stability conditions may be too strong and the bounds of the system response too broad.

The multimodel approach

Assuming that the plant can be described by a few different parameter vectors a controller has to be found that satisfies the design specifications for all these modes of plant operation. That is, the controller has to simultaneously stabilise all these models or even to simultaneously produce a satisfactory closed-loop system response

The parameter space design starts from the state space model of the plant and different tuples ($\mathbf{A}^i$, $\mathbf{B}^i$, $\mathbf{C}^i$) and aims at constructing a state feedback which shifts the closed-loop system poles into a certain region of the complex plane. As this design principle is based on the determination of all rather than a single controller for each of the plant models, robust controllers are received unless no such controllers exist (Section 7.2).

General results on the simultaneous stabilisability are obtained on the basis of the algebraic theory of linear control systems, using a polynom matrix representation of the plant and the controller (Section 7.1).

5 MODELS OF UNCERTAIN SYSTEMS

5.1 The methodology of modelling uncertain systems

The aim of modelling is to set up a mathematical description

$$\hat{\mathbf{y}} = \hat{\mathbf{G}}\mathbf{u} \tag{1}$$

whose I/O-behaviour matches the I/O-behaviour of the real-existing process

$$\mathbf{y} = \mathbf{G}\mathbf{u} \tag{2}$$

as well as possible. Here, we want to deal with systems that can only be incompletely modelled and begin our considerations with the introduction of some important terms, although some of them have already been used in earlier chapters.

Systems are dealt with as **uncertain systems**, if the model leads to great, not neglectable differences between the **original system** output $\mathbf{y}$ and the model output $\hat{\mathbf{y}}$. For such systems eqn (1) can only be referred to as an **approximate model**. The model error

$$\mathbf{f} = \mathbf{y} - \hat{\mathbf{y}} \tag{3}$$

can be symbolically represented by an **error model**

$$\mathbf{f} = \delta\mathbf{G}\mathbf{u} \tag{4}$$

with

$$\delta\mathbf{G} = \mathbf{G} - \hat{\mathbf{G}}. \tag{5}$$

The error model is used to indicate *that* the I/O-behaviour of the original system (2) is not exactly described by the model output $\hat{\mathbf{y}}$. It is marked in Fig. 5.1 by a grey box which should remind us of the uncertainties within this system part. In the simplest form of the uncertain system description the error model (4) lies in parallel with the approximate model (1) and tells us that the output error $\mathbf{f}$ depends on the input $\mathbf{u}$.

The difficulty of modelling uncertain systems consists in finding an appropriate description of the model uncertainties or, equivalently, of the error model (4) in order to indicate *how far* the original behaviour can deviate from the approximate description. Here, we use the **set-theoretic approach**, ie we determine a set of possible representatives of the error model or the overall model, respectively. This can be explained in the best way by describing the dependence of the system performance upon uncertain phenomena within the system as the dependence of the transfer function $\mathbf{G}(p)$ upon a parameter vector $\mathbf{a}$. The parameter vector $\mathbf{a}$ is assumed to belong to the set $\mathcal{A}$. Hence, the original system belongs to the set $\mathcal{G}$

$$\mathcal{G} = \{\mathbf{G}(p, \mathbf{a}) : \mathbf{a} \in \mathcal{A}\}. \tag{6}$$

Each member of $\mathcal{G}$ may represent the original system but we have no information about which member it really does. The only thing we know is that the original system $\mathbf{G}_0$ belongs to $\mathcal{G}$

$$\mathbf{G}_0 \in \mathcal{G}. \tag{7}$$

This set-theoretic approach is motivated by several circumstances of the engineering practice. First, our analysis and design results have to be valid for a uniquely given system. For example, if we intend to design a stabilising controller, then we must be sure that the controller will actually stabilise the given original plant. We are not content with knowing that stability will occur with a certain probability if the controller was applied to many plants all belonging to $\mathcal{G}$. Hence, we have to take into account all possible representations of the original system.

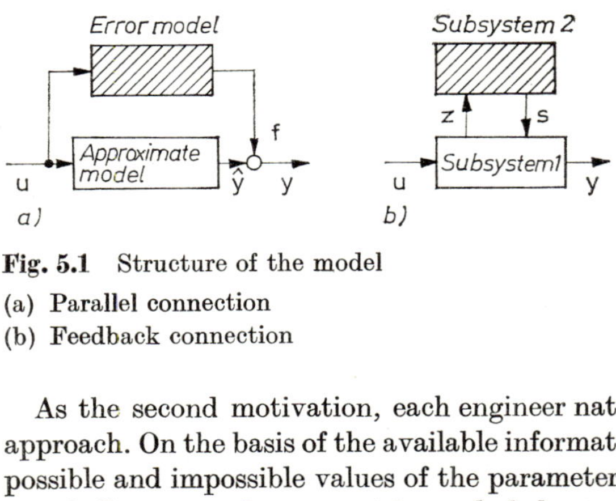

Fig. 5.1 Structure of the model
(a) Parallel connection
(b) Feedback connection

As the second motivation, each engineer naturally uses the set-theoretic approach. On the basis of the available information he distinguishes between possible and impossible values of the parameter vector $\mathbf{a}$. Let us denote the set of all systems that cannot be excluded as representations of the original system by $\bar{\mathcal{G}}$. It is the smallest set of systems that must be considered. To model the uncertain original system means to bring this set $\bar{\mathcal{G}}$ into a standard form $\mathcal{G}$ that can be tackled directly by the analysis and design methods elaborated for models of this kind.

In the alternative probabilistic approach, additional information about the likelihood of the appearance of the possible values is necessary. That is, each member of the set $\mathcal{G}$ is attached with a weighting factor.

For the purpose of robust control, two principal ways of specifying the set $\mathcal{G}$ are used. They will be explained in the next two sections for frequency domain and time domain models of linear systems. More general considerations have shown that set-theoretic models can be extended to time-varying or nonlinear original systems. This should be outlined by some remarks on static nonlinear systems at the end of Section 5.3.

5.2 Specimens of the original system

In the first way of modelling uncertain systems, the set $\mathcal{A}$ consists of q different samples of the uncertain parameter vector $\mathbf{a}$. Hence, the set $\mathcal{G}$ encloses q different models $\mathbf{G}^i = \mathbf{G}(\mathbf{a}^i)$, each of which describes a possible mode of operation of the original system

$$\mathcal{G} = \{\mathbf{G}^1, \mathbf{G}^2, \ldots, \mathbf{G}^q\}. \tag{1}$$

For instance, if the system is described by a state space model

$$\begin{aligned} \dot{\mathbf{x}} &= \mathbf{A}\mathbf{x} + \mathbf{B}\mathbf{u}, \qquad \mathbf{x}(0) = \mathbf{x}_0 \\ \mathbf{y} &= \mathbf{C}\mathbf{x} + \mathbf{D}\mathbf{u} \end{aligned} \tag{2}$$

which has been derived by a linearisation of the more exact nonlinear model

$$\begin{aligned} \dot{\bar{\mathbf{x}}} &= \mathbf{f}(\bar{\mathbf{x}}, \bar{\mathbf{u}}), \qquad \mathbf{x}(0) = \mathbf{x}_0 \\ \bar{\mathbf{y}} &= \mathbf{g}(\bar{\mathbf{x}}, \bar{\mathbf{u}}) \end{aligned}$$

around some operating point $(\mathbf{x}^o, \mathbf{u}^o)$ with $\mathbf{f}(\mathbf{x}^o, \mathbf{u}^o) = \mathbf{0}$, then the matrices $\mathbf{A}$, $\mathbf{B}$, and $\mathbf{C}$, $\mathbf{D}$

$$\mathbf{A} = \left.\frac{\partial \mathbf{f}(\bar{\mathbf{x}}, \bar{\mathbf{u}})}{\partial \bar{\mathbf{x}}}\right|_{\mathbf{x}^o,\mathbf{u}^o} \qquad \mathbf{B} = \left.\frac{\partial \mathbf{f}(\bar{\mathbf{x}}, \bar{\mathbf{u}})}{\partial \bar{\mathbf{u}}}\right|_{\mathbf{x}^o,\mathbf{u}^o}$$

$$\mathbf{C} = \left.\frac{\partial \mathbf{g}(\bar{\mathbf{x}}, \bar{\mathbf{u}})}{\partial \bar{\mathbf{x}}}\right|_{\mathbf{x}^o,\mathbf{u}^o} \qquad \mathbf{D} = \left.\frac{\partial \mathbf{g}(\bar{\mathbf{x}}, \bar{\mathbf{u}})}{\partial \bar{\mathbf{u}}}\right|_{\mathbf{x}^o,\mathbf{u}^o}$$

depend on the point $(\mathbf{x}^o, \mathbf{u}^o)$. If the system is considered under different operating conditions, different sets $(\mathbf{A}^i, \mathbf{B}^i, \mathbf{C}^i, \mathbf{D}^i)$ and, thus, different models (2) result from this linearisation.

In general, all the specimens $\mathbf{G}^i$ of the system behaviour may exhibit completely different dynamical properties. They have no known relationship between each other. Therefore, this form of representation of the system is particularly appropriate to systems with structural uncertainties or large parameter deviations. Besides this, the model (1) is useful for explorative investigations in which only few specimens of the system behaviour should be included. Then the set $\mathcal{G}$ comprises merely extreme rather than all possible modes of plant operation or a few linear models as local substitutes of a globally valid nonlinear model.

5.3 Systems with unknown-but-bounded uncertainties

The second method represents a particular but rather convenient form of the set-theoretic model. The set $\mathcal{A}$ or $\mathcal{G}$ represents a bounded continuum of parameter vectors or models, respectively. To describe the model uncertainties we only have to describe the boundary of $\mathcal{A}$. Then all systems whose parameter vector $\mathbf{a}$ lies within this bound belong to the set $\mathcal{G}$. The set $\mathcal{G}$ contains an infinite number of representations of the original system but they all have a common relationship given by the hull of $\mathcal{A}$.

In a very general form, this philosophy is realised in the state space model

$$\dot{\mathbf{x}} = \mathbf{A}(\mathbf{a})\, \mathbf{x} + \mathbf{B}(\mathbf{a})\, \mathbf{u} \tag{1}$$

$$\mathbf{y} = \mathbf{C}(\mathbf{a})\, \mathbf{x} + \mathbf{D}(\mathbf{a})\, \mathbf{u} \tag{2}$$

where $\mathbf{a} \in \mathcal{A} \subseteq \mathbb{R}^q$ holds with $\mathcal{A}$ being a closed connected subset of the q-dimensional parameter space. In a more convenient but slightly more restricted form, the set $\mathcal{A}$ is described by a nominal value $\hat{\mathbf{a}}$ and some bounds on the deviation $\mathbf{a} - \hat{\mathbf{a}}$. In the following we want to introduce the general model structure first and give several possibilities for describing the bound of $\mathcal{A}$ in a form convenient for analysing such systems. To begin with we restrict our consideration to frequency domain models.

The model structure

In a very useful generalisation of the model structure of Fig. 5.1(a) the original system is represented by two interconnected subsystems (Fig. 5.1(b)). Subsystem 1 (SS 1) describes the completely known part

$$\begin{aligned} \mathbf{y}(p) &= \hat{\mathbf{G}}(p)\, \mathbf{u}(p) + \mathbf{G}_{ys}(p)\, \mathbf{s}(p) \\ \mathbf{z}(p) &= \mathbf{G}_{zu}(p)\, \mathbf{u}(p) + \mathbf{G}_{zs}(p)\, \mathbf{s}(p) \end{aligned} \tag{3}$$

ie the transfer function matrices $\hat{\mathbf{G}}$, $\mathbf{G}_{ys}$, $\mathbf{G}_{zu}$ and $\mathbf{G}_{zs}$ are assumed to be known. $\mathbf{s}$ and $\mathbf{z}$ are interconnection signals. Ignoring the influence of subsystem 2 on subsystem 1 ($\mathbf{s} = \mathbf{0}$), we get

$$\hat{\mathbf{y}}(p) = \hat{\mathbf{G}}(p)\, \mathbf{u}(p)\,. \tag{4}$$

Hence, eqn (3) for $\mathbf{s} = \mathbf{0}$ or, equivalently, eqn (4) represents the approximate model of the original system. (We will sometimes refer to the whole model (3) as the approximate model.)

Subsystem 2 (SS 2) describes the uncertain parts of the original system. If the system is linear, this subsystem can be described by some I/O-relation

$$\mathbf{s}(p) = \mathbf{G}_2(p)\, \mathbf{z}(p) \tag{5}$$

but the transfer function matrix $\mathbf{G}_2(p)$ is not assumed to be known. Eqn (5) is used merely to designate the subsystem 2 as multivariable dynamical system with input $\mathbf{z}$ and output $\mathbf{s}$. For the original system (3), (5) the error model

$$\mathbf{f}(p) = \delta\mathbf{G}(p)\, \mathbf{u}(p) \tag{6}$$

is given with

$$\delta\mathbf{G}(p) = \mathbf{G}_{\mathbf{ys}}(p)\, \mathbf{G}_2(p) \left(\mathbf{I} - \mathbf{G}_{\mathbf{zs}}(p)\, \mathbf{G}_2(p)\right)^{-1} \mathbf{G}_{\mathbf{zu}}(p) \tag{7}$$

provided that the inverse matrix exists.

The main advantage of the model (3), (5) is that it can be easily adapted to the sources of the model uncertainties. If the signals $\mathbf{s}$ and $\mathbf{z}$ and the transfer function matrices $\mathbf{G}_{\mathbf{ys}}$, $\mathbf{G}_{\mathbf{zu}}$ and $\mathbf{G}_{\mathbf{zs}}$ are selected appropriately, eqn (5) comprises the uncertain parts of the original system, and no less and no more. Fig. 5.2 shows special model structures some of which will be used later. In particular, Fig. 5.2(a) and (c) give the often used models in which the uncertainties occur additively or multiplicatively, respectively. In all figures the transfer function matrix occuring in the approximate model (4) is denoted by $\hat{\mathbf{G}}$ elsewhere.

As can be seen from eqn (7), the uncertain subsystem (5) does not directly provide a description of the output error $\mathbf{f}$. To the contrary, the ways in which the signal $\mathbf{z}$ is formed and the signal $\mathbf{s}$ influences the behaviour of subsystem 1, respectively, determine how the model uncertainties $\mathbf{G}_2$ are transformed into the description of the model error. To underline the difference between the model uncertainties as the source of the mismatch of model and original system on the one hand and the output error as the result of the uncertainties on the other hand, eqns (5), (6) will be referred to as **uncertainty model** or error model, respectively (for example see Fig. 5.4).

The uncertainty bound

Now, the question arises how to describe the uncertainty model (5). We want to do this by means of upper bounds of $\mathbf{G}_2(p)$. If eqn (5) represents a single-input single-output system with a scalar transfer function $g_2(p)$, the relation

$$|s(p)| \leqq |g_2(p)|\, |z(p)|$$

can be easily derived. If we know some upper bound $v_2(p)$ of $|g_2(p)|$, ie

$$v_2(p) \geqq |g_2(p)| \qquad \text{for all } p, \tag{8}$$

then $v_2(p)\,|z(p)|$ yields an upper bound of $|s(p)|$

$$|s(p)| \leqq v_2(p)\, |z(p)|\,.$$

This way for describing the uncertainty model (5) can be interpreted as using an auxiliary system

$$r = v_2 w \tag{9}$$

Model structure	Decomposition of the original system $y=Gu$	Description of SS2: $s=G_2z$ Description of SS1:	$\Sigma=\{G=\hat{G}+\delta G: \lvert G_2 \rvert \leq V_2$ and $\delta G=\dots\}$	Uncertainty model G_2
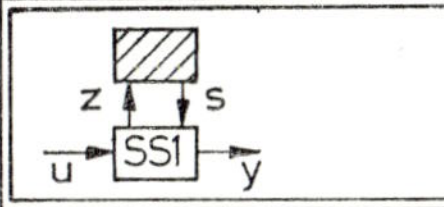	$G=\hat{G}+G_{ys}G_2(I-G_{zs}G_2)^{-1}G_{zu}$	$y=\hat{G}u+G_{ys}s$ $z=G_{zu}u+G_{zs}s$	$\delta G=G_{ys}G_2(I-G_{zs}G_2)^{-1}G_{zu}$	$G_2=(G_{zu}(G-\hat{G})^{-1}G_{ys}+G_{zs})^{-1}$
a) 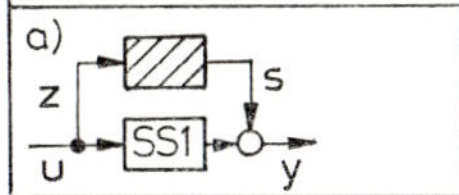	$G=\hat{G}+G_2$	$y=\hat{G}u+s$ $z=u$	$\delta G=G_2$	$G_2=G-\hat{G}$
b) 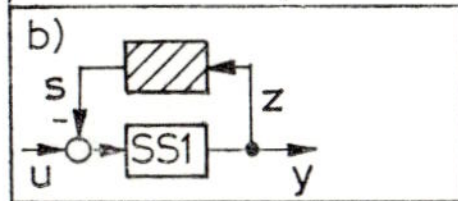	$G=\hat{G}(I+G_2\hat{G})^{-1}$	$y=\hat{G}u-\hat{G}s$ $z=\hat{G}u-\hat{G}s$	$\delta G=-\hat{G}G_2(I+\hat{G}G_2)^{-1}\hat{G}$	$G_2=G^{-1}-\hat{G}^{-1}$
c) 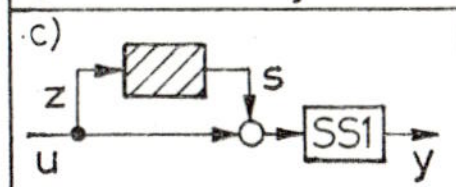	$G=\hat{G}(I+G_2)$	$y=\hat{G}u+\hat{G}s$ $z=u$	$\delta G=\hat{G}G_2$	$G_2=\hat{G}^{-1}G-I$
d) 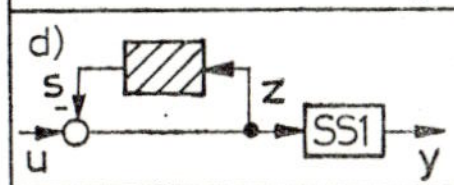	$G=\hat{G}(I+G_2)^{-1}$	$y=\hat{G}u-\hat{G}s$ $z=u-s$	$\delta G=-\hat{G}G_2(I+G_2)^{-1}$	$G_2=G^{-1}\hat{G}-I$
e) 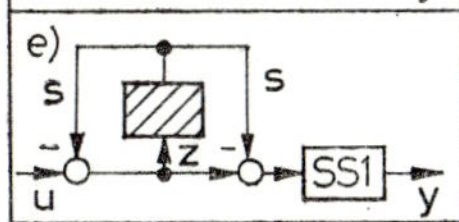	$G=\hat{G}(I-G_2)(I+G_2)^{-1}$	$y=\hat{G}u-2\hat{G}s$ $z=u-s$	$\delta G=-2\hat{G}G_2(I+G_2)^{-1}$	$G_2=(\hat{G}+G)^{-1}(\hat{G}-G)$

SS1 = Subsystem 1

SS2 = Subsystem 2 (Uncertainty model)

Fig. 5.2 General and special structures of the model

that — if applied with $w = |z|$ — majorises the absolute value of the uncertainty model output s

$$v_2 \, |z| = r \geqq |s| = |g_2 z| \, .$$

The system (9) is called **comparison system**. A linear system $r = v_2 w$ is a comparison system of the linear uncertainty model (5) if and only if the transfer function v_2 satisfies the relation (8).

For multi-input multi-output systems we have to explain what $|\mathbf{s}(p)|$, $|\mathbf{z}(p)|$, and $|\mathbf{G}_2(p)|$ mean. Two ways are commonly used. The first is an **element-by-element interpretation**, where $|\mathbf{s}(p)|$, $|\mathbf{z}(p)|$, and $|\mathbf{G}_2(p)|$ mean that each entry of $\mathbf{s}(p)$, $\mathbf{z}(p)$ and $\mathbf{G}_2(p)$ is replaced by its absolute value. Then

$$|\mathbf{s}(p)| \leqq \mathbf{V}_2(p) \, |\mathbf{z}(p)| \tag{10}$$

holds if the matrix $\mathbf{V}_2(p)$ satisfies the relation

$$\mathbf{V}_2(p) \geqq |\mathbf{G}_2(p)| \, . \tag{11}$$

In eqns (10) and (11) the relation $\geqq$ applies to all the corresponding elements of the vectors or matrices, respectively. That is, eqn (11) is equivalent to

$$v_{2ij}(p) \geqq |g_{2ij}(p)| \qquad (i = 1, \ldots, \dim \mathbf{s}\,; j = 1, \ldots, \dim \mathbf{z})$$

for all p, where v_{2ij} and g_{2ij} are the entries of the matrices $\mathbf{V}_2$ and $\mathbf{G}_2$, respectively. The use of element-by-element bounds of $\mathbf{G}_2(p)$ is synonymous with using a multivariable comparison system $\mathbf{r}(p) = \mathbf{V}_2(p) \, \mathbf{w}(p)$ for which

$$\mathbf{V}_2(p) \, |\mathbf{z}(p)| = \mathbf{r}(p) \geqq |\mathbf{s}(p)|$$

holds. Due to eqn (10) the uncertainty model (4) is known to belong to the set

$$\mathbf{G}_2 \in \mathscr{G}_2 = \{\mathbf{G}_2(p) \colon |\mathbf{G}_2(p)| \leqq \mathbf{V}_2(p)\} \, . \tag{12}$$

The second way is a **norm interpretation**. From (5) we have

$$\|\mathbf{s}(p)\| \leqq \|\mathbf{G}_2(p)\|_{\mathrm{M}} \, \|\mathbf{z}(p)\|$$

where $\|.\|$ and $\|.\|_{\mathrm{M}}$ denote a vector norm and the induced matrix norm $\|\mathbf{A}\|_{\mathrm{M}} = \sup\limits_{\|\mathbf{x}\|=1} \|\mathbf{A}\mathbf{x}\|$. Both norms are used for every frequency p. If a scalar function $v_2(p)$ is known that provides an upper bound of $\|\mathbf{G}_2(p)\|_{\mathrm{M}}$

$$v_2(p) \geqq \|\mathbf{G}_2(p)\|_{\mathrm{M}} \tag{13}$$

then an upper bound of $\|\mathbf{s}(p)\|$ is given by

$$\|\mathbf{s}(p)\| \leqq v_2(p) \, \|\mathbf{z}(p)\| \, . \tag{14}$$

The use of norms leads to single-input single-output comparison systems (9) whose transfer function satisfies the relation (13).

Very often, the Euclidean vector norm

$$\|\mathbf{x}\|_2 = \sqrt{\sum_{i=1}^{n} |x_i|^2} \tag{15}$$

is used. The induced matrix norm is the spectral norm

$$\|\mathbf{A}\|_M = \sqrt{\lambda_{\max}[\mathbf{A}^*\mathbf{A}]} = \sigma_{\max}[\mathbf{A}] \tag{16}$$

which corresponds to the maximum singular value of $\mathbf{A}$ (cf Appendix 2). Then $v_2(p)$ has to satisfy the inequality

$$\sigma_{\max}[\mathbf{G}_2(p)] \leqq v_2(p) \tag{17}$$

which leads to the term '*singular value approach*' for the norm-interpretation of $|\mathbf{G}_2|$.

The set of uncertainty models is given by

$$\mathbf{G}_2 \in \mathscr{S}_2 = \{\mathbf{G}_2(p): \|\mathbf{G}_2(p)\|_M \leqq v_2(p)\}\,. \tag{18}$$

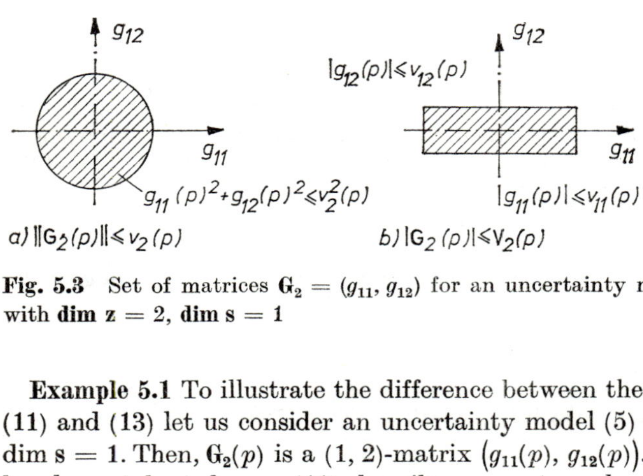

Fig. 5.3 Set of matrices $\mathbf{G}_2 = (g_{11}, g_{12})$ for an uncertainty model with **dim** $\mathbf{z} = 2$, **dim** $\mathbf{s} = 1$

Example 5.1 To illustrate the difference between the uncertainty bounds (11) and (13) let us consider an uncertainty model (5) with dim $\mathbf{z} = 2$ and dim $\mathbf{s} = 1$. Then, $\mathbf{G}_2(p)$ is a (1, 2)-matrix $\big(g_{11}(p), g_{12}(p)\big)$. Using the element-by-element bound, eqn (11) describes a rectangular area $\mathbf{G}_2$ shown in Fig. 5.3(b) since $|g_{11}(p)| \leqq v_{11}(p)$ and $|g_{12}(p)| \leqq v_{12}(p)$. All (1, 2)-matrices lying inside this rectangle are possible realisations of the model uncertainty $\mathbf{G}_2(p)$. On the other hand, eqn (18) yields a circular, elliptical or quadratic area for $\mathbf{G}_2(p)$ whose form depends upon the norm definition. Fig. 5.3(a) shows the circular area $g_{11}^2 + g_{12}^2 \leqq v_2{}^2$ received for the spectral norm (16). Obviously, the model uncertainty $\mathbf{G}_2(p)$ can be described in more detail by eqn (10) than by eqn (13). Whereas the scalar v_2 in eqn (13) mainly depends on the element of $\mathbf{G}_2$ that is worst known, the choice of the matrix $\mathbf{V}_2$ has regard to different ranges of the elements of $\mathbf{G}_2(p)$. □

The model of the original system

The original system is described by eqns (3), (5) and (11) or (13), respectively. That is, the set of candidates is given by

$$\mathscr{G} = \{\mathbf{G} = \hat{\mathbf{G}} + \delta\mathbf{G},\ \delta\mathbf{G} = \mathbf{G}_{ys}\mathbf{G}_2(\mathbf{I} - \mathbf{G}_{zs}\mathbf{G}_2)^{-1}\,\mathbf{G}_{zu} \text{ with } |\mathbf{G}_2| \leqq \mathbf{V}_2\} \tag{19}$$

or

$$\mathscr{G} = \{\mathbf{G} = \hat{\mathbf{G}} + \delta\mathbf{G},\ \delta\mathbf{G} = \mathbf{G}_{ys}\mathbf{G}_2(\mathbf{I} - \mathbf{G}_{zs}\mathbf{G}_2)^{-1}\,\mathbf{G}_{zu} \text{ with } \|\mathbf{G}_2\|_M \leqq v_2\}. \tag{20}$$

These equations show the difference between the uncertainty description (11), (13) and the model error description (7).

To continue with **Example 5.1** consider an original system with two inputs and one output and the (1, 2)-uncertainty model from Fig. 5.3(b). Then

$$\begin{aligned}
\delta\mathbf{G} &= g_{ys}(g_{11}\ g_{12})\left[\mathbf{I} - \begin{pmatrix} g_{zs11} \\ g_{zs21} \end{pmatrix}(g_{11}\ g_{12})\right]^{-1}\begin{pmatrix} g_{zu11} & g_{zu12} \\ g_{zu21} & g_{zu22} \end{pmatrix} \\
&= g_{ys}\left(\frac{g_{11}}{1 - g_{11}g_{zs11} - g_{12}g_{zs21}},\ \frac{g_{12}}{1 - g_{11}g_{zs11} - g_{12}g_{zs21}}\right)\begin{pmatrix} g_{zu11} & g_{zu12} \\ g_{zu21} & g_{zu22} \end{pmatrix} \\
&= \left(g_{ys}\frac{g_{11}g_{zu11} + g_{12}g_{zu21}}{1 - g_{11}g_{zs11} - g_{12}g_{zs21}},\ g_{ys}\frac{g_{11}g_{zu12} + g_{12}g_{zu22}}{1 - g_{11}g_{zs11} - g_{12}g_{zs21}}\right).
\end{aligned} \tag{21}$$

For each value of g_{11} and g_{21} within the rectangular or circular regions of Fig. 5.3, respectively, we get a model error $\delta\mathbf{G}$. The set of all possible (1, 2)-matrices $\delta\mathbf{G} = (\delta g_{11}\ \delta g_{12})$ is depicted in Fig. 5.4. The region of $\delta\mathbf{G}$ is no longer rectangular or circular, the edge of this box is no longer oriented along the coordinate axes, and $\delta\mathbf{G} = \mathbf{0}$ does not lie in the centre of the region. This

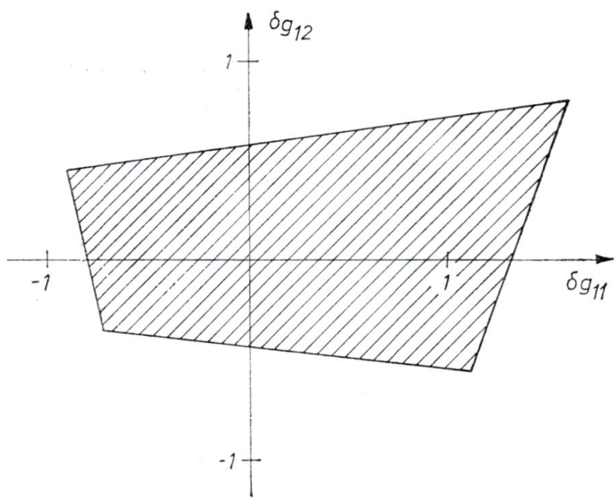

Fig. 5.4 Region of realisations of g_{11}, g_{12} for $v_{11} = 1$, $v_{12} = 0.5$, $g_{zs11} = g_{zs21} = 0.25$, $g_{ys} = 1$, $g_{zu11} = g_{zu22} = 1$, $g_{zu21} = g_{zu12} = 0$

deformation occurs although we have chosen very simple matrices $\mathbf{G}_{ys}$, $\mathbf{G}_{zs}$ and $\mathbf{G}_{zu}$ and have not taken advantage of all the freedom the model (3) and (5) offers. □

Example 5.2 To illustrate the representation of the original system by some approximate model and an uncertainty bound we consider the first-order system $g(p) = 1/(p - \lambda)$. The pole λ is only known to lie within the interval $\lambda_u \leqq \lambda \leqq \lambda_o$. The middle of this interval is used as pole of the approximate model $\hat{g}(p) = 1/(p - \hat{\lambda})$, where $\hat{\lambda} = (\lambda_o + \lambda_u)/2$.
Using the parallel model of Fig. 5.2(a) the uncertainty model is

$$g_2(p) = g(p) - \hat{g}(p) = \frac{\lambda - \hat{\lambda}}{(p - \lambda)(p - \hat{\lambda})}.$$

Note that the uncertainty model has order two although the original system is only of first order. This confirms the general observation that the dynamical order of the uncertainty model is often much larger than the original system order.

To bound the uncertainty we have to find some $v_2(p)$ for which

$$v_2(\mathrm{j}\omega) \geqq \left| \frac{\lambda - \hat{\lambda}}{(\mathrm{j}\omega - \lambda)(\mathrm{j}\omega - \hat{\lambda})} \right|$$

holds. Assume that $0.35 \leqq \lambda \leqq 0.65$ is known. Then

$$v_2(\mathrm{j}\omega) = \left| \frac{0.15}{(\mathrm{j}\omega - 0.35)(\mathrm{j}\omega - 0.5)} \right| = \frac{0.15}{\sqrt{\omega^4 + 0.37\omega^2 + 0.03}}$$

represents such an upper bound. For our simple example, $v_2(\mathrm{j}\omega)$ coincides with $g_2(\mathrm{j}\omega)$ for an extreme value of λ. This is not the case for higher-order systems. The bound given above may be approximated by some rational function with the same or some larger magnitude, but often a graphically given bound is convenient for the further analysis (Fig. 5.5). Note that although $g(p)$ and $\hat{g}(p)$ are unstable $v_2(\mathrm{j}\omega)$ is finite. Moreover, since $g_2(p) \xrightarrow[|p| \to \infty]{} 0$, $v_2(p)$ is known to be strictly proper. □

The time domain model

The time domain model can be derived by analogy. The subsystem (3) is described by a state space model

$$\begin{aligned} \dot{\mathbf{x}} &= \mathbf{A}\mathbf{x} + \mathbf{B}\mathbf{u} + \mathbf{E}\mathbf{s}, \qquad \mathbf{x}(0) = \mathbf{x}_0 \\ \mathbf{y} &= \mathbf{C}\mathbf{x} + \mathbf{D}\mathbf{u} + \mathbf{F}\mathbf{s} \\ \mathbf{z} &= \mathbf{C}_z\mathbf{x} + \mathbf{D}_z\mathbf{u} + \mathbf{F}_z\mathbf{s} \end{aligned} \tag{22}$$

or, alternatively, by a convolution model

$$\begin{aligned} \mathbf{y}(t) &= \hat{\mathbf{G}}(t) * \mathbf{u}(t) + \mathbf{G}_{ys}(t) * \mathbf{s}(t) \\ \mathbf{z}(t) &= \mathbf{G}_{zu}(t) * \mathbf{u}(t) + \mathbf{G}_{zs}(t) * \mathbf{s}(t). \end{aligned} \tag{23}$$

The approximate model has the form

$$\dot{\mathbf{x}} = \mathbf{A}\mathbf{x} + \mathbf{B}\mathbf{u}, \qquad \mathbf{x}(0) = \mathbf{x}_0$$
$$\hat{\mathbf{y}} = \mathbf{C}\mathbf{x} + \mathbf{D}\mathbf{u} \tag{24}$$

or

$$\hat{\mathbf{y}}(t) = \hat{\mathbf{G}}(t) * \mathbf{u}(t), \tag{25}$$

respectively, whereas the uncertainty model is given by

$$\dot{\mathbf{x}}_e = \mathbf{A}_e\mathbf{x}_e + \mathbf{B}_e\mathbf{z}$$
$$\mathbf{s} = \mathbf{C}_e\mathbf{x}_e + \mathbf{D}_e\mathbf{z} \tag{26}$$

or

$$\mathbf{s}(t) = \mathbf{G}_2(t) * \mathbf{z}(t). \tag{27}$$

The analogues of eqns (6) and (7) are

$$\mathbf{f}(t) = \delta\mathbf{G}(t) * \mathbf{u}(t) \tag{28}$$

and

$$\delta\mathbf{G}(t) = \mathbf{G}_{ys}(t) * \overline{\mathbf{G}}(t) * \mathbf{G}_{zu}(t) \tag{29}$$

with

$$\overline{\mathbf{G}}(t) = \mathbf{G}_2(t) + \mathbf{G}_2(t) * \mathbf{G}_{zs}(t) * \overline{\mathbf{G}}(t)$$

where the impulse response matrix $\overline{\mathbf{G}}(t)$ has been introduced since the inverse operation of the convolution is not in general use.

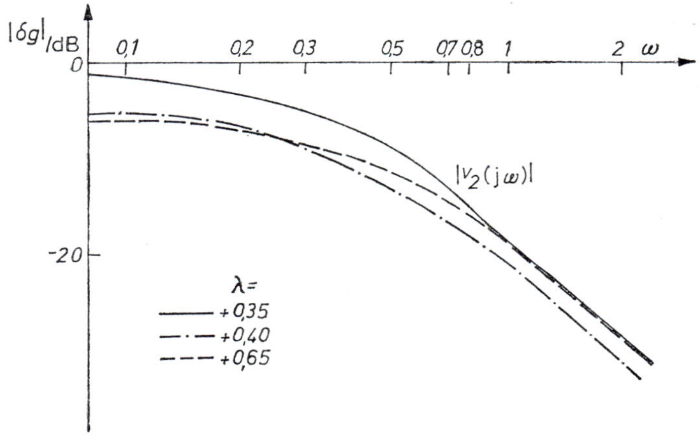

Fig. 5.5 Bode plot of $g(j\omega)$ from Example 5.2

To define the uncertainty bounds we again use the symbols $|.|$ and $||.||$ to indicate the modulus (element-by-element) and the norm of the vectors and matrices. As the convolution operation includes matrix-vector multiplication and integration, eqn (27) yields

$$|\mathbf{s}(t)| \leqq |\mathbf{G}_2(t)| * |\mathbf{z}(t)|$$

or

$$||\mathbf{s}(t)|| \leqq ||\mathbf{G}_2(t)||_M * ||\mathbf{z}(t)||.$$

Hence, if

$$\mathbf{V}_2(t) \geqq |\mathbf{G}_2(t)| \tag{30}$$

holds, then we have

$$|\mathbf{s}(t)| \leqq \mathbf{V}_2(t) * |\mathbf{z}(t)|. \tag{31}$$

Likewise, if v_2 is known to satisfy

$$v_2(t) \geqq ||\mathbf{G}_2(t)||_M \tag{32}$$

then

$$||\mathbf{s}(t)|| \leqq v_2(t) * ||\mathbf{z}(t)|| \tag{33}$$

provides the upper bound. Eqns (31) and (33) can be interpreted as using comparison systems $r(t) = v_2(t) * w(t)$ or $\mathbf{r}(t) = \mathbf{V}_2(t) * \mathbf{w}(t)$ that satisfy $v_2 * ||\mathbf{z}|| \geqq ||\mathbf{s}||$ or $\mathbf{V}_2 * |\mathbf{z}| \geqq |\mathbf{s}|$, respectively. The time domain counterparts of the set-theoretic models (19) and (20) of the original system

$$\mathbf{y}(t) = \mathbf{G}(t) * \mathbf{u}(t)$$

are

$$\mathcal{G} = \{\mathbf{G}(t) = \hat{\mathbf{G}}(t) + \delta\mathbf{G}(t),\ \delta\mathbf{G} = \mathbf{G}_{\mathbf{ys}} * \bar{\mathbf{G}} * \mathbf{G}_{\mathbf{zu}} \text{ with } |\mathbf{G}_2(t)| \leqq \mathbf{V}_2(t)\} \tag{34}$$

or

$$\mathcal{G} = \{\mathbf{G}(t) = \hat{\mathbf{G}}(t) + \delta\mathbf{G}(t),\ \delta\mathbf{G} = \mathbf{G}_{\mathbf{ys}} * \bar{\mathbf{G}} * \mathbf{G}_{\mathbf{zu}} \text{ with } ||\mathbf{G}_2(t)|| \leqq v_2(t)\}, \tag{35}$$

respectively, where $\bar{\mathbf{G}}(t)$ is defined in eqn (29). In this model the impulse response $\hat{\mathbf{G}}(t)$ of the approximate model (25) is often given in the state-space form (24).

Example 5.3 Consider two vehicles moving one behind the other along a straight line. Use the velocity of the second vehicle as state variable x_1 and the distance between the vehicles as state variable x_2. Then we have the following equation (Fig. 5.6 (a))

$$\frac{\mathrm{d}}{\mathrm{d}t}\begin{pmatrix}x_1\\x_2\end{pmatrix}=\begin{pmatrix}-\frac{1}{m} & 0\\ -1 & 0\end{pmatrix}\begin{pmatrix}x_1\\x_2\end{pmatrix}+\begin{pmatrix}\frac{1}{m}\\0\end{pmatrix}u+\begin{pmatrix}0\\1\end{pmatrix}d$$

where m denotes the mass of the second vehicle, u the accelerating force as control input and d the velocity of the first vehicle as disturbance. The problem is to model this system for different value of the mass m. With $\hat{m}=1$ we have

$$x_1(t)=\frac{1}{m}\exp\left(-\frac{t}{m}\right)*u\,,\qquad \hat{x}_1(t)=\exp(-t)*u$$

$$x_1(t)-\hat{x}_1(t)=\left(\frac{1}{m}\exp\left(-\frac{t}{m}\right)-\exp(-t)\right)*u\,.$$

Hence, the model can be decomposed into

$$\dot{\hat{x}}_1=-\hat{x}_1+u$$

$$x_1=\hat{x}_1+s$$

$$s=\left(\frac{1}{m}\exp\left(-\frac{t}{m}\right)-\exp(-t)\right)*u$$

$$|s(t)|\leqq\max_m\left|\frac{1}{m}\exp\left(-\frac{t}{m}\right)-\exp(-t)\right|*|u|\,.$$

Then a model of the form (22), (31) is given by

$$\frac{\mathrm{d}}{\mathrm{d}t}\begin{pmatrix}\hat{x}_1\\x_2\end{pmatrix}=\begin{pmatrix}-1 & 0\\ -1 & 0\end{pmatrix}\begin{pmatrix}\hat{x}_1\\x_2\end{pmatrix}+\begin{pmatrix}1\\0\end{pmatrix}u+\begin{pmatrix}0\\1\end{pmatrix}d+\begin{pmatrix}0\\-1\end{pmatrix}s$$

$$\mathbf{x}=\begin{pmatrix}\hat{x}_1\\x_2\end{pmatrix}+\begin{pmatrix}1\\0\end{pmatrix}s$$

$$z=u$$

$$|s|\leqq v_2*|u|$$

with

$$v_2(t)=\max_m\left|\frac{1}{m}\exp\left(-\frac{t}{m}\right)-\exp(-t)\right|.$$

The bound $v_2(t)$ is drawn in Fig. 5.6(b) for $0.8\leqq m\leqq 1.2$. Note that we have chosen a model structure which takes into account that the model error is caused by the mass uncertainty which directly affects only the first state (Fig. 5.6(c)). □

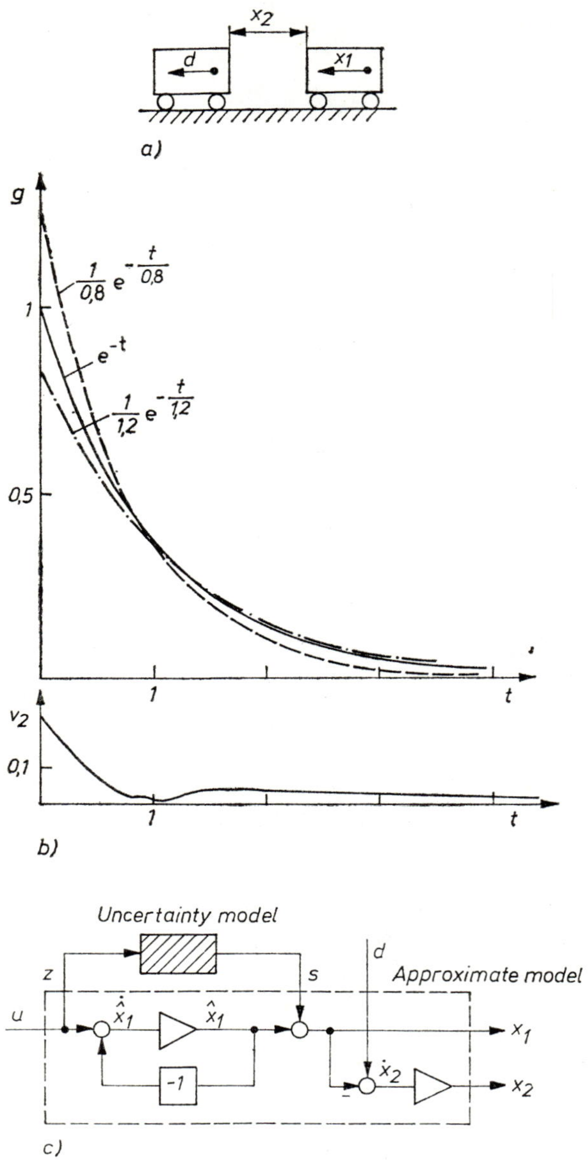

Fig. 5.6 Modelling the vehicle behaviour for uncertain mass

(a) System to be modelled

(b) Determination of $v_2(t)$

(c) Model of vehicle 2

Extensions to nonlinear original systems

As seen so far, the description of systems subject to bounded uncertainties necessitates merely a reasonable decomposition of the model and the use of a comparison system as a means to get an upper uncertainty bound. This general way does not presuppose the linearity assumption made so far. We want to use it now for static nonlinear original systems to outline how straightforward extensions to nonlinear or time-varying systems look like.

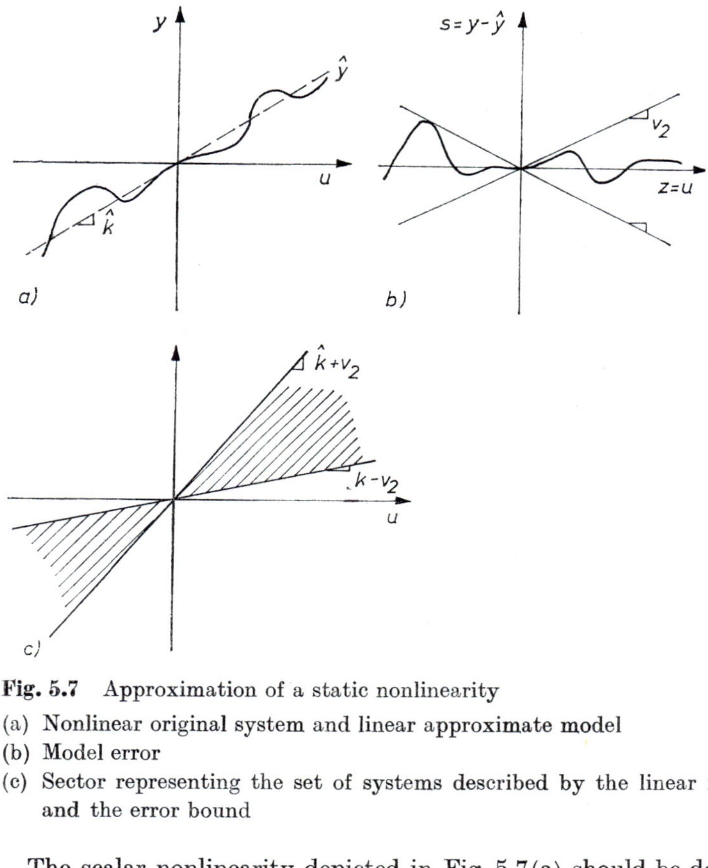

Fig. 5.7 Approximation of a static nonlinearity
(a) Nonlinear original system and linear approximate model
(b) Model error
(c) Sector representing the set of systems described by the linear model and the error bound

The scalar nonlinearity depicted in Fig. 5.7(a) should be described by a linear approximate model

$$\hat{y} = \hat{k}u \tag{36}$$

and some error bound. The scalar $\hat{k}$ is chosen so that the graph of (36) lies 'in the middle' of the graph of the given nonlinearity. Using the parallel model of Fig. 5.2(a), subsystem 1 has the form

$$\begin{aligned} y &= \hat{k}u + s \\ z &= u\,. \end{aligned} \tag{37}$$

The uncertainty model

$$s = n(z) \tag{38}$$

is some nonlinear function $n(z)$ whose graph (Fig. 5.7(b)) corresponds to the difference between the solid and the dashed lines in Fig. 5.7(a). It can be majorised by some static linear comparison system $r = v_2 w$ which yields the upper bound

$$|s| \leqq v_2 \, |z|, \tag{39}$$

although the error model (38) is nonlinear (Fig. 5.7(b)). Then the set $\mathcal{S}$ comprises all linear and nonlinear systems $y = \bar{n}(u)$ that are described by eqns (37) and (39). It is graphically represented by the sector with boundary ascents $\hat{k} + v_2$ and $\hat{k} - v_2$ (Fig. 5.7(c)).

This generalisation towards systems with static nonlinearities relies on the fact that many different kinds of model uncertainty can be bounded by a linear comparison system. The range between the bounds may be filled out by linear, nonlinear or even time-varying original systems.

For multidimensional nonlinearities

$$\mathbf{s} = \mathbf{N}(\mathbf{z}) \tag{40}$$

eqn (39) has to be replaced by a multidimensional bound

$$|\mathbf{s}| \leqq \mathbf{V}_2 \, |\mathbf{z}|. \tag{41}$$

The matrix $\mathbf{V}_2$ must be determined so as to satisfy the inequality

$$|\mathbf{N}(\mathbf{z})| \leqq \mathbf{V}_2 \, |\mathbf{z}| \tag{42}$$

for all possible $\mathbf{z}$.

Conic sectors

A description of the uncertain original system which modifies the model (20) can be received by constructing a conic sector. If we use the parallel model of Fig. 5.2(a), the model (20) has been received by bounding the model error $\mathbf{y} - \hat{\mathbf{y}} = (\mathbf{G} - \hat{\mathbf{G}})\, \mathbf{u}$ in the form

$$\|\mathbf{y}(p) - \hat{\mathbf{y}}(p)\| \leqq v_2 \, \|\mathbf{u}(p)\| \tag{43}$$

with

$$v_2(p) \geqq \|\mathbf{G}(p) - \hat{\mathbf{G}}(p)\|_{\mathrm{M}}. \tag{44}$$

In both equations we may use an arbitrary vector norm $\|.\|$ and the corresponding matrix norm $\|.\|_{\mathrm{M}}$, which both do not depend on the complex frequency p. A more general inequality (43) can be derived, however, if we introduce stable frequency-dependent weighting matrices $\mathbf{S}(p)$ and $\mathbf{R}(p)$ instead of $v_2(p)$. Then we have

$$\big\|\mathbf{S}(p)\,\big(\mathbf{y}(p) - \hat{\mathbf{y}}(p)\big)\big\| \leqq \|\mathbf{R}(p)\, \mathbf{u}(p)\|. \tag{45}$$

With an explicit use of the approximate model $\hat{\mathbf{y}} = \hat{\mathbf{G}}\mathbf{u}$ we get

$$\left\|\mathbf{S}(p)\left(\mathbf{y}(p) - \hat{\mathbf{G}}(p)\,\mathbf{u}(p)\right)\right\| \leqq \|\mathbf{R}(p)\,\mathbf{u}(p)\|. \tag{46}$$

Definition 5.1 The original system $\mathbf{G}(p)$ is said to be inside the *conic sector* $(\hat{\mathbf{G}}, \mathbf{S}, \mathbf{R})$ if equation (45) holds for all $\mathbf{u}(p)$ where $\|.\|$ denotes a vector norm.

The term 'conic sector' (or 'cone') will be explained later. By using $\mathbf{u}(p) = \mathbf{R}(p)^{-1}\,\tilde{\mathbf{u}}(p)$, eqn (46) yields

$$\left\|\mathbf{S}(p)\left(\mathbf{G}(p) - \hat{\mathbf{G}}(p)\right)\mathbf{R}(p)^{-1}\,\tilde{\mathbf{u}}(p)\right\| \leqq \|\tilde{\mathbf{u}}(p)\|$$

and

$$\left\|\mathbf{S}(p)\left(\mathbf{G}(p) - \hat{\mathbf{G}}(p)\right)\mathbf{R}(p)^{-1}\right\|_{\mathrm{M}} \leqq 1. \tag{47}$$

Hence, if the model error $\mathbf{G}(p) - \hat{\mathbf{G}}(p)$ is stable and the inequality (47) is satisfied for $p = j\omega$ then $\mathbf{G}(p)$ is inside the cone $(\hat{\mathbf{G}}, \mathbf{S}, \mathbf{R})$. The conic sector represents a set of candidates of the original system like $\mathscr{S}$ in eqn (20).

A cone of the static nonlinearity

$$\mathbf{y} = \mathbf{N}(\mathbf{u}) \tag{48}$$

can be defined in a similar way. As an obvious approximate model is given by the static relation

$$\mathbf{y} = \hat{\mathbf{K}}\mathbf{u} \tag{49}$$

and constant matrices $\mathbf{S}$ and $\mathbf{R}$ can be used, we get the following definition.

Definition 5.2 The original system (48) is said to be inside the conic sector $(\hat{\mathbf{K}}, \mathbf{S}, \mathbf{R})$ if

$$\left\|\mathbf{S}(\mathbf{N}(\mathbf{u}) - \hat{\mathbf{K}}\mathbf{u})\right\| \leqq \|\mathbf{R}\mathbf{u}\| \tag{50}$$

holds for all $\mathbf{u}$.

If we use $\mathbf{S} = \mathbf{I}$, $\hat{\mathbf{K}} = \frac{1}{2}(k_2 + k_1)\,\mathbf{I}$, $\mathbf{R} = \frac{1}{2}(k_2 - k_1)\,\mathbf{I}$, $-\infty \leqq k_1 < 1 < k_2 \leqq \infty$, eqn (50) has the form

$$\frac{\left\|\mathbf{N}(\mathbf{u}) - \frac{1}{2}(k_1 + k_2)\,\mathbf{u}\right\|}{\|\mathbf{u}\|} \leqq \frac{1}{2}(k_2 - k_1).$$

For single-input single-output systems this inequality describes all systems (48) whose graph lies within the sector $k_1 \leqq \frac{n(u)}{u} \leqq k_2$. For this reason, the set of systems satisfying eqn (46) or (50) is called conic sectors.

Comparing the model (46) with those given in eqns (19) and (20) we see that conic sectors provide a generalisation of the norm-bounded uncertainty representation for the parallel model of Fig. 5.2(a). In a more general sense, these cones can be used for a much more general class as linear or static nonlinear systems. This, however, necessitates a functional analytic extension of (46) or (50) which the interested reader can find in the references. We will come back to sector bounds of static nonlinearities in Section 6.1 and will use the norm-bounded or element-by-element bounded uncertainty description of eqns (19) and (20) for linear systems.

Guidelines for modelling systems with unknown-but-bounded uncertainties

In modelling systems with unknown-but-bounded uncertainties we aim at determing a set $\mathcal{S}$ which is as small as possible. To this end we have to take advantage of the freedom in the choice of the model structure and to use the smallest possible error bound. In more detail, we have to go two steps:

1. Choice of the model structure: According to the circumstances that lead to the model uncertainties the signals $\mathbf{s}$ and $\mathbf{z}$ and the elements $\mathbf{G}_{ys}$, $\mathbf{G}_{zs}$ and $\mathbf{G}_{zu}$ have to be chosen. These elements determine how the signals $\mathbf{z}$ and $\mathbf{s}$ are connected with the approximate model $\hat{\mathbf{G}}$.
2. Choice of the approximate model and the error bound: $\mathbf{V}_2$ and $\hat{\mathbf{G}}$ have to be chosen so as to end up with the smallest possible set $\mathcal{S}$.

Example 5.4 This example demonstrates how the freedom in the choice of the model structure can be exploited. We consider a system whose uncertain part is known to be stable. But we also know that this part can, in connection with the other parts of the system, cause instability of the whole system. If we used the parallel model of Fig. 5.2(a) the uncertainty model and, thus, the uncertainty bound $v_2(p)$ would have to be unstable. However, by using, for example, the model structure shown in Fig. 5.2(d) the instability of the original system can be presented by means of a stable uncertainty model.

For instance, if the uncertainty model is of first order

$$g_2(p) = \frac{-k\lambda}{p - \lambda}$$

it is stable for all $\lambda < 0$. If λ and k are known to lie within the intervals $\lambda_u \leqq \lambda \leqq \lambda_o < 0$ and $0 < k_u \leqq k \leqq k_o < 1$ then

$$|g_2(j\omega)| = |k| \left| \frac{-\lambda}{(j\omega - \lambda)} \right|$$

assumes its maximum for $\lambda = \lambda_u$ and $k = k_o$. Consequently,

$$v_2(j\omega) = \left| \frac{k_o \lambda_u}{(j\omega - \lambda_u)} \right|$$

provides an upper bound of $|g_2(j\omega)|$ for all parameter values. The original model

$$g(p) = \frac{\hat{g}(p)}{1 - g_2(p)} = \frac{p - \lambda}{(p - (1 - k)\,\lambda)}\,\hat{g}(p)$$

is unstable for all λ and k under consideration. This possibility to model unstable original systems by means of stable uncertainty models will turn out to be important for the stability analysis of the closed-loop original system (cf Chapter 8). □

Assessment of this form of the model

The form (3), (5) and (11), or (13), respectively, has been introduced as a formalisation of the description of systems with unknown-but-bounded uncertainties. The main attributes of this model are the decomposition of the model into two separate subsystems, the first representing the approximate model and the second being the uncertainty model, as well as the description of the uncertainties by some upper bound. In assessing this form of the model we have to keep in mind that the modelling task is complicated by the necessary tradeoff between engineering convenience and physical correctness. The contradictory requirements are the following:

- The model must be applicable to engineering problems, ie it must actually be possible to determine the structure and parameters of the model by means of the information available about the uncertain original system.
- The set $\mathcal{S}$ must have a form which is convenient for the solution of analysis and synthesis problems.
- The set $\mathcal{S}$ must be as small as possible, ie it should enclose only those systems that can actually be a representative of the original system.

The first and the second requirements are met by our model. A suitable decomposition of the model naturally results from the elaboration of the sources of the model uncertainties and is compatible with efficient tools for analysing the set $\mathcal{S}$. The possibility of clearly defining the entry point of the uncertainties is of great importance (cf Section 8.3). From the engineer's point of view the description of the uncertainties by upper bounds is very convenient because it corresponds with the commonly used method of characterising errors by upper bounds. This is particularly true for parameter uncertainties. In the analysis, such upper bounds can be easily handled because the range of possible model errors is described in the compact form (11) or (13).

On the other hand, we must be aware of some restrictions concerning the possibilities to adapt the model $\mathcal{S}$ of the set $\bar{\mathcal{S}}$ of systems that, due to the available information, cannot be excluded as possible representatives of the original system. At first, the uncertainty model (5) is only limited by some upper bound but is otherwise *unstructured*. This may be a rather coarse characterisation, if not an interval or dense subset of the parameter space but only specific parameter values are possible candidates of the system description. In this case, the model form (5.2-1) is more appropriate. Furthermore, since the upper bounds refer only to the absolute values of the elements of the transfer function matrix or the impulse response matrix, respectively, no phase information can be included into the uncertainty bound.

Second, the prescribed form of the model decomposition restricts the forms of uncertainties that can be described by narrow bounds. From eqn (19) we have

$$\text{rank}\, \delta \mathbf{G} \leqq \text{rank}\, \mathbf{G}_2 .$$

This inequality provides a relation between the kind and the dimension of the model uncertainties. If, for instance, only one scalar factor k is unknown, the best possible way to bound the model uncertainties is to use a one-dimensional subsystem $2\ s = k_2 z$. Due to eqn (7) this implies that all model errors $\delta\mathbf{G}$ must lie on a one-dimensional hyperplane in the space of $m \times r$-matrices. That is the reason why eg the set $\bar{\mathcal{G}} = \{\hat{\mathbf{G}} + k\delta\tilde{\mathbf{G}}, |k| \leqq \bar{k}\}$ with rank $\delta\tilde{\mathbf{G}} > 1$ has no equivalent description of the form (19). Here, some greater set $\mathcal{G}$, eg

$$\mathcal{G} = \{\mathbf{G} = \hat{\mathbf{G}} + \delta\mathbf{G};\ \delta\mathbf{G} = \text{diag}\, k_i \delta\tilde{\mathbf{G}},\ |k_i| \leqq \bar{k}\} \supset \bar{\mathcal{G}},$$

must be used to bring the model in the standard form.

Bibliographical notes

As the elaboration of suitable models for describing uncertain systems has never been considered as a task of its own, these models emerged in the literature on the analysis of uncertain feedback systems. Therefore, we mention only a few papers that made general considerations on the description of uncertain systems and refer the reader to the bibliographical notes of Chapters 6—8 for more details.

The unknown-but-bounded uncertainty approach to multivariable systems began with the consideration of norm-uncertain systems (Donati and Carlucci, 1975) where the norm of the system operator (eg see eqn (2.1-25)) was bounded from above. In our style of writing they used a single scalar constant v_2 which satisfies $\sup_p \|\mathbf{G}_2(p)\|_M \leqq v_2$. The method to bound the matrix norm or maximum singular value of $\mathbf{G}_2(p)$ was introduced in the singular value approach (Doyle, 1979; Safonov, 1979 etc.). A detailed discussion of this form of uncertainty representation can be found in (Postlethwaite and Foo, 1984). To provide a means for a more structured representation of the model uncertainties, the general model structure and element-by-element bounds were proposed by Lunze (1980a; 1982a; 1986), and Kosut (1982) or Lunze (1982a, 1984a, b), Kantor and Andres (1983), Owens and Chotai (1984), respectively. Corresponding time domain methods have been described by El-Atter and Vidyasagar (1978), Owens (1979) and Lunze (1981a, b, 1982c). A common basis of all these methods is provided by the general majorisation principle derived by Lunze (1983c, 1984c). It extends this form of uncertainty description to nonlinear and time-varying, and to discrete-time systems.

Conic sectors are extensively used in the stability analysis of composite systems as a global description of the subsystems and the interconnections. One of the first papers in this field was published by Zames (1966). In the robustness analysis of feedback systems conic sectors have been used eg by Safonov (1978, 1980), Kosut (1982), and Daniel and Kouvaritakis (1985) for presenting general methods of robustness analysis, Safonov and Athans (1977), Molander and Willems (1980), and Lehtomaki et al. (1981b) for investigating and improving the stability margin of LQ- and LQG-regulators, or Noldus (1982) for designing linear feedback controllers for nonlinear plants. Thompson et al. (1983) have described conic sectors for sampled-data systems.

6 QUALITATIVE ROBUSTNESS ANALYSIS AND DESIGN

Qualitative robustness concerns properties that are maintained throughout some open neighbourhood of the nominal plant parameter $\hat{\mathbf{a}}$. For example, if the original system is described by the state-space model

$$\dot{\mathbf{x}} = \mathbf{A}\mathbf{x} + \mathbf{B}\mathbf{u}$$

$$\mathbf{y} = \mathbf{C}\mathbf{x}$$

with $\mathbf{A} = \hat{\mathbf{A}} + \delta\mathbf{A}$, $\mathbf{B} = \hat{\mathbf{B}} + \delta\mathbf{B}$, $\mathbf{C} = \hat{\mathbf{C}} + \delta\mathbf{C}$, the closed-loop system is said to be robustly stabilisable, if there exist a feedback controller $\mathbf{u} = \mathbf{K}\mathbf{y}$ and uncertainty bounds $d_{\mathbf{A}}$, $d_{\mathbf{B}}$ and $d_{\mathbf{C}}$ such that the stability of $\bar{\mathbf{A}} = \hat{\mathbf{A}} + \hat{\mathbf{B}}\mathbf{K}\hat{\mathbf{C}}$ implies the stability of $\tilde{\mathbf{A}}(\delta\mathbf{A}, \delta\mathbf{B}, \delta\mathbf{C}) = \bar{\mathbf{A}} + \delta\mathbf{A} + (\hat{\mathbf{B}} + \delta\mathbf{B})\,\mathbf{K}(\hat{\mathbf{C}} + \delta\mathbf{C})$ for all perturbations with $\|\delta\mathbf{A}\|_{\mathrm{M}} \leqq d_{\mathbf{A}}$, $\|\delta\mathbf{B}\|_{\mathrm{M}} \leqq d_{\mathbf{B}}$ and $\|\delta\mathbf{C}\|_{\mathrm{M}} \leqq d_{\mathbf{C}}$.

Qualitative robustness investigations are mainly directed to structural problems. Typical examples are the selection of properties that can be robustly produced by feedback control, and the choice of the controller structure for ensuring such robustness specifications.

Most of the results are stability-oriented. In Sections 6.1 and 6.2 we consider the stability margin and the stability degree as robustness indicators and ways for improving them by means of feedback. Sections 6.3 and 6.4 present methods resulting from the singular perturbation analysis and high-gain feedback considerations. Finally, topological aspects of robust multivariable control are outlined as a field of current research.

6.1 Stability margin

The robustness of the closed-loop stability under plant perturbations can be characterised by means of the stability margin. Consider the single-loop unity-feedback system of Fig. 6.1, where $\hat{g}(p)$ denotes the approximate open-loop transfer function and $g_2(p)$ presents the plant perturbation. Both $\hat{g}(p)$ and $\hat{g}(p)\,g_2(p)$ are assumed to be stable. Then the stability margin is defined to be the size of the smallest perturbation g_2 of the open-loop transfer function that produces closed-loop instability. More precisely, the stability margin is defined as the gain or phase margin, respectively, as follows.

Definition 6.1 Refer to the unity-feedback system of Fig. 6.1 and assume that the open-loop system is stable for all $g_2(p)$ under consideration. If $g_2(j\omega) = k$, the smallest value of $k > 1$ and the largest value of $k < 1$ that cause closed-loop instability are called *gain margin* (or upward gain margin) and gain reduction margin (or downward gain margin), respectively. If $g_2(j\omega) = e^{j\phi}$, the smallest positive or negative values of ϕ for which the closed loop becomes unstable is called positive or negative *phase margin*, respectively.

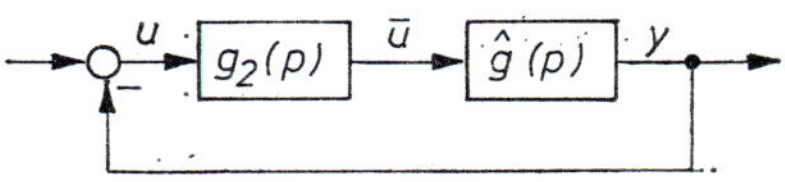

Fig. 6.1 Unity feedback system including plant perturbation $g_2(p)$

The stability margin gained widespread acceptance as indicator of robustness because its size is displayed by various graphical means which are used in classical frequency domain design techniques. In Fig. 6.2 the stability margin is shown in the Nyquist and Bode plots of the approximate open-loop transfer function $\hat{g}(p)$. Furthermore, the extension of the notion of

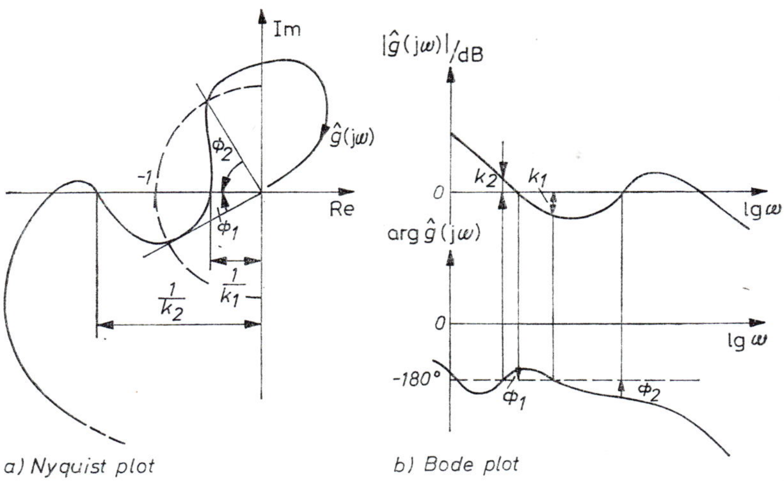

Fig. 6.2 Stability margin

k_1 — Gain margin
k_2 — Gain reduction margin
Φ_1 — Negative phase margin
Φ_2 — Positive phase margin

stability margin to nonlinear perturbations g_2 becomes obvious by considering the generalised rather than the 'ordinary' circle criterion (cf Section 2.2). If a nonlinearity $\bar{u} = g_2(u)$ with

$$f_u \leqq \frac{\bar{u}}{u} \leqq f_o, \qquad 0 < f_u < 1 < f_o \tag{1}$$

is introduced into the loop, stability is maintained as long as the Nyquist plot of $\hat{g}(p)$ does not intersect the disk centred on the real axis and passing through the points $-\frac{1}{f_u}$ and $-\frac{1}{f_o}$. Pairs of f_u and f_o for which this condition holds determine gain margins for nonlinear perturbations (Fig. 2.4).

A straightforward generalisation of the notion of stability margin to multi-input multi-output systems seems to be possible if the feedback paths are analysed independently of each other while treating the loop under consideration as single-input single-output system. This can be done if in the multivariable extension of Fig. 6.1 the perturbation $\mathbf{G}_2(p)$ is restricted to have the form

$$\mathbf{G}_2(p) = \operatorname{diag} g_{2i}(p) \tag{2}$$

where all but one element $g_{2i}(p)$ $(i = 1, 2, \ldots; i \neq j)$ equal unity and the closed-loop stability is considered in terms of the remaining element $g_{2j}(p)$. However, in this way we fail to check the robustness with respect to model uncertainties that occur simultaneously in several feedback paths. Since the plant perturbations are unlikely to confine themselves to a single loop unless they are caused by actuator failures we should deal with simultaneous perturbations of all loops.

Definition 6.2 Consider the multivariable unity-feedback system with the structure of Fig. 6.1, where the open-loop system is assumed to be stable for all $\mathbf{G}_2(p)$ and the perturbations are restricted to be of form (2) but may simultaneously occur. If the stability of the closed-loop system is maintained for all perturbations with $\operatorname{Re}[g_{2i}(j\omega)] < k \leqq 1$ or $\operatorname{Re}[g_{2i}(j\omega)] > k$ with $k > 1$ $(i = 1, 2, \ldots, m)$ then the closed loop is said to possess, respectively, gain reduction tolerance or *gain margin k* independently in each control loop. If the stability occurs for all perturbation $g_{2i}(j\omega) = e^{j\phi_i(\omega)}$ with $|\phi_i(\omega)| \leqq \bar{\phi}$ then the system is said to have $\pm\bar{\phi}$ *phase margin* independently in all control loops.

This definition can be easily extended to general perturbations $\mathbf{G}_2(p)$ since the open-loop transfer function matrix $\mathbf{G}_2(p)\,\hat{\mathbf{G}}(p)$ can be transformed by means of two matrices $\mathbf{V}(p)$ and $\mathbf{W}(p)$ with $\mathbf{V}(p)\,\mathbf{W}(p) = \mathbf{I}$ and $\mathbf{V}(p)\,\mathbf{G}_2(p)\,\mathbf{W}(p) = \operatorname{diag} g_{2i}(p)$. Then the characteristic gains $g_{2i}(p)$ appear in the open-loop matrix

$$\mathbf{V}(p)\,\mathbf{G}_2(p)\,\hat{\mathbf{G}}(p)\,\mathbf{W}(p) = \operatorname{diag} g_{2i}(p)\,\mathbf{V}(p)\,\hat{\mathbf{G}}(p)\,\mathbf{W}(p)\,. \tag{3}$$

As $\mathbf{G}_2(p)$ is assumed to be stable, $\mathbf{V}(p)$ and $\mathbf{W}(p)$ are bounded for all frequencies p with Re $[p] > 0$, and the stability of the closed-loop system can be investigated by means of the transfer function matrix on the right-hand side of eqn (3) applying Definition 6.2.

The matrices $\mathbf{V}(p)$ and $\mathbf{W}(p)$ in eqn (3) determine in which way the uncertainties $g_{2i}(p)$ are distributed within the whole system. If $\mathbf{G}_2(\mathrm{j}\omega)$ is real symmetric for all ω or unitary (ie $\mathbf{G}_2{}^*\mathbf{G}_2 = \mathbf{I}$) then the characteristic gains $g_{2i}(\mathrm{j}\omega)$ are real or equal $\mathrm{e}^{\mathrm{j}\phi_i}$, respectively, as claimed in Definition 6.2.

For an extension to nonlinear systems the perturbation $\mathbf{G}_2$ is replaced by a cone-bounded $(m \times m)$-nonlinearity

$$\tilde{\mathbf{u}} = \mathbf{N}(\mathbf{u}) \tag{4}$$

with

$$\sup_{\mathbf{u}} \big(\mathbf{N}(\mathbf{u}) - k_1\mathbf{u}\big)' \big(\mathbf{N}(\mathbf{u}) - k_2\mathbf{u}\big) < 0 \tag{5}$$

$$-\infty \leqq k_1 \leqq 1 \leqq k_2 \leqq \infty$$

(cf Section 5.3).

Definition 6.3 Consider a unity feedback system whose open loop consists of the linear part $\hat{\mathbf{G}}(p)$ and some cone-bounded nonlinearity (4). The closed-loop system is said to have a *robustness sector* (k_1, k_2) if it is globally asymptotically stable for all nonlinearities (4) that satisfy the condition (5).

Stability margin of LQ-regulators

As an application of the stability margin we consider the stability robustness of the feedback system that consists of the plant with approximate model

$$\begin{aligned} \dot{\mathbf{x}} &= \mathbf{A}\mathbf{x} + \mathbf{B}\mathbf{u}, \qquad \mathbf{x}(0) = \mathbf{x}_0 \\ \mathbf{y} &= \mathbf{C}\mathbf{x} \end{aligned} \tag{6}$$

and the LQ-feedback

$$\mathbf{u}(t) = -\mathbf{K}\mathbf{x}(t) \tag{7}$$

with $\mathbf{K}$ from eqn (2.4-4). In addition to the sensitivity reduction property derived in Section 3.4 this system can be proved to have a substantial stability margin. We want to discuss this advantageous property of LQ-regulation and simultaneously point to the danger of overestimating the importance of the stability margin as robustness indicator.

For single-input systems the stability margin follows directly from Theorem 2.1. As the transfer function $l(\mathrm{j}\omega) = \mathbf{k}'(\mathrm{j}\omega\mathbf{I} - \mathbf{A})^{-1}\mathbf{b}$ of the open loop (with cut at the input signal $u(t)$) satisfies the inequality

$$|1 + l(\mathrm{j}\omega)| \geqq 1 \qquad \text{for all } \omega$$

the Nyquist plot of $l(p)$ avoids the unit circle around the point $-1 + j0$ in the complex plane. Consequently, a comparison of Fig. 6.3 with Figs. 6.2 and 2.4 yields the following result.

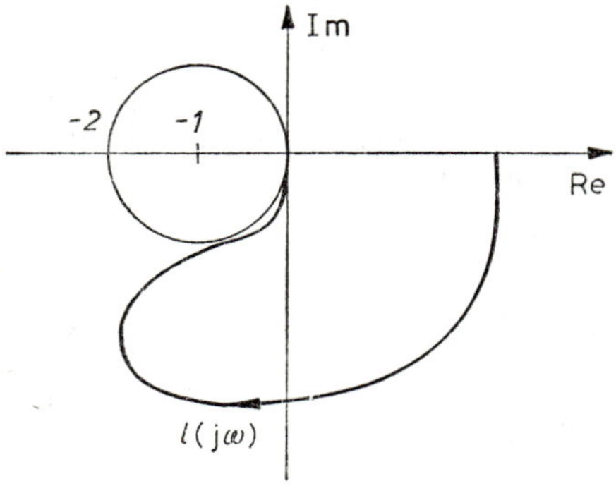

Fig. 6.3 Nyquist plot of the open-loop transfer function $l(j\omega)$ of a single-input plant with LQ-regulator

Theorem 6.1 (*Stability margin of LQ-regulation for single-input plants*) For linear original systems the **LQ**-regulator guarantees a phase margin of at least $\pm 60°$, a gain reduction tolerance of at least 50% (equivalent to 6 dB) and an infinite upwards gain margin. Furthermore, the closed-loop system is stable despite nonlinear perturbations satisfying the sector condition (1) with $f_u = \frac{1}{2}$ and $f_o = \infty$.

For multi-input plants the problem of determing the stability margin of LQ-regulators is more involved. No doubt, eqn (3.4-15) leads to

$$\left|\Pi\left(1 + \lambda_i[\mathbf{L}(j\omega)]\right)\right| \geqq 1 \tag{8}$$

and proves that the characteristic loci have a certain distance from the point $-1 + j0$. But no general assertion concerning the stability margin is possible. In fact, an inappropriate choice of the weighting matrices $\mathbf{Q}$ and $\mathbf{R}$ may lead to a poor stability margin. If, however, $\mathbf{Q}$ and $\mathbf{R}$ are chosen to be positive definite and $\mathbf{R}$ to be a diagonal matrix and, furthermore, the uncertainties are restricted to occur independently in the input channels as described by eqn (2) for linear perturbations, then similar stability margins as in the single-input case can be guaranteed.

Theorem 6.2 (*Stability margin of LQ-regulation for multi-input plants*) If the weighting matrix $\mathbf{Q}$ is positive definite and $\mathbf{R}$ presents a positive definite diagonal matrix then the closed-loop system has stability margins as described in Theorem 6.1 for simultaneous but independent perturbations of the input channels.

At first glance, the stability margin of LQ-regulators seems to be excellent. As the upward gain margin is infinite, a better gain reduction margin can be easily produced by using $\frac{1}{2d}\mathbf{K}$ with $d > 0$ instead of $\mathbf{K}$ as feedback matrix in (7). Then Theorem 6.1 proves a gain reduction margin of d while the infinite upward gain margin remains unaltered. However, examined closely, the stability margin of LQ-regulators may be too small as to ensure stability under practically occuring perturbations. First, since the gain reduction margin is only 50% the results of Theorem 6.2 do not prove stability in case of actuator failures that correspond to a gain reduction to zero. Secondly, $\pm 60°$ phase margin can be used up by an uncertainty that is similar to a first-order element, which in turn may produce up to 90° phase lag in the critical frequency range. Thirdly, the results of Theorems 6.1 and 6.2 refer to separate gain or phase uncertainties but not to a mixture of both.

A second point of discussion should focus our attention again on eqns (3.4-14) and (3.4-15). Low sensitivity over the whole frequency range can be produced only if the open-loop system has a pole excess of at most two (cf Section 2.3). This open-loop property is produced here by the use of full state feedback in connection with the LQ-design of the feedback matrix. The LQ-regulator (7) introduces new zeros so that the n-th order open-loop system possesses as much as $n - 1$ zeros.

The results of Theorems 6.1 and 6.2 are tied to the condition that the LQ-regulator is actually applied as a state feedback. If only some outputs of the plant can be measured the feedback (7) can only be applied using the reconstructed states $\hat{\mathbf{x}}$. The signal $\hat{\mathbf{x}}$ is produced by a Kalman filter or a Luenberger observer which, due to the separation theorem, can be designed independently of the LQ-state feedback. However, although both components of the controller have excellent stability margins the overall system exhibits often a poor stability robustness with respect to uncertainties in the actuators. The reason for this is given by the fact that the separation theorem does not longer hold if plant perturbations occur because the Kalman filter or Luenberger observer, respectively, have been designed on the basis of the approximate plant model and are not updated to the new plant performances. The chain connection of the plant and the filter or observer has a least pole excess of two since both components are nonanticipatory systems. Hence, the introduction of the filter or observer, respectively, decreases the stability margin.

For plants without right-half plane zeros the stability margin of the LQ-regulator can be asymptotically recovered if very high gains in the observer part of the controller are used and the matrices $\mathbf{Q}$ and $\mathbf{R}$ are particularly chosen. However, the use of high loop gains may cause the stability margin to be used up by very small plant perturbations. Such problems with high-gain feedback will be discussed in Section 6.4.

A final remark should point out that for optimal discrete-time regulators no a-priori phase and gain margins exist. Except for trivial cases the corresponding return difference $1 + \mathbf{k}'(z\mathbf{I} - \mathbf{A})^{-1}$ is not always larger than unity for z on the unit circle. This inconsistency between the continuous and the discrete time system should be a further reason for being careful with the robustness estimates that are received for optimal regulators in terms of the stability margin.

6.2 Stability degree

The stability degree can be considered as the time domain counterpart of the stability margin because it represents another indicator of robust stability and can be easily interpreted in terms of the state-space model. For the plant

$$\begin{aligned} \dot{\mathbf{x}} &= \mathbf{A}\mathbf{x} + \mathbf{B}\mathbf{u}, \qquad \mathbf{x}(0) = \mathbf{x}_0 \\ \mathbf{y} &= \mathbf{C}\mathbf{x} \end{aligned} \tag{1}$$

and the controller

$$\mathbf{u} = -\mathbf{K}\mathbf{y} \tag{2}$$

the eigenvalues of the closed-loop system matrix $\bar{\mathbf{A}} = \mathbf{A} + \mathbf{BKC}$ determine the stability property. As all eigenvalues have to lie in the left-half complex plane the minimum distance of the eigenvalues from the imaginary axis as given by

$$d_s = -\max_i \operatorname{Re}\left[\lambda_i[\mathbf{A}]\right] \tag{3}$$

can be used as a measure of stability robustness.

This has been done in connection with pole assignment techniques by demanding the closed-loop poles to lie to the left of the parallel line to the imaginary axis through a given point $-d_s + \mathrm{j}0$. In LQ-regulation, the modification of the performance criterion towards

$$\mathbf{I} = \int_0^\infty \mathrm{e}^{2dt}(\mathbf{x}'\mathbf{Q}\mathbf{x} + \mathbf{u}'\mathbf{R}\mathbf{u})\,\mathrm{d}t \to \min_{\mathbf{u}(t)} \tag{4}$$

for given scalar $d > 0$ and positive definite $\mathbf{Q}$ and $\mathbf{R}$ leads to a state feedback

$$\mathbf{u} = -\mathbf{K}\mathbf{x} \tag{5}$$

that ensures a stability degree of the closed-loop system (1), (5) of a least d. Using the transformation

$$\tilde{\mathbf{x}} = \mathrm{e}^{dt}\mathbf{x}, \qquad \tilde{\mathbf{u}} = \mathrm{e}^{dt}\mathbf{u}$$

the LQ-problem (4), (5) can be reformulated as minimising

$$\tilde{\mathbf{I}} = \int_0^\infty (\tilde{\mathbf{x}}'\mathbf{Q}\tilde{\mathbf{x}} + \tilde{\mathbf{u}}'\mathbf{R}\tilde{\mathbf{u}})\, dt \to \min_{\tilde{\mathbf{u}}}$$

subject to

$$\dot{\tilde{\mathbf{x}}} = (\mathbf{A} + d\mathbf{I})\, \tilde{\mathbf{x}} + \mathbf{B}\tilde{\mathbf{u}}\,.$$

The solution to this problem is as simple as the original one. The controller matrix is received as

$$\mathbf{K} = \mathbf{R}^{-1}\mathbf{B}'\tilde{\mathbf{P}} \tag{6}$$

with $\tilde{\mathbf{P}}$ being the positive definite solution of the equation

$$(\mathbf{A}' + d\mathbf{I})\, \tilde{\mathbf{P}} + \tilde{\mathbf{P}}(\mathbf{A} + d\mathbf{I}) - \tilde{\mathbf{P}}\mathbf{B}\mathbf{R}^{-1}\mathbf{B}'\tilde{\mathbf{P}} + \mathbf{Q} = \mathbf{0}\,. \tag{7}$$

Like the stability margin, the stability degree is only a qualitative measure of uncertainty tolerance. It does not tell us which perturbations do disturb the closed-loop stability. The lack of this information becomes obvious by comparing the stability degree with the norm $r(\hat{\mathbf{A}})$ of the minimal perturbation $\delta\mathbf{A}$ that makes $\hat{\mathbf{A}}$ unstable

$$r(\hat{\mathbf{A}}) = \min_{\delta\mathbf{A} \in \bar{\mathcal{A}}} \|\delta\mathbf{A}\| \tag{8}$$

where $\bar{\mathcal{A}}$ is the set of perturbations $\delta\mathbf{A}$ for which $\hat{\mathbf{A}} + \delta\mathbf{A}$ is unstable. If $\hat{\mathbf{A}}$ is stable with eigenvalues $\lambda_i = \delta_i + \mathrm{j}\omega_i$ ($i = 1, \ldots, n$; $\delta_1 \leqq \delta_2 \cdots \leqq \delta_n$) then $(\delta_n + \mathrm{j}\omega_n)\,\mathbf{I} - \hat{\mathbf{A}}$ is singular. Hence, we have

$$r(\hat{\mathbf{A}}) \leqq |\delta_n| \tag{9}$$

where δ_n is the stability degree. That is, the norm of the smallest destabilising perturbations $\delta\mathbf{A}$ is, generally, less than the stability degree. The equality sign holds only for normal matrices. If $\hat{\mathbf{A}}$ is not normal the stability degree can be a very misleading indicator of stability robustness.

The following example illustrates this assertion. Consider the matrix $\hat{\mathbf{A}} = \begin{pmatrix} -a & a^3 \\ 0 & -a \end{pmatrix}$ which is stable for $a > 0$. Obviously, $\delta\mathbf{A} = \begin{pmatrix} 0 & 0 \\ \frac{1}{a} & 0 \end{pmatrix}$ destabilises $\hat{\mathbf{A}}$. If $a \to \infty$ then the stability degree of $\hat{\mathbf{A}}$ grows unlimited. However, the norm of the destabilising perturbation $\delta\mathbf{A}$ approaches zero.

6.3 Robustness of two-time scale systems

This section is devoted to the feedback design for singularly perturbed systems. As this design is carried out by means of approximate models the applicability of the resulting control law necessitates a substantial closed-loop robustness.

If the plant incorporates state variables with widely distinct speed it is convenient to decompose the whole system into reduced-order slow and fast subsystems. This is often implicitely done by neglecting fast 'parasitic' elements such as the dynamics of actuators or measurement devices, whose effects on the overall system performance is considered as not essential. More generally, many high-dimensional systems consist of a mixture of fast and slow parts which can be separated by a time-scale decomposition.

From a theoretical point of view, systems of such kind are referred to as *singularly perturbed systems*. Assuming that the fast and the slow models have already been separated, the original system has the state equations

$$\begin{aligned} \dot{\mathbf{x}}_1 &= \mathbf{A}_{11}\mathbf{x}_1 + \mathbf{A}_{12}\mathbf{x}_2 + \mathbf{B}_1\mathbf{u} \\ \varepsilon\dot{\mathbf{x}}_2 &= \mathbf{A}_{21}\mathbf{x}_1 + \mathbf{A}_{22}\mathbf{x}_2 + \mathbf{B}_2\mathbf{u} \qquad (1) \\ \mathbf{y} &= \mathbf{C}_1\mathbf{x}_1 + \mathbf{C}_2\mathbf{x}_2 \end{aligned}$$

where the overall state vector $\mathbf{x}$ of dimension n consists of two parts of dimensions n_1 and n_2: $\mathbf{x} = (\mathbf{x}_1{}'\,\mathbf{x}_2{}')$. ε is a small positive scalar. $\mathbf{A}_{22}$ is assumed to be nonsingular.

The first equation refers to the slow subsystem and the second to the fast one since the eigenvalues of $\mathbf{A}_{11}$ and $\mathbf{A}_{22}/\varepsilon$ form two separate clusters in the complex plane if ε is sufficiently small. A reasonable approximation of the system description is based on the assumption that both subsystems do not interact dynamically. If the original system is considered for the slowly varying part $\hat{\mathbf{u}}$ of the input signal and for a long time horizon the second equation is used with $\varepsilon = 0$. Then the fast subsystem is represented by

$$\hat{\mathbf{x}}_2 = -\mathbf{A}_{22}^{-1}\mathbf{A}_{21}\hat{\mathbf{x}}_1 - \mathbf{A}_{22}^{-1}\mathbf{B}_2\hat{\mathbf{u}} \qquad (2)$$

which describes the static couplings of $\mathbf{x}_1$ and $\mathbf{u}$ towards $\mathbf{x}_2$. Using eqn (2) and the slow subsystem

$$\begin{aligned} \dot{\hat{\mathbf{x}}}_1 &= (\mathbf{A}_{11} - \mathbf{A}_{12}\mathbf{A}_{22}^{-1}\mathbf{A}_{21})\,\hat{\mathbf{x}}_1 + (\mathbf{B}_1 - \mathbf{A}_{12}\mathbf{A}_{22}^{-1}\mathbf{B}_2)\,\hat{\mathbf{u}} \\ \hat{\mathbf{y}} &= (\mathbf{C}_1 - \mathbf{C}_2\mathbf{A}_{22}^{-1}\mathbf{A}_{21})\,\hat{\mathbf{x}}_1 - \mathbf{C}_2\mathbf{A}_{22}^{-1}\mathbf{B}_2\hat{\mathbf{u}} \end{aligned} \qquad (3)$$

provides an approximate model of the original system (1) for long time investigations. It is abbreviated by

$$\begin{aligned} \dot{\hat{\mathbf{x}}} &= \hat{\mathbf{A}}\hat{\mathbf{x}} + \hat{\mathbf{B}}\hat{\mathbf{u}} \\ \hat{\mathbf{y}} &= \hat{\mathbf{C}}\hat{\mathbf{x}} + \hat{\mathbf{D}}\hat{\mathbf{u}}\,. \end{aligned} \qquad (4)$$

Note that $\hat{\mathbf{y}}$ depends directly on the input $\hat{\mathbf{u}}$ although the original system (1) has not direct throughput.

For short time investigations the slow state vector $\mathbf{x}_1$ and the slowly varying input $\hat{\mathbf{u}}$ are assumed to remain constant during fast transients of $\mathbf{x}_2$ towards the static value $\hat{\mathbf{x}}_2$ given by (2). From eqns (1) and (2) we obtain

$$\begin{aligned} \varepsilon\dot{\mathbf{x}}_2 &= \mathbf{A}_{21}\hat{\mathbf{x}}_1 + \mathbf{A}_{22}\mathbf{x}_2 + \mathbf{B}_2\mathbf{u} \\ &= \mathbf{A}_{22}(\mathbf{x}_2 - \hat{\mathbf{x}}_2) + \mathbf{B}_2(\mathbf{u} - \hat{\mathbf{u}}) \end{aligned} \tag{5}$$

and

$$\begin{aligned} \varepsilon \frac{\mathrm{d}}{\mathrm{d}t}(\mathbf{x}_2 - \hat{\mathbf{x}}_2) &= \mathbf{A}_{22}(\mathbf{x}_2 - \hat{\mathbf{x}}_2) + \mathbf{B}_2(\mathbf{u} - \hat{\mathbf{u}}) \\ \mathbf{y} - \hat{\mathbf{y}} &= \mathbf{C}_2(\mathbf{x}_2 - \hat{\mathbf{x}}_2) \end{aligned} \tag{6}$$

since $\dot{\hat{\mathbf{x}}}_2 = \mathbf{0}$. The second equation provides an estimate $\hat{\mathbf{f}}$ rather than the correct model error $\mathbf{f} = \mathbf{y} - \hat{\mathbf{y}}$ that occurs during the fast transients of $\mathbf{x}_2$. Eqn. (6) can be abbreviated by

$$\begin{aligned} \varepsilon\dot{\bar{\mathbf{x}}}_2 &= \mathbf{A}_{22}\bar{\mathbf{x}}_2 + \mathbf{B}_2\mathbf{u}_f \\ \hat{\mathbf{f}} &= \mathbf{C}_2\bar{\mathbf{x}}_2 \end{aligned} \tag{7}$$

where

$$\mathbf{u}_f = \mathbf{u} - \hat{\mathbf{u}} \tag{8}$$

represents the fast varying part of the input.

Corresponding to this model decomposition the controller

$$\mathbf{u} = -\mathbf{K}\mathbf{y} \tag{9}$$

is divided into two parts

$$\hat{\mathbf{u}} = -\mathbf{K}_1\hat{\mathbf{y}} \tag{10}$$

$$\mathbf{u}_f = -\mathbf{K}_2\hat{\mathbf{f}} \tag{11}$$

which determine the slowly and the quickly varying components of $\mathbf{u}$ independently of one another. These controller parts are designed separately for the approximate models (4) and (7), respectively. Both together form the controller (9) (cf eqn (8)).

The question arises under which conditions the performance of the closed-loop original system (1), (9) with small ε is actually close to the performance that can be predicted by means of the closed-loop approximate models (4), (10) and (7), (11). In other words, the robustness of the controller against the model simplification, whose degree depends on ε, must be proved. We want to tackle this problem here for controllers that are designed for the slow subsystem, neglecting the fast transients completely. Therefore, the model (7) is used with $\mathbf{u} = \hat{\mathbf{u}}$. The following investigation of two different situations should demonstrate the merit and the pitfalls of approximations which result from the singular perturbation approach.

The following lemma provides the basis for further investigations.

Lemma 6.1 Consider the autonomous singularly perturbed system

$$\dot{\mathbf{x}}_1 = \mathbf{A}_1\mathbf{x}_1 + \mathbf{A}_2\mathbf{x}_2$$

$$\varepsilon\dot{\mathbf{x}}_2 = \mathbf{A}_3\mathbf{x}_1 + \mathbf{A}_4\mathbf{x}_2$$

where $\mathbf{A}_4$ is a stable matrix. If the system is stable for $\varepsilon = 0$, ie the matrix

$$\mathbf{R} = \mathbf{A}_1 - \mathbf{A}_2\mathbf{A}_4^{-1}\mathbf{A}_3$$

is stable, then there is a positive constant $\bar{\varepsilon}$ such that the system is stable for all ε with

$$0 \leqq \varepsilon \leqq \bar{\varepsilon}.$$

If the controller represents a feedback of the slow modes of the plant

$$\mathbf{u} = -\mathbf{K}_1\mathbf{x}_1 \tag{12}$$

the approximate closed-loop system (4), (12) with $\mathbf{x}_1 = \hat{\mathbf{x}}$ is given by

$$\dot{\hat{\mathbf{x}}} = (\hat{\mathbf{A}} - \hat{\mathbf{B}}\mathbf{K})\,\hat{\mathbf{x}}$$

$$\hat{\mathbf{y}} = (\hat{\mathbf{C}} - \hat{\mathbf{D}}\mathbf{K})\,\hat{\mathbf{x}}.$$

If the pair $(\hat{\mathbf{A}}, \hat{\mathbf{B}})$ is controllable the controller matrix can be chosen so as to make the matrix

$$\tilde{\mathbf{A}} = \hat{\mathbf{A}} - \hat{\mathbf{B}}\mathbf{K} \tag{13}$$

stable. The closed-loop original system (1), (12) is described by

$$\begin{aligned} \dot{\mathbf{x}}_1 &= (\mathbf{A}_{11} - \mathbf{B}_1\mathbf{K})\,\mathbf{x}_1 + \mathbf{A}_{12}\mathbf{x}_2 \\ \varepsilon\dot{\mathbf{x}}_2 &= (\mathbf{A}_{21} - \mathbf{B}_2\mathbf{K})\,\mathbf{x}_1 + \mathbf{A}_{22}\mathbf{x}_2 \\ \mathbf{y} &= \mathbf{C}_1\mathbf{x}_1 + \mathbf{C}_2\mathbf{x}_2. \end{aligned} \tag{14}$$

If $\mathbf{A}_{22}$ is stable the stability of this system is ensured for sufficiently small ε due to Lemma 6.1 since the matrix $\mathbf{R}$ in the lemma applied to eqn (14) is equivalent to $\tilde{\mathbf{A}}$ and, thus, stable

$$\mathbf{R} = (\mathbf{A}_{11} - \mathbf{B}_1\mathbf{K}) - \mathbf{A}_{12}\mathbf{A}_{22}^{-1}(\mathbf{A}_{21} - \mathbf{B}_2\mathbf{K}) = \tilde{\mathbf{A}}.$$

Theorem 6.3 Consider a singularly perturbed system (1) with stable matrix $\mathbf{A}_{22}$. Then for every controller (12) that stabilises the approximate model (4) there exists a positive bound $\bar{\varepsilon}$ such that the original closed-loop system (1), (12) is stable for all ε with $0 < \varepsilon \leqq \bar{\varepsilon}$.

This theorem justifies the neglection of parasitic elements when designing state feedback controllers.

The advantage of designing the controller for the approximate model (4) is twofold. Firstly, only the n_1-th order approximate model (4) instead of the $(n_1 + n_2)$-th order system (1) has to be considered. For the realisation of the control law (12) only the slow states rather than the whole state vector have to be measured. Secondly, only the matrices of the static reinforcement of the fast subsystem appearing in eqn (2) must be known.

Now, let us turn our attention to the design of an output feedback

$$\mathbf{u} = -\mathbf{K}\mathbf{y}\,. \tag{15}$$

With $\mathbf{u} = \hat{\mathbf{u}}$ and $\mathbf{y} = \hat{\mathbf{y}}$ eqns (4), (15) yield

$$\hat{\mathbf{u}} = -(\mathbf{I} + \mathbf{K}\hat{\mathbf{D}})^{-1}\,\mathbf{K}\hat{\mathbf{C}}\hat{\mathbf{x}}$$

and

$$\begin{aligned} \dot{\hat{\mathbf{x}}} &= \left(\hat{\mathbf{A}} - \hat{\mathbf{B}}(\mathbf{I} + \mathbf{K}\hat{\mathbf{D}})^{-1}\,\mathbf{K}\hat{\mathbf{C}}\right)\hat{\mathbf{x}} \\ \hat{\mathbf{y}} &= (\mathbf{I} + \hat{\mathbf{D}}\mathbf{K})^{-1}\,\hat{\mathbf{C}}\hat{\mathbf{x}} \end{aligned} \tag{16}$$

provided that $(\mathbf{I} + \hat{\mathbf{D}}\mathbf{K})$ is invertible. The existence of a stabilising output feedback matrix depends on $\hat{\mathbf{A}}, \hat{\mathbf{B}}, \hat{\mathbf{C}}, \hat{\mathbf{D}}$. We assume to have found such a matrix $\mathbf{K}$. Then the closed-loop original system is described by

$$\begin{aligned} \dot{\mathbf{x}}_1 &= (\mathbf{A}_{11} - \mathbf{B}_1\mathbf{K}\mathbf{C}_1)\,\mathbf{x}_1 + (\mathbf{A}_{12} - \mathbf{B}_1\mathbf{K}\mathbf{C}_2)\,\mathbf{x}_2 \\ \varepsilon\dot{\mathbf{x}}_2 &= (\mathbf{A}_{21} - \mathbf{B}_2\mathbf{K}\mathbf{C}_1)\,\mathbf{x}_1 + (\mathbf{A}_{22} - \mathbf{B}_2\mathbf{K}\mathbf{C}_2)\,\mathbf{x}_2 \\ \mathbf{y} &= \mathbf{C}_1\mathbf{x}_1 + \mathbf{C}_2\mathbf{x}_2\,. \end{aligned} \tag{17}$$

Since the term $-\mathbf{B}_2\mathbf{K}\mathbf{C}_2$ can make the system matrix of the fast subsystem unstable, the original closed-loop system may be unstable for all ε, although the controller stabilises the approximate model.

To underline this danger of the singular perturbation method consider the following example. The original system is given by

$$\begin{aligned} \dot{x}_1 &= x_1 + (1 \quad 0)\,\mathbf{x}_2 + 2u \\ \varepsilon\dot{\mathbf{x}}_2 &= \begin{pmatrix} 0 \\ 1 \end{pmatrix} x_1 + \begin{pmatrix} -1 & 0 \\ -1 & -1 \end{pmatrix} \mathbf{x}_2 + \begin{pmatrix} -1 \\ -1 \end{pmatrix} u \\ y &= x_1 + (0 \quad 1)\,\mathbf{x}_2\,. \end{aligned}$$

Obviously, $\mathbf{A}_{22}$ is stable, whereas the slow subsystem has to be stabilised. For the approximate model

$$\begin{aligned} \dot{\hat{x}}_1 &= \hat{x}_1 + u \\ \hat{y} &= 2\hat{x} \end{aligned}$$

the stability is ensured by an output feedback (15) with $k > 1/2$. For the closed-loop original system we get the description

$$\begin{aligned} \dot{x}_1 &= (1 - 2k)\,x_1 + (1 \quad -2k)\,\mathbf{x}_2 \\ \varepsilon\dot{\mathbf{x}}_2 &= \begin{pmatrix} +k \\ 1 \ +k \end{pmatrix} x_1 + \begin{pmatrix} -1 & k \\ -1 & -1 + k \end{pmatrix} \mathbf{x}_2\,. \end{aligned}$$

Dividing the second equation by ε and investigating the trace of the system matrix of the overall system we can derive

$$(1 - 2k)\,\varepsilon + k > 2$$

as sufficient condition for the instability of the closed-loop original system. Hence, if the approximate model is stabilised by some controller with $k > 2$ (which is a very reasonable controller for the approximate model) the overall system is unstable however small ε may be. There is no stability robustness against 'parasitic' elements at all. This example demonstrates that the 'continuity assumption' which is usually made by control engineers and suggests that sufficiently small uncertainties can give rise only to neglectably small perturbations of the closed loop must not be used without proving its applicability.

Instability may not occur if $\mathbf{B}_2 = \mathbf{0}$ or $\mathbf{C}_2 = \mathbf{0}$ hold. In particular, if the fast modes are due to actuator or measurement dynamics, the slow and the fast subsystems have cascade connection. Then one of these restrictions is known to be satisfied and a result similar to Theorem 6.3 can be proved for output feedback.

6.4 High-gain feedback

The use of high-gain feedback for improving the robustness of feedback systems is suggested by several facts. As mentioned in Section 3.4, high-gain feedback is likely to decrease the sensitivity matrix and to improve the stability robustness of the closed-loop system. On the other hand, fundamental results on root loci state that the poles of the closed loop asymptotically approach the zeros of the open-loop system. If the system is minimum-phase or, equivalently, all zeros have negative real part then high-gain feedback leads the poles to go to the stable positions prescribed by the system zeros. In this way even unstable open-loop systems can be stabilised.

Both motivating facts do not depend on specific dynamical properties of the plant but hold for all minimum-phase systems. We want to perform these ideas further in a general form first and then look at details of an example.

Consider the one-degree-of-freedom system of Fig. 2.5(c1) concerning the command input $\mathbf{v}$ and the plant disturbance $\mathbf{d}$. It is described by

$$\mathbf{y}(p) = \mathbf{T}(p)\,\mathbf{v}(p) + \mathbf{S}(p)\,\mathbf{d}(p)$$

with

$$\mathbf{S}(p) = \big(\mathbf{I} + \mathbf{G}(p)\,\mathbf{K}(p)\big)^{-1}$$

$$\mathbf{T}(p) = \big(\mathbf{I} + \mathbf{G}(p)\,\mathbf{K}(p)\big)^{-1}\,\mathbf{G}(p)\,\mathbf{K}(p) = \mathbf{I} - \mathbf{S}(p)\,.$$

There are several definitions of the loop gain $g(\mathbf{GK})$, for instance

$$g(\mathbf{GK}) = \|\mathbf{G}(p)\,\mathbf{K}(p)\|_{\mathbf{M}} \quad \text{or} \quad g(\mathbf{GK}) = \sigma_{\min}[\mathbf{G}(p)\,\mathbf{K}(p)]$$

$$\text{or} \quad g(\mathbf{GK}) = \sup_p \|\mathbf{G}(p)\,\mathbf{K}(p)\|_{\mathbf{M}}$$

where $\sigma_{\min}[.]$ denotes the minimum singular value and the supremum is taken over p in the right-half complex plane. Using the second definition and

$$\sigma_{\max}[(\mathbf{I} + \mathbf{GK})^{-1}] = \sigma_{\min}[(\mathbf{I} + \mathbf{GK})]$$

$$\sigma_{\min}[\mathbf{GK}] - 1 \leqq \sigma_{\min}[(\mathbf{I} + \mathbf{GK})] \leqq \sigma_{\min}[\mathbf{GK}] + 1$$

we have

$$\begin{aligned} \mathbf{S}(p) &= \big(\mathbf{I} + \mathbf{G}(p)\,\mathbf{K}(p)\big)^{-1} \xrightarrow[\sigma_{\min}[\mathbf{GK}] \to \infty]{} \mathbf{0} \\ \mathbf{T}(p) &= \mathbf{I} - \mathbf{S}(p) \xrightarrow[\sigma_{\min}[\mathbf{GK}] \to \infty]{} \mathbf{I}. \end{aligned} \tag{1}$$

Hence,

$$\mathbf{y}(p) \approx \mathbf{v}(p).$$

Such a high loop gain can be produced by a controller

$$\mathbf{K}(p) = k\tilde{\mathbf{K}}(p) \tag{2}$$

with large scalar k provided that $\mathbf{G}\tilde{\mathbf{K}}$ has full rank.

Tight-loop systems provide ideal tracking properties including the complete decoupling of the v_i—y_i-connections, disturbance rejection, and insensitivity. These properties hold independently of $\mathbf{G}(p)$, which might be highly uncertain.

However, the problem is to ensure that the closed-loop system remains stable. We have to investigate which plants $\mathbf{G}(p)$ are stable under an appropriately chosen high-gain feedback controller $\mathbf{K}(p)$ in order to mark off the field of application of such controllers. The second question concerns how to choose such controllers.

In this simplest way, high-gain feedback is produced by the controller (2) where $\tilde{\mathbf{K}}(p)$ is a prespecified matrix and the loop gain depends merely on a scalar k. Then the stability of the closed-loop system can be investigated by means of **root locus** results. For single-input single-output plants the transfer function $l(p)$ of the open loop consisting of the plant $g(p)$ and the controller (2) can be brought into the form

$$l(p) = k\,\frac{\prod_{i=1}^{\bar{n}} (p - p_{0i})}{\prod_{i=1}^{n} (p - p_i)}.$$

Then the root locus describes the values of p for which $1 + l(p) = 0$ holds. The following result is recalled from root locus theory.

Lemma 6.2 The n branches of the root locus of the closed-loop system which results from the open loop with transfer function $l(p)$ begin for $k = 0$ in the poles p_i. $\bar{n}$ branches end for $k \to \infty$ in the zeros p_{0i} of the open loop while the remaining $n - \bar{n}$ branches approach infinity ($|p| = \infty$) along asymptotes which cross the real axis of the complex plane at

$$\bar{p} = \frac{\sum_{i=1}^{\bar{n}} p_{0i} - \sum_{i=1}^{n} p_i}{\bar{n} - n} \tag{3}$$

and whose angles with the real axis are

$$\phi_i = \frac{\pi}{\bar{n} - n}\, i \qquad (i = 0, \pm 1, \pm 2, \ldots). \tag{4}$$

A similar result for m-input r-output systems states that the finite eigenvalues of the closed-loop system approaches the transmission zeros of the open loop whereas the asymptotes of the remaining poles are again symmetrically distributed within the complex plane. (For the definition of multivariable transmission zeros see Section 2.1).

Since for uncertain plants right-half plane zeros are not exactly known and cannot be compensated for by unstable compensator poles the plant zeros must lie in the left-half complex plane. Furthermore, as the asymptotes (3), (4) can be adjusted by means of the controller so as to point to stable pole positions only if there are at most two of them, the open-loop system may have at most two poles more than zeros.

Theorem 6.4 The following conditions are necessary for the stability of the closed-loop system with high-gain feedback and uncertain plant:

(i) All plant transmission zeros have negative real part, ie the plant be minimum-phase.
(ii) The plant has at most two poles more than zeros.

This theorem provides very restrictive conditions for the use of high-gain feedback. As the plant is uncertain, unmodelled dynamics may easily violate the condition (ii).

The question how to choose the controller matrix $\mathbf{K}(p)$ so as to ensure closed-loop stability in spite of high loop gains can be answered in different ways. The first possibility is to use *LQ-regulation* with very small weighting matrix $\mathbf{R}$. If $\mathbf{R}$ in the objective function (2.4-2) is used in the form $\varrho\tilde{\mathbf{R}}$ where ϱ is a small positive scalar and $\tilde{\mathbf{R}}$ is symmetric positive definite, and if $\mathbf{Q}$ has the form $\mathbf{Q} = \mathbf{H}'\mathbf{H}$ for an arbitrary (m, n)-matrix $\mathbf{H}$, then the closed-loop system remains stable for $\varrho \to 0$ provided that it is minimum-phase. With the reduction of the input power penalty in the cost criterion the resulting state feedback matrix $\mathbf{K}$ in eqn (2.4-4) has increasing gain. $n - m$

eigenvalues of the closed-loop system (2.4-6) approaches the zeros of the system $(\mathbf{A}, \mathbf{B}, \mathbf{H})$ (see definition of zeros in Section 2.1 with $\mathbf{C} = \mathbf{H}$ and $\mathbf{D} = \mathbf{0}$) whereas for the other eigenvalues

$$\lambda_i(\varrho) \to \frac{\bar{\lambda}_i}{\sqrt{\varrho}} \quad \text{with} \quad |\bar{\lambda}_i| < \infty, \qquad \mathrm{Re}\,[\bar{\lambda}_i] < 0 \tag{5}$$

holds. This result shows the possibility to produce stable high-gain feedback systems, which have the properties of tight loops described above. Since the LQ-state-feedback system satisfies condition (ii) of Theorem 6.4, the stability is ensured for all minimum-phase systems. However, the application of the LQ-regulator necessitates the knowledge of the precise state-space model of the plant.

The second, more practical example deals with the determination of *high-gain PI-controllers* for uncertain minimum-phase plants. The plant is assumed to be quadratic ($\dim \mathbf{y} = \dim \mathbf{u} = m$) and to have a pole excess of at most one. It should be linear so that it can be described by the model

$$\begin{aligned} \dot{\mathbf{x}} &= \mathbf{A}\mathbf{x} + \mathbf{B}\mathbf{u} \\ \mathbf{y} &= \mathbf{C}\mathbf{x}, \end{aligned} \tag{6}$$

although the knowledge of the matrices $\mathbf{A}$, $\mathbf{B}$, and $\mathbf{C}$ will not be supposed. The transfer function matrix $\mathbf{G}(p)$ is given by

$$\mathbf{G}(p) = \mathbf{C}(p\mathbf{I} - \mathbf{A})^{-1}\,\mathbf{B} = \frac{1}{p}\sum_{i=0}^{\infty}\frac{1}{p^i}\,\mathbf{C}\mathbf{A}^i\mathbf{B}. \tag{7}$$

Its inverse $\mathbf{G}(p)^{-1}$ might be decomposed into three terms

$$\mathbf{G}(p)^{-1} = p\mathbf{F}_0 + \mathbf{F}_1 + \mathbf{F}_0\mathbf{H}(p) \tag{8}$$

where $\mathbf{H}(p)$ is a stable, proper real rational matrix with

$$\mathbf{H}(0) = \mathbf{0}. \tag{9}$$

$\mathbf{F}_0$, $\mathbf{F}_1$ and $\mathbf{H}(p)$ are received from $\mathbf{G}(p)^{-1}\,\mathbf{G}(p) = \mathbf{I}$ and

$$\begin{aligned} \mathbf{G}(p)^{-1}\,\mathbf{G}(p) &= \big(p\mathbf{F}_0 + \mathbf{F}_1 + \mathbf{F}_0\mathbf{H}(p)\big)\,\frac{1}{p}\sum_{i=0}^{\infty}\frac{1}{p^i}\,\mathbf{C}\mathbf{A}^i\mathbf{B} \\ &= \mathbf{F}_0\mathbf{C}\mathbf{B} + \sum_{i=1}^{\infty}\frac{1}{p^i}\,(\mathbf{F}_0\mathbf{C}\mathbf{A}^i\mathbf{B} + \mathbf{F}_1\mathbf{C}\mathbf{A}^{i-1}\mathbf{B}) \\ &\quad + \sum_{i=1}^{\infty}\frac{1}{p^i}\,\mathbf{F}_0\mathbf{H}(p)\,\mathbf{C}\mathbf{A}^{i-1}\mathbf{B}. \end{aligned}$$

Hence, we get

$$F_0 = (\mathbf{C}\mathbf{B})^{-1} = \big(p\mathbf{G}(p)\big)^{-1}\big|_{|p|\to\infty}, \tag{10}$$

where the inverse exists since the plant is assumed to have at most one pole more than zeros. The sum of the second and the third term, or both terms in the equation above have to vanish. In both cases

$$\mathbf{F}_1 = -\mathbf{F}_0\mathbf{CAB}(\mathbf{CB})^{-1} = \mathbf{G}(0)^{-1} \tag{11}$$

results. Hence, $\mathbf{F}_0$ and $\mathbf{F}_1$ can be determined by measuring the ascent and the final values of the step responses of the plant (cf eqns (2.1-10), (2.1-11)). Because of measurement errors, approximates $\hat{\mathbf{F}}_0$ and $\hat{\mathbf{F}}_1$ rather than precise matrices $\mathbf{F}_0$ and $\mathbf{F}_1$ are available.

Instead of the complete model $\mathbf{G}(p)^{-1}$ only the approximate model

$$\hat{\mathbf{G}}(p)^{-1} = p\hat{\mathbf{F}}_0 + \hat{\mathbf{F}}_1 \tag{12}$$

is used for the design of the PI-controller

$$\mathbf{u}(p) = \mathbf{K}(p)\,\big(\mathbf{v}(p) - \mathbf{y}(p)\big)$$

where the controller matrix is chosen to have the form

$$\mathbf{K}(p) = \hat{\mathbf{F}}_0 \operatorname{diag}\left(k_i + c_i + \frac{k_i c_i}{p}\right) - \hat{\mathbf{F}}_1 \tag{13}$$

with k_i, c_i being positive scalars. That is, we use the model structure of Fig. 5.2(b) with $\mathbf{L} = \mathbf{GK}$ and $\hat{\mathbf{L}} = \hat{\mathbf{G}}\mathbf{K}$ instead of $\mathbf{G}$ and $\hat{\mathbf{G}}$ (Fig. 6.4). The uncertainty model has the transfer function matrix $(\mathbf{GK})^{-1} - (\hat{\mathbf{G}}\mathbf{K})^{-1}$.

Ignoring the uncertainty model, Fig. 6.4 shows the closed-loop approximate model

$$\begin{aligned}\hat{\mathbf{y}} &= (\mathbf{I} + \hat{\mathbf{G}}\mathbf{K})^{-1}\,\hat{\mathbf{G}}\mathbf{K}\mathbf{v} \\ &= (\hat{\mathbf{G}}^{-1} + \mathbf{K})^{-1}\,\mathbf{K}\mathbf{v} \\ &= \operatorname{diag}\frac{1}{(p + k_i)\,(p + c_i)}\operatorname{diag}\big((k_i + c_i)\,p + k_i c_i\big) - p\mathbf{F}_0^{-1}\mathbf{F}_1\big).\end{aligned} \tag{14}$$

It is stable provided that the controller parameters k_i and c_i are positive and nonnegative, respectively. The question arises under which condition this controller stabilises the original system $\mathbf{G}(p)$ too.

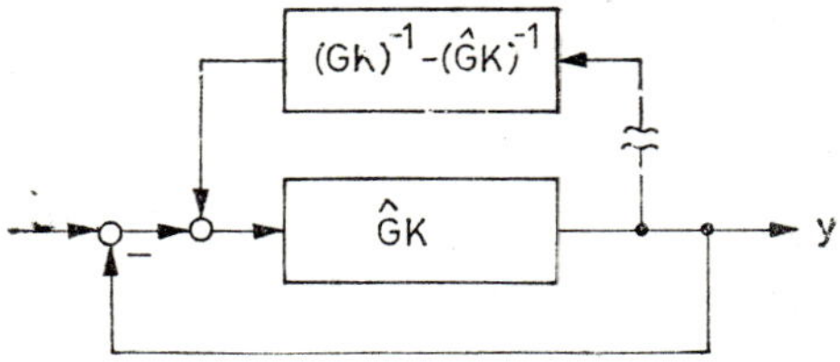

Fig. 6.4 The closed-loop original system

Anticipating a method for proving the stability, which will be explained in more detail in Section 8.3, this problem can be solved by means of a sufficient condition for the closed-loop stability. For the loop cut shown in Fig. 6.4 the return difference matrix is given by

$$\mathbf{F}(p) = \mathbf{I} - \tilde{\mathbf{L}}(p) \tag{15}$$

where

$$\tilde{\mathbf{L}}(p) = (\mathbf{I} + \hat{\mathbf{G}}\mathbf{K})^{-1}\,\hat{\mathbf{G}}\mathbf{K}\big((\mathbf{G}\mathbf{K})^{-1} - (\hat{\mathbf{G}}\mathbf{K})^{-1}\big)$$
$$= (\hat{\mathbf{G}}^{-1} + \mathbf{K})^{-1}\,(\mathbf{G}^{-1} - \hat{\mathbf{G}}^{-1})$$
$$= \left(p\hat{\mathbf{F}}_0 + \hat{\mathbf{F}}_1 + \hat{\mathbf{F}}_0 \operatorname{diag}\left(k_i + c_i + \frac{k_i c_i}{p}\right) - \hat{\mathbf{F}}_1\right)^{-1}$$
$$\times \big(p\mathbf{F}_0 + \mathbf{F}_1 + \mathbf{F}_0\mathbf{H}(p) - p\hat{\mathbf{F}}_0 - \hat{\mathbf{F}}_1\big)$$
$$= \operatorname{diag} \frac{p}{(p + k_i)\,(p + c_i)}\,[\hat{\mathbf{F}}_0^{-1}\mathbf{F}_0\mathbf{H}(p) + p\hat{\mathbf{F}}_0^{-1}(\mathbf{F}_0 - \hat{\mathbf{F}}_0)$$
$$+ \hat{\mathbf{F}}_0^{-1}(\mathbf{F}_1 - \hat{\mathbf{F}}_1)]. \tag{16}$$

According to the Nyquist criterion (Theorem 2.2) the stable open loop $\tilde{\mathbf{L}}(p)$ yields a stable closed-loop system if det $\mathbf{F}(p)$ does not encircle the origin of the complex plane for p on the Nyquist contour $\mathscr{D}$. Since

$$\det\big(\mathbf{F}(p)\big) = \det\big(\mathbf{I} - \tilde{\mathbf{L}}(p)\big) = \prod_{i=1}^{m}\big(1 + \lambda_i[\tilde{\mathbf{L}}(p)]\big) \geqq \big(1 - \|\tilde{\mathbf{L}}(p)\|_{\mathrm{M}}\big)^m > 0$$

provided that

$$\|\tilde{\mathbf{L}}(p)\|_{\mathrm{M}} < 1 \tag{17}$$

eqn (17) for $p \in \mathscr{D}$ represents a sufficient stability condition. This condition is satisfied for $k_i \geqq \bar{k}$ with sufficiently high $\bar{k}$ if

$$\|\hat{\mathbf{F}}_0^{-1}(\mathbf{F}_0 - \hat{\mathbf{F}}_0)\|_{\mathrm{M}} < 1 \tag{18}$$

because $\mathbf{H}(p)$ is assumed to be proper and, consequently $\|\tilde{\mathbf{L}}(p)\|$ takes its maximum value on the semicircle $|p| = R$ of the contour $\mathscr{D}$.

> **Theorem 6.5** Consider the minimum-phase system (6) with pole excess of at most one. Assume that an approximate model (12) has been determined. Then the controller (13) ensures closed-loop stability for non-negative parameters c_i and sufficiently high positive parameters k_i if the model uncertainty satisfies eqn (18).

Note that the plant may be unstable. For stable plants (6) the information necessary to set up the approximate model (12) can be received by simple experiments (cf eqns (10), (11)). Condition (18) claims that the 'relative error' of the measured ascent $\mathbf{F}_0$ of the step response matrix of the plant (6) is less than unity. For single-input single-output systems eqn (18) reads as $\left|\dfrac{f_0 - \hat{f}_0}{\hat{f}_0}\right| < 1$.

High-gain feedback has been used in many theoretical investigations to prove the reachability of the design objectives. The possibilities to produce high stability robustness and to approach a perfect tracking and disturbance rejection belong to the merits of this method. As shown by the design example of a PI-controller, high-gain feedback may be used to stabilise unstable plants even if the causes of instability are uncertain.

For practical applications the suitability of this method depends on how large the gain must be actually made in order to achieve the desired properties. Besides the fact that high gains cannot be realised over an arbitrarily large frequency range or may make the actuator signals exceed their technically given bounds, there are methodological problems which make its application problematic.

At first, all arguments hold only if the closed-loop system remains stable. Theorem 6.4 and practical experience show that indiscriminately increasing the gain may not be compatible with stability. Secondly, the consideration of the robustness alone is not typical for control engineering. Other objectives as bounded actuator signals, acceptable measurement noise attenuation or suitable overshoot and settling time may prohibit high gains.

6.5 Algebraic aspects of robustness

In the previous sections, we investigated which perturbations $\delta\mathbf{G}(p)$ of $\hat{\mathbf{G}}(p)$ are 'possible' in the sense that we can construct a single controller $\mathbf{K}(p)$ that stabilises both $\hat{\mathbf{G}}(p)$ and $\mathbf{G}(p) = \hat{\mathbf{G}}(p) + \delta\mathbf{G}(p)$ for small enough $\delta\mathbf{G}(p)$. We concentrated on particular classes of uncertainties. For example, in singularly perturbed systems the amount of uncertainties $\delta\mathbf{G}(p)$ was given by the value of ε, whereas the stability margins were defined for perturbations that can be interpreted as variations of the gain or the phase of the open loop.

Now, we want to look at the robustness problem in a more general way. The perturbations $\delta\mathbf{G}(p)$ need not depend on a single scalar ε but may encompass a much broader class including unstable errors. We will outline some recent results concerning the question what 'neighbourhood' and 'proximity' of plants $\hat{\mathbf{G}}(p)$ subject to general perturbations $\delta\mathbf{G}(p)$ means. A topology for the class of plants $\mathbf{G}(p)$ and a metric to measure the distance between $\hat{\mathbf{G}}(p)$ and $\hat{\mathbf{G}}(p) + \delta\mathbf{G}(p)$ will be defined that enables us to develop general assertions concerning the robustness of feedback systems.

We assume that all systems under consideration can be described by some transfer function matrix $\mathbf{G}(p) \in \mathscr{S}(p)^{m\times m}$ (for definition of $\mathscr{S}(p)$ see Section 2.1.3). $\mathbf{G}(p)$ is used in its right-coprime factorisation

$$\mathbf{G}(p) = \mathbf{N}(p)\,\mathbf{D}^{-1}(p) \tag{1}$$

which is assumed to be normalised, ie

$$\mathbf{D}'(-p)\,\mathbf{D}(p) + \mathbf{N}'(-p)\,\mathbf{N}(p) = \mathbf{I}. \tag{2}$$

$\mathbf{D}(p)$ and $\mathbf{N}(p)$ are unique up to right multiplication by an orthogonal matrix. Hence, each plant can be represented by $\begin{pmatrix}\mathbf{D}(p)\\ \mathbf{N}(p)\end{pmatrix} \in \mathscr{S}(p)^{2m\times m}$.

To consider 'small enough' perturbations in an algebraic setting, sequences $\{\delta \mathbf{G}^i\}$ or $\{\mathbf{G}^i\}$ are considered that converge to $\mathbf{0}$ or $\hat{\mathbf{G}}(p)$, respectively. The one-degree-of-freedom controller $\mathbf{K}(p)$ (Fig. 2.5(c)) has to stabilise $\hat{\mathbf{G}}(p)$ and $\mathbf{G}^i = \hat{\mathbf{G}}(p) + \delta\mathbf{G}^i(p)$ for large enough i. Moreover, we claim that the I/O-behaviour of the perturbed closed-loop system approaches that of the nominal system. This can be described by the sensitivity matrix

$$\mathbf{S}(\mathbf{G}^i, \mathbf{K}) = \big(\mathbf{I} + \mathbf{G}^i(p)\, \mathbf{K}(\mathrm{p})\big)^{-1} \tag{3}$$

and its complement

$$\mathbf{T}(\mathbf{G}^i, \mathbf{K}) = \big(\mathbf{I} + \mathbf{G}^i(\mathrm{p})\, \mathbf{K}(\mathrm{p})\big)^{-1}\, \mathbf{G}^i(\mathrm{p})\, \mathbf{K}(\mathrm{p})\,. \tag{4}$$

Robustness means

$$\mathbf{S}(\mathbf{G}^i, \mathbf{K}) \xrightarrow[i\to\infty]{} \mathbf{S}(\hat{\mathbf{G}}, \mathbf{K}) \tag{5}$$

$$\mathbf{T}(\mathbf{G}^i, \mathbf{K}) \xrightarrow[i\to\infty]{} \mathbf{T}(\hat{\mathbf{G}}, \mathbf{K})\,. \tag{6}$$

At first, let us define a topology on the plants $\mathbf{G}(p) \in \mathscr{R}(p)^{m\times m}$.

Definition 6.4 A sequence of plants $\{\mathbf{G}^i(p)\}$ with $\mathbf{G}^i(p) \in \mathscr{R}(p)^{m\times m}$ converges to $\hat{\mathbf{G}}(p)$ in *graph topology* if there exist right-coprime factorisations $(\mathbf{N}^i, \mathbf{D}^i)$ of $\mathbf{G}^i$ and $(\hat{\mathbf{N}}, \hat{\mathbf{D}})$ of $\hat{\mathbf{G}}$ such that $\mathbf{N}^i \to \hat{\mathbf{N}}$ and $\mathbf{D}^i \to \hat{\mathbf{D}}$ in $\mathscr{S}(p)^{m\times m}$.

In this definition, the convergence of unstable plants are examined in terms of the convergence of the stable transfer functions $\mathbf{D}^i(p)$ and $\mathbf{N}^i(p)$ towards $\hat{\mathbf{D}}(p)$ and $\hat{\mathbf{N}}(p)$. This can be proved by means of the norm (2.1-23) of $\mathscr{S}(p)^{m\times m}$. The name 'graph topology' originates from a graph-theoretical interpretation of the I/O-pairs of the plant $\mathbf{G}(p)$ in the space of bounded functions. As the following result shows, this topology defines perturbations against which feedback control is robust.

Theorem 6.6 Suppose that $\{\mathbf{G}^i(p)\}$ converges to $\hat{\mathbf{G}}(p)$ in graph topology. Then the following holds:

(i) Every controller $\mathbf{K}(p)$ that stabilises $\hat{\mathbf{G}}(p)$ does stabilise $\mathbf{G}^i(p)$ for large enough i.

(ii) For every controller $\mathbf{K}(p)$ stabilising $\hat{\mathbf{G}}(p)$ eqns (5) and (6) hold in $\mathscr{S}(p)^{m\times m}$.

The converse is also true: Consider a sequence $\{\mathbf{G}^i\}$ for which there exists a controller that stabilises $\hat{\mathbf{G}}(p)$ and $\mathbf{G}^i(p)$ for large enough i and for which (5) and (6) are true. Then this sequence converges in graph topology. For this reason, the graph topology is the weakest topology of $\mathscr{R}(p)^{m\times m}$ for which feedback is robust against plant perturbations.

To demonstrate the usefulness of the algebraic investigations let us come back to the control of singularly perturbed systems (6.3-1). As opposed to the results presented in Section 6.3, no a-priori restrictions are imposed on the controller $\mathbf{K}(p)$ here. Therefore, the following investigations concern linear controllers of any dynamical order and parameters. Moreover, robustness does not only concern stability but the I/O-behaviour as well. The result is summarised in the following theorem.

Theorem 6.7 Consider the system (6.3-1) with stable matrix $\mathbf{A}_{22}$. Suppose that the approximate model (6.3-3) is stabilisable and detectable. Then the plant $\mathbf{G}^\varepsilon(p)$ given by (6.3-1) approaches $\hat{\mathbf{G}}(p)$ in eqn (6.3-3) for $\varepsilon \to 0$ in graph topology if

$$\mathbf{C}_2(p\mathbf{I} - \mathbf{A}_{22})^{-1}\mathbf{B}_2 = 0. \tag{7}$$

Because of the assumptions on $\hat{\mathbf{G}}(p)$ there exist controllers $\mathbf{K}(p)$ stabilising $\hat{\mathbf{G}}(p)$. Then, according to Theorem 6.6 every such controller is robust. It stabilises $\mathbf{G}^\varepsilon(p)$ for small enough ε and possesses the property (5), (6) for $\varepsilon \to 0$. This result holds for all controller structures that might be used to stabilise $\hat{\mathbf{G}}(p)$. Obviously, Theorem 6.7 extends the results concerning the robust stability of singularly perturbed systems discussed at the end of Section 6.3.

The algebraic investigations can be extended to obtain quantitative robustness results. A metric that induces the graph topology can be defined as follows.

Definition 6.5 The *graph metric* $d(\mathbf{G}^1, \mathbf{G}^2)$ is defined by

$$d(\mathbf{G}^1, \mathbf{G}^2) = \max\{\delta(\mathbf{G}^1, \mathbf{G}^2),\ \delta(\mathbf{G}^2, \mathbf{G}^1)\} \tag{8}$$

where

$$\delta(\mathbf{G}^1, \mathbf{G}^2) = \inf_{\substack{\|\mathbf{U}\| \leq 1 \\ \mathbf{U} \in \mathscr{S}(p)^{m \times 2m}}} \left\| \begin{pmatrix} \mathbf{D}^1(p) \\ \mathbf{N}^1(p) \end{pmatrix} - \begin{pmatrix} \mathbf{D}^2(p) \\ \mathbf{N}^2(p) \end{pmatrix} \mathbf{U}(p) \right\| \tag{9}$$

with $\mathbf{N}^i$, $\mathbf{D}^i$ being the normalised right-coprime factors of $\mathbf{G}^i(p)$ over $\mathscr{S}(p)^{m \times m}$.

Eqn (9) refers to the norm (2.1-23) of the set $\mathscr{S}(p)^{m \times 2m}$ of stable transfer function matrices. $d(\mathbf{G}^1, \mathbf{G}^2)$ assumes real values in the interval [0, 1].

On this basis bounds on $d(\hat{\mathbf{G}}, \hat{\mathbf{G}} + \delta\mathbf{G})$ can be derived so as to ensure the robustness of the closed-loop system under the perturbation $\delta\mathbf{G}$. The following preliminary result illustrates that powerful quantitative results can be expected to be found in near future by algebraic investigations.

Lemma 6.3. Assume that the closed-loop system consisting of the controller $\mathbf{K}(p)$ and the plant model $\hat{\mathbf{G}}(p)$ is stable. Then the system remains stable under perturbations $\delta\mathbf{G}(p)$ if the distance between $\hat{\mathbf{G}}(p)$ and $\mathbf{G}(p) = \hat{\mathbf{G}}(p) + \delta\mathbf{G}(p)$ as measured by the graph metric remains smaller than $1/(2 + 4r)$

$$d(\hat{\mathbf{G}}, \mathbf{G}) < \frac{1}{2 + 4r} \tag{10}$$

where

$$r = \left(1 + \left\| \begin{matrix} (\mathbf{I} + \mathbf{GK})^{-1} & -\mathbf{G}(\mathbf{I} + \mathbf{KG})^{-1} \\ \mathbf{K}(\mathbf{I} + \mathbf{GK})^{-1} & (\mathbf{I} + \mathbf{KG})^{-1} \end{matrix} \right\|^2 \right)^{1/2}. \tag{11}$$

In comparison with the results that will be presented in Section 8.3 this lemma does not presuppose that $\hat{\mathbf{G}}(p)$ and $\mathbf{G}(p)$ have the same number of right-half plane poles or zeros.

This short section on algebraic robustness investigations should be concluded with two remarks. The first concerns the advantage of the algebraic approach itself. Algebraic investigations do not necessitate a-priori information about the plant and the controller structures (dynamical order, number of right-half plane poles or zeros etc) and yield the specification of the controller structure as a result. Therefore, very general assertions can be made.

The second remark deals with the choice of the plant model. As pointed out by Definition 6.4 and Theorem 6.4, I/O-models have the property that

$$\{\mathbf{G}^i\} \to \hat{\mathbf{G}} \quad \text{implies} \quad \{\mathbf{y}^i = \mathbf{G}^i\mathbf{u}\} \to \mathbf{y} = \hat{\mathbf{G}}\mathbf{u}.$$

A corresponding relation is not true for the state space model since $\{(\mathbf{A}^i, \mathbf{B}^i, \mathbf{C}^i, \mathbf{D}^i)\} \to (\mathbf{A}, \mathbf{B}, \mathbf{C}, \mathbf{D})$ does not imply and is not implied by

$$\{\mathbf{G}^i = \mathbf{C}^i(p\mathbf{I} - \mathbf{A}^i)^{-1}\,\mathbf{B}^i + \mathbf{D}^i\} \to \hat{\mathbf{G}} \quad \text{or} \quad \{\mathbf{y}^i\} \to \hat{\mathbf{y}}.$$

Therefore, the I/O-models should be preferred if system perturbations are considered. We will stick to this advice, at least for the description of the uncertainty model.

Summary

This chapter was devoted to qualitative robustness considerations. The results proved, but in several cases refuted, the intuitively confirmed statement that a controller $\mathbf{K}(p)$ that stabilises the approximate model does stabilise the original plant $\mathbf{G}(p) = \hat{\mathbf{G}}(p) + \delta\mathbf{G}(p)$ as well provided that $\hat{\mathbf{G}}(p)$ is 'close enough' to $\mathbf{G}(p)$. The results emphasise the possibility of robust control by pointing out for which classes of systems this statement is really true.

However, no quantitative results concerning the 'allowable' distance $\delta\mathbf{G}$ between $\mathbf{G}$ and $\hat{\mathbf{G}}$ are made and most of the results regard stability as the only property to be preserved. To answer the question 'Is the given plant uncertainty tolerable?' it is necessary to get quantitative robustness results as those that will be given in the next chapters.

Bibliographical notes

There are several definitions of the stability margin of multi-input multi-output systems. Those given in Definition 6.1 through 6.3 correspond with the notions introduced by Safonov and Athans (1977), and Molander and Willems (1980). The stability margin of LQ-regulators has been considered by Anderson and Moore (1971) for single-input systems on the basis of Kalman's proof of the return difference to be greater than unity at all frequencies (1964) (Theorem 6.1). Extensions to multi-input plants have been presented by Safonov and Athans (1977) for nonlinear or stable linear perturbations and by Lehtomaki et al. (1981) for unstable perturbations (Theorem 6.2). Lehtomaki et al. also published an example showing poor stability robustness of LQ-regulators if the weighting matrix **R** is not diagonal, and considered the stability of LQ-regulation that is applied by means of Kalman filters. The asymptotic recovery of the stability margin has been described by Doyle (1978) and Doyle and Stein (1979). Critical remarks on the results of this asymptotic recovery method and examples demonstrating poor stability robustness are given eg by Shaked and Soroka (1985) and Grimble and Owens (1986).

Methods for searching specific weighting matrices in order to improve the robustness of the LQ-regulators beyond the a-priori margins described in Theorems 6.1 and 6.2 have been proposed eg by Molander and Willems (1980) and Joshi (1984). Extensions of the stability margin to interconnected systems and decentralised controllers are presented by Sezer and Siljak (1981).

The stability margin of discrete-time LQ-regulators has been investigated by Kwakernaak and Sivan (1972).

The notion of the stability degree has been used by Anderson and Moore (1971) to improve the stability robustness of LQ-regulators. Extensions to decentralised controllers can be found in (Hassan and Singh, 1979; Sezer and Siljak, 1980). Hinrichsen and Prichard (1985) called the evidence of the stability degree as distance of a matrix from instability into question by proving that the norm of the smallest perturbation that may cause some eigenvalue to drift into the right-half complex plane is generally smaller than the stability degree (eqn (6.2-9)). They described the example which concludes Section 6.2.

A thorough survey on the singular perturbation approach to the modelling of control systems has been given by Kokotovic et al. (1976). Its application to controller design has been dealt with eg in (Chow and Kokotovic, 1978). Theorem 6.3 has been proved by Porter (1974). The example of Section 6.3 was taken from (Khalil, 1981). Most of the recent work along the lines of singularly perturbed systems is devoted to multi-modelling and control of large-scale interconnected systems (Khalil and Kokotovic, 1979).

The high-gain philosophy has been proposed by Bode (1945) and elaborated by Horowitz (1963) as a means to produce low-sensitivity single-input single-output systems. The root locus result presented in Lemma 6.2 is well-known (see eg (Solodownikow, 1971)). Extensions to multi-input multi-output systems have been described by Kouvaritakis and Shaked (1976) and Davison and Wang (1977). Kimura (1984) considered high-gain feedback for producing perfect regulation while Kwakernaak (1982), and Owens and Chotai (1982) used this method for stabilising uncertain systems. Owens and Chotai gave the original proof of Theorem 6.5. A generalisation to infinite-dimensional systems has been given by Logemann and Owens (1987). Advantages and shortcomings of the method are discussed in detail by Doyle and Stein (1981) and Tolle (1983).

The development of the algebraic theory of feedback systems started with the work by Rosenbrock (1974) and Youla et al. (1976), who proposed to use polynomial matrices as description of linear systems and elaborated a global parametrisation of the class of stabilising compensators. Recently, this theory has been used by Francis and Vidyasagar (1983), Vidyasagar (1984), and others to approach the robust stability problem from an algebraic point of view in order to get a general description of the kind and the amount of uncertainties that may be tolerated by feedback systems. The results presented in Section 6.5 are due to Vidyasagar (1984).

7 SIMULTANEOUS STABILISATION AND POLE ASSIGNMENT

In this first chapter on quantitative robustness investigations we assume to have a discrete set $\mathscr{S}$ of completely specified models at our disposal. The deviations between these models may be large. The problem is to stabilise all the systems of $\mathscr{S}$ or even to assign certain poles by means of a single controller.

The main difficulties that arise in the solution of this problem are caused by the absence of a simple relationship between all or some elements of the set $\mathscr{S}$. Even if the different models result from a linearisation of a nonlinear model at different operating points, there is in general no obvious relationship among the different modes of operation (although in reality such a law may exist). As the specimens do not lie in the vicinity of one another, there is even no reasonable continuity assumption which would enable us to extrapolate the properties of the second closed-loop system from those of the first.

That is the reason why only relatively few results and constructive methods are available concerning simultaneous stabilisation and pole assignment. We will review the results on simultaneous stabilisability in Section 7.1 and discuss the parameter space design method for pole region assignment in Section 7.2.

7.1 Simultaneous stabilisability

With the restriction to the least robustness requirement claiming that the closed-loop system be stable in all modes of operation, we investigate the question whether a controller does exist that stabilises all plants of a given set $\mathscr{S}$. First, we consider two of the many simple examples that illustrate the hardship of this problem.

Example 7.1 Consider the set of single-variable systems

$$g^i(p) = \frac{a^i}{(p - \lambda^i)} \tag{1}$$

$(i = 1, 2, \ldots, q)$ under proportional feedback

$$u(p) = -ky(p)\,. \tag{2}$$

The closed-loop system is stable if and only if

$$\lambda^i < ka^i \qquad (i = 1, 2, \ldots, q) \tag{3}$$

holds for some constant k. That is, the pairs (a^i, λ^i) have to lie on one side of a straight line passing the origin and being not parallel to the λ-axis (Fig. 7.1). The systems (1) and (2) in Fig. 7.1 are simultaneously stabilisable by a proportional feedback with $k > \lambda^1/a^1$, whereas there is no common controller for the systems (1), (3).

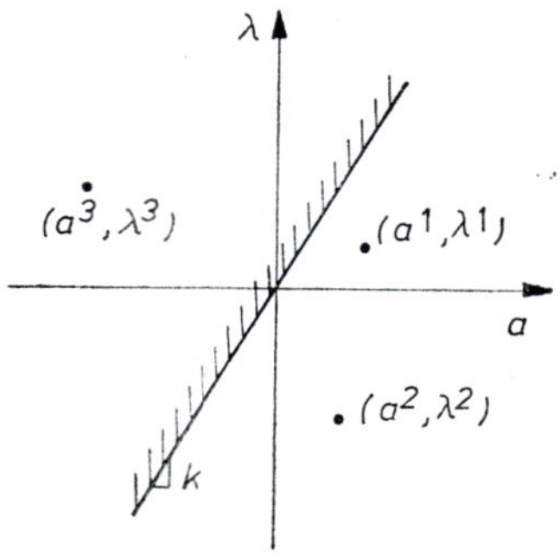

Fig. 7.1 Location of simultaneously stabilisable first-order plants in the parameter space

If a proper stable feedback controller

$$u(p) = -k \frac{\prod_{j=1}^{m_c} (p - p_{oj})}{\prod_{j=1}^{n_c} (p - p_j)} y(p), \qquad m_c < n_c \tag{4}$$

is used then the closed-loop characteristic equation is given by

$$p^{n_c+1} + p^{n_c}\left[\sum_{j=1}^{n_c}(-p_j) - \lambda^i\right] + \cdots + \left[\left(-\lambda^i \prod_{j=1}^{n_c}(-p_j) - ka^i \prod_{j=1}^{m_c}(-p_{oj})\right)\right] = 0.$$

Since all coefficients must be positive we have

$$\lambda^i \prod_{j=1}^{n_c}(-p_j) < ka^i \prod_{j=1}^{m_c}(-p_{oj})$$

which requires the same relationship (3) between the plant models as the proportional controller (2). Therefore, the introduction of controller dynamics does not weaken the simultaneous stabilisability condition on the system (1). □

For higher-order or multivariable systems the introduction of controller dynamics may make the stabilisation possible whereas the set $\mathcal{S}$ cannot be simultaneously stabilised by a proportional controller. This renders the problem more severe because it emphasises the necessity to choose the controller structure appropriately (although we have few results and methods for doing this).

Example 7.2 Consider the plants

$$\dot{\mathbf{x}} = \begin{pmatrix} 0 & 1 \\ -\mathrm{a}_1 & -\mathrm{a}_2 \end{pmatrix} \mathbf{x} + \begin{pmatrix} 0 \\ 1 \end{pmatrix} u$$

$$y = (1 \quad 0)\, \mathbf{x}$$

with $a_1{}^1 = 5$, $a_2{}^1 = -2$, and $a_1{}^2 = 8$, $a_2{}^2 = -3$. They are not simultaneously stabilisable by the proportional controller (2), but the dynamical controller

$$\dot{x}_{\mathrm{r}} = -4x_{\mathrm{r}} + 4x_1$$

$$u = 44x_{\mathrm{r}} - 4x_1$$

stabilises both plants. □

A general answer to the question of simultaneous stabilisability necessitates a characterisation of all (or a subset of all) stabilising controllers of a single linear plant. Applied to different plant models such a characterisation leads to different sets of controllers whose intersection represents the set of simultaneously stabilising controllers. If the intersection is not empty the plants are simultaneously stabilisable.

Unfortunately, the problem of describing the set of all stabilising controllers for a given plant with fixed parameters has not attracted considerable attention in linear control theory or remained unsolved as, for example, the problem of stabilisability by means of a constant-gain output feedback. The following results are derived by means of a characterisation of all stabilising state-feedback controllers or dynamic compensators of arbitrary dynamical order, respectively.

Simultaneous stabilisability by state-feedback

We consider the controllable plant

$$\dot{\mathbf{x}} = \mathbf{A}\mathbf{x} + \mathbf{B}\mathbf{u} \tag{5}$$

under state feedback

$$\mathbf{u} = -\mathbf{K}\mathbf{x}. \tag{6}$$

Lemma 7.1 The set of all controllers (6) for which the closed-loop system (5), (6) is stable is given by

$$\mathbf{K} = \mathbf{K}^0 + \mathbf{L}\mathbf{B}'\mathbf{P} \tag{7}$$

where $\mathbf{K}^0$ is an arbitrary real (m, n)-matrix and $\mathbf{P}$ is the positive definite solution of the matrix-Riccati-equation

$$(\mathbf{A} - \mathbf{B}\mathbf{K}^0)'\,\mathbf{P} + \mathbf{P}(\mathbf{A} - \mathbf{B}\mathbf{K}^0) - 2\mathbf{P}\mathbf{B}\mathbf{L}\mathbf{B}'\mathbf{P} + \mathbf{S} = \mathbf{0}. \tag{8}$$

$\mathbf{L}$ and $\mathbf{S}$ are arbitrary, symmetric positive definite matrices.

The proof of this lemma is based on the fact that the matrix $\mathbf{P}$ represents the solution of the Lyapunov equation

$$(\mathbf{A} - \mathbf{B}\mathbf{K})'\,\mathbf{P} + \mathbf{P}(\mathbf{A} - \mathbf{B}\mathbf{K}) = -\mathbf{S} \tag{9}$$

of the closed-loop system (5), (6). If different parameter sets $(\mathbf{A}^i, \mathbf{B}^i)$ are given we use the controller

$$\mathbf{K} = \mathbf{K}^0 + \sum_{i=1}^{q} \mathbf{L}^i \mathbf{B}^{i\prime} \mathbf{P}^i \tag{10}$$

which with $\mathbf{K}^{0i} = \mathbf{K}^0 + \sum_{j \neq i} \mathbf{L}^j \mathbf{B}^{j\prime} \mathbf{P}^j$ has the form (7) if $\mathbf{P}^j$ is the solution of eqn (8) for the system $(\mathbf{A}^i, \mathbf{B}^i)$.

Theorem 7.1 The systems (5) with parameter sets $(\mathbf{A}^i, \mathbf{B}^i)$ $(i = 1, 2, \ldots, q)$ is simultaneously stabilisable by a state feedback (6) if and only if the Riccati equations

$$\Big(\mathbf{A}^i - \mathbf{B}^i\mathbf{K}^0 - \mathbf{B}^i \sum_{j \neq i} \mathbf{L}^j \mathbf{B}^{j\prime} \mathbf{P}^j\Big)'\,\mathbf{P}^i + \mathbf{P}^i\Big(\mathbf{A}^i - \mathbf{B}^i\mathbf{K}^0 - \mathbf{B}^i \sum_{j \neq i} \mathbf{L}^j \mathbf{B}^{j\prime} \mathbf{P}^j\Big)$$
$$- 2\mathbf{P}^i\mathbf{B}^i\mathbf{L}^i\mathbf{B}^{i\prime}\mathbf{P}^i + \mathbf{S}^i = \mathbf{0} \tag{11}$$

$(i = 1, \ldots, q)$ with arbitrary real matrix $\mathbf{K}^0$ and arbitrary positive definite matrices $\mathbf{S}^i$ and $\mathbf{L}^i$ have positive definite solutions $\mathbf{P}^i$. If this condition is satisfied, a stabilising controller is given by eqn (10).

Note that the equations (11) are coupled and can only be solved iteratively.

Example 7.3 As an illustration apply Theorem 7.1 to Example 7.1 with $a^1 = 1$, $\lambda^1 = -1$ and $a^2 = -2$, $\lambda^2 = 1$. Using $s^1 = s^2 = l^1 = l^2 = 1$ and $k^0 = 0$ the Riccati equations (11) read as $2p^1(-1 + 2p^2) - 2(p^1)^2 + 1 = 0$ and $2p^2(1 + 2p^1) - 8(p^2)^2 + 1 = 0$, which actually have a positive solution $p^1 = -\sqrt{2}$ and $p^2 = 1.207$. Hence, both systems can be stabilised by the feedback (6) with $k = 1.71 + (-2)\,1.21 = -0.71$ (Eqn (3) gives $-1 \leq k \leq -0.5$). □

Eqn (10) suggests to use a combination of the individual LQ-regulators $\mathbf{K}^i = (\mathbf{R}^i)^{-1}\,\mathbf{B}^{i\prime}\mathbf{P}^i$, which are received for the given plant parameter sets $(\mathbf{A}^i, \mathbf{B}^i)$ and arbitrarily fixed weighting matrices $(\mathbf{Q}^i, \mathbf{R}^i)$ via the Riccati equation, as the common controller. For instance, $\mathbf{K}$ can be chosen to be an 'average controller'

$$\mathbf{K} = \sum_{i=1}^{q} \varepsilon^i \mathbf{K}^i = \sum_{i=1}^{q} \varepsilon^i (\mathbf{R}^i)^{-1}\,\mathbf{B}^{i\prime}\mathbf{P}^i. \tag{12}$$

Then the existence of a solution to eqn (11) can be checked with ε^i as weighting factors to be fixed so as to satisfy Theorem 7.1.

Simultaneous stabilisability by dynamic controllers

As the introduction of dynamics into the controller may make the problem solvable, the existence of simultaneous stabilising controllers has to be tackled in a broader framework than previously. We will use now results from algebraic system theory to present a more general result. The m-input m-output plant

$$\mathbf{y}(p) = \mathbf{G}(p)\,\mathbf{u}(p) \tag{13}$$

is assumed to have a strictly proper rational transfer function matrix $\mathbf{G}(p)$ whose right and left coprime factorisations over $\mathcal{S}(p)^{m\times m}$ are given by

$$\mathbf{G}(p) = \mathbf{N}(p)\,\mathbf{D}(p)^{-1} = \tilde{\mathbf{D}}(p)^{-1}\,\tilde{\mathbf{N}}(p)\,. \tag{14}$$

Then there exist matrices $\mathbf{P}, \mathbf{Q}, \tilde{\mathbf{P}}$ and $\tilde{\mathbf{Q}}$ such that

$$\mathbf{PN} + \mathbf{QD} = \mathbf{I}, \qquad \tilde{\mathbf{N}}\tilde{\mathbf{P}} + \tilde{\mathbf{D}}\tilde{\mathbf{Q}} = \mathbf{I} \tag{15}$$

hold (cf Section 2.1). The plant (13) is to be stabilised by a dynamic controller

$$\mathbf{u}(p) = \mathbf{K}(p)\,\big(\mathbf{v}(p) - \mathbf{y}(p)\big)\,. \tag{16}$$

Lemma 7.2 The set of controllers (16) for which the closed-loop system (13), (16) is stable is given by

$$\mathbf{K}(p) = (\mathbf{Q} - \mathbf{R}\tilde{\mathbf{N}})^{-1}\,(\mathbf{P} + \mathbf{R}\tilde{\mathbf{D}}) \tag{17}$$

with $\mathbf{P}$, $\mathbf{Q}$, $\tilde{\mathbf{N}}$ and $\tilde{\mathbf{D}}$ from eqns (14) and (15). $\mathbf{R}(p)$ is an arbitrary matrix from $\mathcal{S}(p)^{m\times m}$.

If two plants (13) whose matrices $\mathbf{G}^1(p)$ and $\mathbf{G}^2(p)$ are factorised by $\mathbf{N}^i$, $\mathbf{D}$ etc have to be stabilised the individual sets of stabilising controllers must have at least one common element. Hence,

$$(\mathbf{Q}^1 - \mathbf{R}^1\tilde{\mathbf{N}}^1)^{-1}\,(\mathbf{P}^1 + \mathbf{R}^1\tilde{\mathbf{D}}^1) = (\mathbf{Q}^2 - \mathbf{R}^2\tilde{\mathbf{N}}^2)^{-1}\,(\mathbf{P}^2 + \mathbf{R}^2\tilde{\mathbf{D}}^2) \tag{18}$$

must hold for some $\mathbf{R}^1, \mathbf{R}^2 \in \mathcal{S}(p)^{m\times m}$. This condition can be reformulated by means of two newly defined matrices $\mathbf{A}^0$ and $\mathbf{B}^0$

$$\begin{pmatrix}\mathbf{A}^0\\ \mathbf{B}^0\end{pmatrix} = \begin{pmatrix}\mathbf{Q}^1 & \mathbf{P}^1\\ -\tilde{\mathbf{N}}^1 & \tilde{\mathbf{D}}^1\end{pmatrix}\begin{pmatrix}\mathbf{D}^2\\ \mathbf{N}^2\end{pmatrix} = \begin{pmatrix}\mathbf{Q}^1\mathbf{D}^2 + \mathbf{P}^1\mathbf{N}^2\\ -\tilde{\mathbf{N}}^1\mathbf{D}^2 + \tilde{\mathbf{D}}^1\mathbf{N}^2\end{pmatrix} \tag{19}$$

in the following way:

Theorem 7.2 Two systems (13) with strictly proper rational transfer function matrices $\mathbf{G}^1(p)$ and $\mathbf{G}^2(p)$ can be stabilised by means of the same controller (16) if and only if the system (13) with $\mathbf{G}(p) = \mathbf{B}^0(p)\,\mathbf{A}^0(p)^{-1}$ can be stabilised by a stable controller (16). If this condition is satisfied, simultaneously stabilising controllers are given by

$$\mathbf{K}(p) = (\mathbf{Q}^1 - \mathbf{R}\tilde{\mathbf{N}})^{-1}\,(\mathbf{P}^1 + \mathbf{R}\tilde{\mathbf{D}}^1)$$

for all $\mathbf{R} \in \mathcal{S}(p)^{m\times m}$ for which $\mathbf{A}^0 + \mathbf{R}\mathbf{B}^0$ is unimodular.

That is, the stabilisability of two systems $\mathbf{G}^1(p)$ and $\mathbf{G}^2(p)$ can be checked by means of the auxiliary system $\mathbf{B}^0(p)\,\mathbf{A}^0(p)^{-1}$, which must be stabilisable by a *stable* controller $\mathbf{K}(p)$ ('strongly stabilisable'). For single-variable plants (13) the existence of a stable stabilising controller (16) depends merely on the distribution of the *real* right-half plane poles and zeros.

Lemma 7.3 Consider a single-input single-output system (13). Denote the total number of real poles to the right of the different real right-half plane zeros p_{0i} of the plant transfer function $g(p)$ by ν_i. There exists a stable stabilising controller (16) if and only if all ν_i are odd (or even).

This property of the plant is referred to as the '*parity interlacing property*'.

If one of the two plants, say $\mathbf{G}_1$, is stable then the auxiliary system $\mathbf{B}_0\mathbf{A}_0^{-1}$ has an obvious interpretation. Since $\mathbf{N}^1 = \tilde{\mathbf{N}}^1 = \mathbf{G}^1$, $\mathbf{D}^1 = \tilde{\mathbf{D}}^1 = \mathbf{I}$ represents a coprime factorisation over $\mathscr{S}(p)^{m\times m}$, we get $\mathbf{P}^1 = \tilde{\mathbf{P}}^1 = \mathbf{0}$, $\mathbf{Q}^1 = \tilde{\mathbf{Q}}^1 = \mathbf{I}$ and, finally,

$$\mathbf{B}^0(\mathbf{A}^0)^{-1} = \mathbf{G}^2 - \mathbf{G}^1. \tag{20}$$

The stable system $\mathbf{G}^1(p)$ and the possibly unstable system $\mathbf{G}^2(p)$ are simultaneously stabilisable, if and only if the difference between both systems is strongly stabilisable.

Example 7.4 For Example 7.1 with the parameters given in Example 7.3 we have $g^2 - g^1 = -\dfrac{3p+1}{(p+1)\,(p-1)}$ which has no right-half plane zero and is, by Lemma 7.3, stabilisable by a stable controller (16). Hence, both systems are simultaneously stabilisable. □

The general problem of simultaneous stabilisability of $q \geqq 2$ plants can be investigated by Theorem 7.2 if $\mathbf{G}^2, \mathbf{G}^3, \ldots, \mathbf{G}^q$ are combined with $\mathbf{G}^1$. We get $q - 1$ systems of the form $\mathbf{B}^0(\mathbf{A}^0)^{-1}$ which have to be stabilisable by stable compensators. This condition, however, is not computationally verifiable for $q > 2$.

Theorem 7.2 has regard to the stabilisation by a dynamic feedback (16) of arbitrary dynamical order. The question which minimal dynamical order is necessary to accomplish the stability task is quite open. Examples have shown that there is no a-priori upper bound of the controller order in terms of the plant order.

7.2 Simultaneous pole region assignment

In this section we describe a design procedure for robust controllers which represents an extension of the well-known pole assignment principle to the multi-model design problem. As the sharp requirement that all closed-loop systems have exactly the same poles can only be met for specific plant properties or plant uncertainties the design objective is formulated as placement of the closed-loop eigenvalues into a given region.

Accordingly, for the single-input plant

$$\dot{\mathbf{x}} = \mathbf{A}\mathbf{x} + \mathbf{b}u \tag{1}$$

a state-feedback controller

$$u = -\mathbf{k}'\mathbf{x} \tag{2}$$

has to be found so that the eigenvalues of the closed-loop system (1), (2) lie within a prescribed region $\mathscr{L}$ of the complex plane for all given plant parameter sets. The closed-loop system characteristic polynomial is given by

$$f(p, \mathbf{c}^i) = \det(p\mathbf{I} - \mathbf{A}^i + \mathbf{b}^i\mathbf{k}') = p^n + c_1{}^i p^{n-1} + \cdots + c_n{}^i \tag{3}$$

with $\mathbf{c}^i = (c_n{}^i, c_{n-1}^i, \ldots, c_1{}^i)'$. The region $\mathscr{L}$ of 'nice stability' depicted in Fig. 7.2(a) requires the closed-loop eigenvalues to lie to the left of a hyberbola (B_1)

$$\omega^2 = \delta^2 - d_1{}^2 \tag{4}$$

and to the right of a circle (B_2)

$$\omega^2 + \delta^2 = d_2{}^2 \tag{5}$$

and, hence, describes closed-loop systems with substantial stability degree and limited bandwidth.

The method of simultaneous pole region assignment proceeds in two steps:

(1) The eigenvalue constraints shown in Fig. 7.2(a) are transformed into equivalent constraints on the coefficients of the characteristic polynomial (3). As a result, we bound a subset $\mathscr{C}_{\mathscr{L}}$ of admissible polynomial coefficients within the real coefficient space $\mathbb{R}^n$.

(2) For each plant parameters ($\mathbf{A}^i$, $\mathbf{b}^i$) the set $\mathscr{C}_{\mathscr{L}}$ can be related to a corresponding subset $\mathscr{K}_{\mathscr{L}}{}^i$ of the space $\mathbb{R}^n$ of real state-feedback parameters. The intersection $\mathscr{K}_{\mathscr{L}}$ of these subsets yields the set of controllers (2) which simultaneously assign the closed-loop system eigenvalues into $\mathscr{L}$:

$$\mathscr{K}_{\mathscr{L}} = \bigcap_{i=1,\ldots,q} \mathscr{K}_{\mathscr{L}}{}^i. \tag{6}$$

The first step can be accomplished without reference to the plant parameters. If the eigenvalues are varied continuously with a real eigenvalue λ or a conjugate-complex pair of eigenvalues $\delta \pm j\omega$ on the boundaries of $\mathscr{L}$, the characteristic polynomial (3) can be decomposed according to

$$f(p, \mathbf{c}) = (p - \lambda)\,\tilde{\tilde{f}}(p, \tilde{\tilde{\mathbf{c}}}) \tag{7}$$

or

$$f(p, \mathbf{c}) = (p - \delta - j\omega)(p - \delta + j\omega)\,\tilde{f}(p, \tilde{\mathbf{c}}) \tag{8}$$

where $\tilde{\tilde{f}}(p, \tilde{\tilde{\mathbf{c}}})$ and $\tilde{f}(p, \tilde{\mathbf{c}})$ are polynomials of order $n - 1$ and $n - 2$, respectively. Then the characteristic equation $f(p, \mathbf{c}) = 0$ with $f(p, c)$ from eqns (7) or (8) describes the corresponding boundaries within the coefficient space $\mathbb{R}^n$.

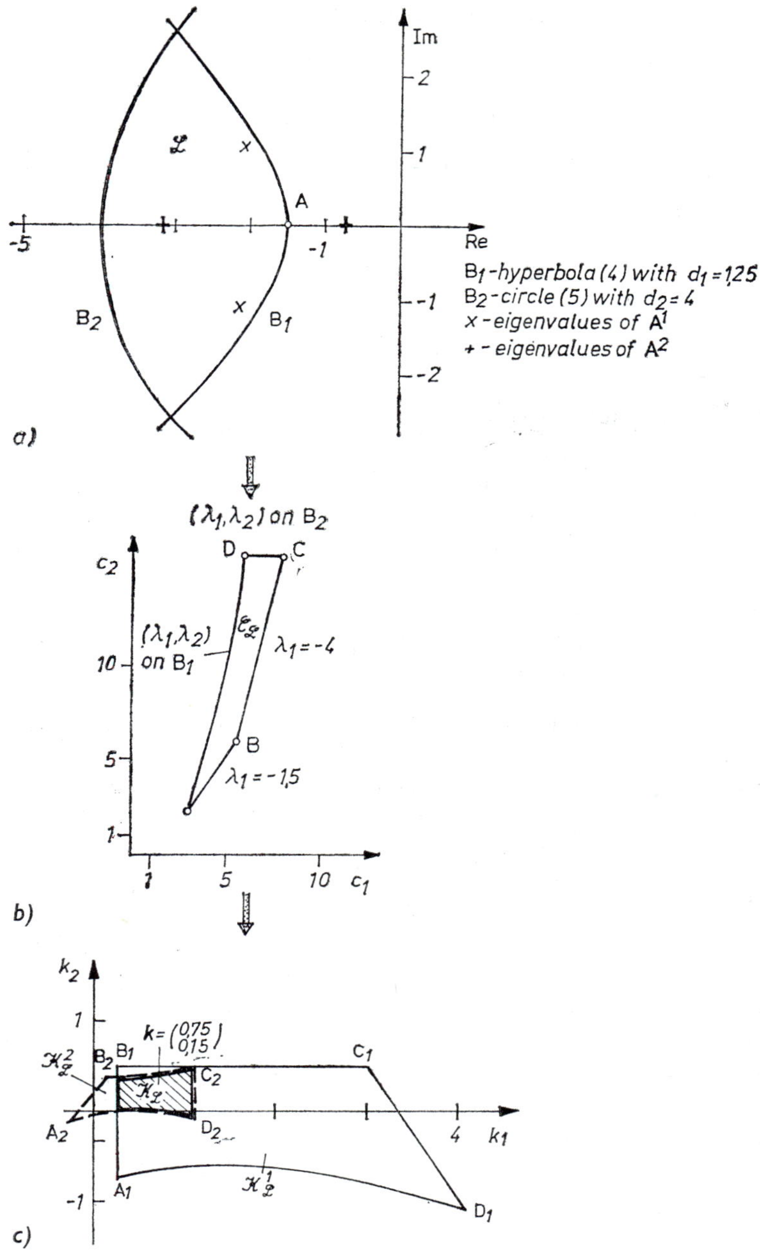

Fig. 7.2 Pole region assignment

(a) Admissible region $\mathscr{L}$ of the closed-loop system eigenvalues

(b) Corresponding region $\mathscr{C}_{\mathscr{L}}$ of the polynomial coefficients

(c) Overlapping sets $\mathscr{K}_{\mathscr{L}}{}^{i}$ of the individual controllers and set $\mathscr{K}_{\mathscr{L}}$ of robust controllers

For $\mathscr{L}$ in Fig. 7.2(a) we get for a second-order plant (1) from eqn (7)

$$f(p, \mathbf{c}) = (p + 1.5)(p - \lambda_2) = p^2 + (1.5 - \lambda_2)\, p - 1.5\lambda_2 \tag{9}$$

and

$$f(p, \mathbf{c}) = (p + 4)(p - \lambda_2) = p^2 + (4 - \lambda_2)\, p - 4\lambda_2, \tag{10}$$

and from eqn (8)

$$f(p, \mathbf{c}) = p^2 - 2\delta p + 2\delta^2 - 2.25, \qquad -3 \leqq \delta \leqq 1.5 \tag{11}$$

and

$$f(p, \mathbf{c}) = p^2 - 2\delta p + 16, \qquad -4 \leqq \delta \leqq -3. \tag{12}$$

Eqns (9)—(12) describe the pairs (c_1, c_2) that form the boundary of the subset $\mathscr{C}_{\mathscr{L}}$ within the coefficient space $\mathbb{R}^2$ (Fig. 7.2(b)).

In the second step, the controller $\mathbf{k}'$ is determined so as to produce polynomial coefficients $\mathbf{c} \in \mathscr{C}_{\mathscr{L}}$. A direct relation between $\mathbf{k}'$ and $\mathbf{c}$ is given for the plant (1) by Ackermann's formula

$$\mathbf{k}' = (c_n, c_{n-1}, \ldots, c_1, 1) \begin{pmatrix} \mathbf{e}' \\ \mathbf{e}'\mathbf{A} \\ \mathbf{e}'\mathbf{A}^n \end{pmatrix} = (\mathbf{c}' \quad 1)\, \mathbf{E} \tag{13}$$

with $\mathbf{e}' = (0, 0, \ldots, 0, 1)(\mathbf{b}, \mathbf{Ab}, \ldots, \mathbf{A}^{n-1}\mathbf{b})^{-1}$. It can be used for every parameter set $(\mathbf{A}^i, \mathbf{b}^i)$ to determine $\mathscr{K}_{\mathscr{L}}{}^i$. As eqn (13) defines an affine mapping $\mathbb{R}^n \to \mathbb{R}^n$, straight lines and hyperplanes bounding $\mathscr{C}_{\mathscr{L}}$ are mapped into straight lines and hyperplanes, respectively, that bound $\mathscr{K}_{\mathscr{L}}{}^i$. Then the set $\mathscr{K}_{\mathscr{L}}$ of robust controllers is given by the intersection (6).

If we use $\mathbf{A}^1 = \begin{pmatrix} -1 & 0.5 \\ 1 & -3 \end{pmatrix}$, $\mathbf{b}^1 = \begin{pmatrix} 1 \\ 2 \end{pmatrix}$ and $\mathbf{A}^2 = \begin{pmatrix} -1 & 1 \\ -2 & -3 \end{pmatrix}$, $\mathbf{b}^2 = \begin{pmatrix} 2 \\ 4 \end{pmatrix}$ for our example, two overlapping regions $\mathscr{K}_{\mathscr{L}}{}^1$ and $\mathscr{K}_{\mathscr{L}}{}^2$ are received (Fig. 7.2(c)). All controllers whose parameters lie within the shaded region in Fig. 7.2(c) simultaneously move the eigenvalues of the closed-loop system into the given region $\mathscr{L}$. For example, if we choose the point $\mathbf{k}' = (0.75 \;\; 0.35)$ in the middle of the admissible region, the eigenvalues of the closed-loop systems with $(\mathbf{A}^1, \mathbf{b}^1)$ and $(\mathbf{A}^2, \mathbf{b}^2)$ are -3.2, -2.02 and $-3.45 \pm j0.7$, respectively. Note that both plant models differ severely since they have quite different eigenvalues (see Fig. 7.2(a)) and different static couplings between u and $\mathbf{x}$. Nevertheless, simultaneous pole region assignment is possible.

For systems with dynamical order greater than 2 or 3 the visualisation of the sets $\mathscr{K}_{\mathscr{L}}{}^i$ and $\mathscr{K}_{\mathscr{L}}$ represents the most difficult problem of this design technique. It can be overcome by fixing $n - 2$ eigenvalues and controller gains, and considering the regions of the remaining two controller parameters in dependence upon the remaining two eigenvalues. The characteristic polynomial for $n - 2$ fixed eigenvalues is given by

$$f(p, \mathbf{c}) = \tilde{f}(p, \tilde{\mathbf{c}})\,(p^2 + \bar{c}_1 p + \bar{c}_2) \tag{14}$$

where the coefficients of the $(n - 2)$-th order polynomial $\tilde{f}(p, \tilde{\mathbf{c}})$ are determined by the fixed closed-loop poles. Moving the free eigenvalues on the boundary of $\mathscr{L}$, $f(p, \mathbf{c}) = 0$ with $f(p, \mathbf{c})$ from (14) yields the admissible region $\mathscr{C}_{\mathscr{L}}$ of $(\bar{c}_1, \bar{c}_2)$.

If we assume, for the sake of simplicity, that the first two controller parameters are to be chosen, then eqn (13) can be transformed into a relation between k_1, k_2 and $\bar{c}_1$, $\bar{c}_2$ which is similar to (13). Using

$$(\mathbf{c}' \quad 1) = (\tilde{\mathbf{c}}' \quad 1)\begin{pmatrix} \bar{c}_2 & \bar{c}_1 & 1 & 0 & \dots & & 0 \\ 0 & \bar{c}_2 & \bar{c}_1 & 1 & & & 0 \\ \vdots & \vdots & \vdots & \vdots & & & \vdots \\ 0 & 0 & 0 & 0 & \dots \bar{c}_2 & \bar{c}_1 & 1 \end{pmatrix}$$

and denoting the last row of the explicitly given matrix by $\mathbf{t}'$ and the remaining $(n - 2, n + 1)$-matrix by $\mathbf{S}$ we get

$$(k_1, k_2, \mathbf{k}_f') = (\tilde{\mathbf{c}}' \quad 1)\begin{pmatrix} \mathbf{S} \\ \mathbf{t}' \end{pmatrix}(\mathbf{e}_1\ \mathbf{e}_2\ \mathbf{E}_f)$$

where the matrix $\mathbf{E}$ has been decomposed appropriately. Hence, we have

$$\tilde{\mathbf{c}}' = (\mathbf{k}_f' - \mathbf{t}'\mathbf{E}_f)\,(\mathbf{S}\mathbf{E}_f)^{-1} \tag{15}$$

and

$$(k_1\ k_2) = (\mathbf{c}' \quad 1)\,(\mathbf{e}_1, \mathbf{e}_2)\,. \tag{16}$$

These equations describe the region for the two free controller parameters k_1 and k_2 in dependence upon $(\bar{c}_1, \bar{c}_2) \in \mathscr{C}_{\mathscr{L}}$ included in $\mathbf{t}$ and $\mathbf{S}$.

In the same way as we got the boundaries of k_1 and k_2 for our second-order example, boundaries of k_1 and k_2 can be received for fixed $n - 2$ eigenvalues of the n-th order plant. Using this procedure step-by-step for different pairs of controller parameters all elements of the vector $\mathbf{k}'$ can be determined successively. The same way enables us to design output feedback controllers, where some elements of $\mathbf{k}'$ are fixed at zero.

By means of powerful computer graphics the same procedure can be used with $n - 3$ eigenvalues and controller parameters fixed. None the less, the complexity of the design grows significantly with increasing number of controller parameters. This is the reason why an extension to multi-input plants remains to be thoroughly elaborated.

This design method has the obvious advantage to produce all robust controllers of the given structure. Therefore, the existence of a robust controller (2) becomes evident during the design process although no a-priori existence test is available. The choice of the controller $\mathbf{k}'$ from the set $\mathscr{K}_{\mathscr{L}}$ of admissible parameters can be made with regard to further design objectives such as bounds on $|u|$ or $|\dot{u}|$ or $||\mathbf{k}||$. The design method has been successfully applied eg to a flight and a loading bridge control problem (see references).

Bibliographical notes

The problem of simultaneous stabilisability has been tackled by Vidyasagar and Viswanadham (1982) who derived the result presented in Theorem 7.2 on the basis of the characterisation of stabilising controllers (Lemma 7.2) by Desoer et al. (1980). Lemma 7.3 is due to Youla et al. (1974). Smith and Sondergeld (1987) have shown very recently that no a-priori bound on the dynamical order of a strongly stabilising controller exists in terms of the plant order.

More specific results have been described by Emre (1983), Ghosh (1983), and Ghosh and Byrnes (1983) who investigated the simultaneous stabilisation and pole assignment problems by algebraic methods and derived necessary or sufficient stabilisability conditions. Alosh (1983) derived a sufficient condition for the stability of the closed-loop system under actuator or measurement device failures by considering the integrity of the closed-loop system as a problem of simultaneous stabilisation.

The results reported in Lemma 7.1 and Theorem 7.1 are due to Bux (1975).

Ackermann's method of simultaneous pole region assignment is described in detail in the paper (Ackermann 1980) and the textbook (Ackermann 1983). The problems that arise by fixing $n-2$ eigenvalues in order to reduce the design task has been tackled by Ackermann and Türk (1982), and Ackermann and Kaesbauer (1981) who gave numerically efficient algorithms and a **D**-decomposition for determining the resulting hyperplane in the $\mathcal{K}$-space. Application examples have been extensively discussed by Franklin and Ackermann (1981), and Ackermann (1983) for a flight and a loading bridge control problem, respectively. Robustness against sensor failures has been investigated as a problem of pole region assignment by Ackermann (1984).

Other results along the lines of pole region assignment have been published eg by Frank et al. (1982) and Heger (1984) who investigated which pole regions are likely to be simultaneously assignable. Based on the sensitivity analysis of the state-feedback system, guidelines have been proposed for deriving reasonable regions for the desirable closed-loop eigenvalues. The controller is designed for fixed eigenvalues within this region and for one plant parameter set, and afterwards investigated in connection with all specimens of the plant. Looze (1983) extended the LQ-design principle to the multi-model design problem.

8 QUANTITATIVE ROBUSTNESS OF SYSTEMS WITH BOUNDED UNCERTAINTIES

This chapter is devoted to systems whose uncertainties are described by some upper bounds. The set $\mathcal{S}$ of possible representatives of the original system embodies a bounded continuum, ie the plant perturbations can vary continuously. For such systems, the transformation of the uncertainty bounds of the plant into bounds on the closed-loop uncertainties turns out to be the basic problem of robustness assessment.

Section 8.1 explains the quantitative feedback design method for single-loop systems which provides a good insight into the problem of choosing the controller with regard to a continuous set of plants. Sections 8.2—8.5 as the main part of this chapter describe how the transformation of the open-loop into the closed-loop uncertainties can be carried out for multi-input multi-output systems once a controller with fixed parameters has been given. These sections present the methods for robustness analysis which constitute the major steps of the design strategy explained in Section 8.6. As norm-bounded as well as element-by-element-bounded uncertainties play important roles in the current literature and will be used in the application studies of the next chapters, the results received for both kinds of the uncertainty description will be derived in parallel. The last section of this chapter deals with the evaluation of tolerable uncertainties and criteria for robust stability if the uncertainties are given in terms of the parameters of the state-space model.

8.1 The quantitative feedback design

The 'quantitative feedback design method' of Horowitz has been proposed to make the relation between the plant uncertainties, the controller parameters, and the resulting variations of the closed-loop performance explicit. It enables the designer to relate the amount of feedback used to the amount of uncertainties encountered in single-input single-output systems.

It is assumed that quantitative bounds of the plant uncertainties and of the desired closed-loop system response are given. Using a two-degrees-of-freedom control configuration (Fig. 2.5(b)) compensator and prefilter functions are to be found that guarantee a satisfactory I/O-behaviour over the given range of the plant uncertainties. More specifically, the plant is described by the frequency response $g(j\omega, \mathbf{a})$ where the uncertain parameter vector is known to belong to a given set $\mathcal{A}$, eg to the interval $\bar{\bar{\mathbf{a}}} \leqq \mathbf{a} \leqq \bar{\mathbf{a}}$. As the plant is assumed to be minimum-phase the magnitude $|t(j\omega)|_{dB}$ of the closed-loop frequency response completely characterises the closed-loop performance. A given tolerance band describes the design objectives concerning command tracking (Fig. 8.1, cf Section 1.1). $|t|_{dB}$ is used as abbreviation of $20 \lg |t|$. As a further design specification, sensor noise has to be attenuated according to

$$\left| \frac{l(j\omega, \mathbf{a})}{1 + l(j\omega, \mathbf{a})} \right|_{dB} \leqq d \tag{1}$$

where the term on the left-hand side describes the transmission of the disturbance through the closed loop with $l(j\omega, \mathbf{a})$ denoting the open-loop frequency response.

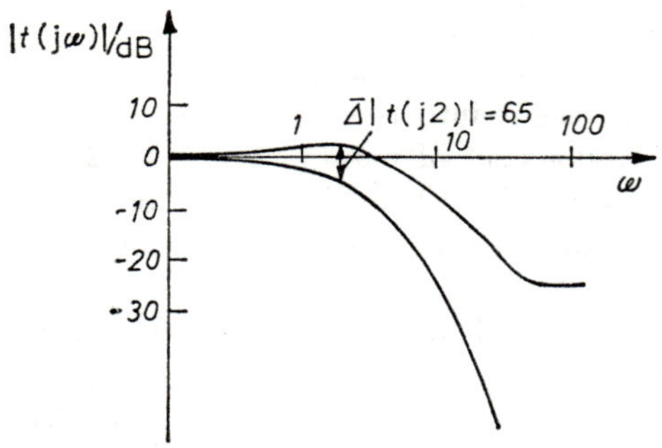

Fig. 8.1 Design specification: tolerance band for $t(j\omega)$

Preliminary considerations

As we use a two-degrees-of-freedom configuration as shown in Fig. 2.5(b1) the design task can be divided into two subtasks. At first, the feedback compensator $k(j\omega)$ has to be chosen, so that the uncertainties $\Delta\,|\tilde{t}(j\omega)|_{dB}$ of the frequency response

$$\tilde{t}(j\omega, \mathbf{a}) = \frac{g(j\omega, \mathbf{a})\, k(j\omega)}{1 + g(j\omega, \mathbf{a})\, k(j\omega)} \tag{2}$$

of the closed-loop system without prefilter is smaller than the width $\bar{\Delta}\,|t(j\omega)|_{dB}$ of the given tolerance band in Fig. 8.1

$$\Delta\,|\tilde{t}(j\omega)|_{dB} = \max_{\mathbf{a}\in\mathcal{A}} |\tilde{t}(j\omega, \mathbf{a})|_{dB} - \min_{\mathbf{a}\in\mathcal{A}} |\tilde{t}(j\omega, \mathbf{a})|_{dB} \leqq \bar{\Delta}\,|t(j\omega)|_{dB} \tag{3}$$

and that eqn (1) is satisfied. Then $\Delta\,|t(j\omega)|$ is small enough because the width of the tolerance band remains unaltered by the prefilter. This can be seen from

$$t(j\omega, \mathbf{a}) = f(j\omega)\, \tilde{t}(j\omega, \mathbf{a}) \tag{4}$$

and

$$\begin{aligned}\Delta\, |t(j\omega)|_{dB} &= |t(j\omega, \tilde{\mathbf{a}})|_{dB} - |t(j\omega, \tilde{\tilde{\mathbf{a}}})|_{dB} \\ &= [|f(j\omega)|_{dB} + |\tilde{t}(j\omega, \tilde{\mathbf{a}})|_{dB}] - [|f(j\omega)|_{dB} + |\tilde{t}(j\omega, \tilde{\tilde{\mathbf{a}}})|_{dB}] \\ &= \Delta\, |\tilde{t}(j\omega)|_{dB} \leqq \bar{\Delta}\, |t(j\omega)|_{dB}\end{aligned}$$

where $\tilde{\mathbf{a}}$ and $\tilde{\tilde{\mathbf{a}}}$ denote the parameter vectors for which $|t(j\omega, \mathbf{a})|$ assumes its maximum or minimum value for the given frequency ω.

At second, the prefilter $f(j\omega)$ has to be designed so that the form of the tolerance band of $|t(j\omega)|$ is as required. As the width of the tolerance band has already been adjusted this subtask can be accomplished by designing $f(j\omega)$ for the nominal parameter vector $\hat{\mathbf{a}}$.

The second preliminary consideration concerns the representation of $l(j\omega, \mathbf{a}) = g(j\omega, \mathbf{a})\, k(j\omega)$ in the *Nichols chart*. The Nichols chart is a graphical means for determining the closed-loop frequency response $t(j\omega)$ for a given frequency response $l(j\omega)$ of the open loop. It was commonly used by control engineers before computers became available on a large scale and has fallen into oblivion afterwards. Appendix 3 provides a short refresher of it. From $l(j\omega, \mathbf{a}) = g(j\omega, \mathbf{a})\, k(j\omega)$ we have

$$|l(j\omega, \mathbf{a})|_{dB} = |g(j\omega, \mathbf{a})|_{dB} + |k(j\omega)|_{dB} \tag{5}$$

$$\phi(\omega) = \arg l(j\omega, \mathbf{a}) = \arg g(j\omega, \mathbf{a}) + \arg k(j\omega)\,. \tag{6}$$

Hence, the open-loop point $l(j\omega, \mathbf{a})$ in the Nichols chart is received by shifting the point $g(j\omega, \mathbf{a})$ of the plant the distance $\arg k(j\omega)$ to the right and $|k(j\omega)|_{dB}$ to the top.

Design steps

The set $\mathcal{S}$ of plant representatives correponds to a region in the Nichols chart. This region has to be shifted into a position for which the conditions (1) and (3) are satisfied. The conditions can be checked by means of the parameters of the contour lines that go through the new region. The horizontal and vertical distances between the shifted and the original regions determine the feedback $k(j\omega)$ which is necessary for the uncertainty attenuation. Finally, the prefilter has to be designed for the system with a fixed parameter vector $\hat{\mathbf{a}}$. These design steps will be explained in more detail now by means of the example

$$g(j\omega, \mathbf{a}) = \frac{a_1 a_2}{j\omega(j\omega + a_2)}, \qquad \begin{array}{l} 1 \leqq a_1 \leqq 10 \\ 1 \leqq a_2 \leqq 10\,. \end{array} \tag{7}$$

1. *Determination of the region of* $g(j\omega, \mathbf{a})$ in the Nichols chart for given frequency points $\omega_1, \omega_2, \ldots$ The uncertainties superimpose in the following way

$$|g(j\omega, \mathbf{a})|_{dB} = |a_1|_{dB} + \left|\frac{a_2}{\sqrt{a_2^2 + \omega^2}}\right|_{dB} \tag{8}$$

$$\arg g(j\omega, \mathbf{a}) = \arc\tan\frac{a_2}{\omega} - \frac{\pi}{2}. \tag{9}$$

At $\overline{\omega} = 2$ the first term in eqn (8) yields 0...20 dB for $a_1 = 1 \ldots 10$ and the second term -13 dB...-6.2 dB for $a_2 = 1 \ldots 10$. Both terms depend monotonically on a_1 or a_2, respectively. Eqn (9) leads to $-153.4° \ldots 101.3°$ for $a_2 = 1 \ldots 10$. The region for $g(j\omega, \mathbf{a})$ is shown in Fig. 8.2 where the points A, B, C, and D correspond to the parameter combinations (a_1, a_2) indicated there. The boundaries AB, BC, CD, and DA are given by eqns (8), (9) for $a_2 = 1$, $a_1 = 10$, $a_2 = 10$ and $a_1 = 1$, respectively. For $\omega = 1$ or 10 similar

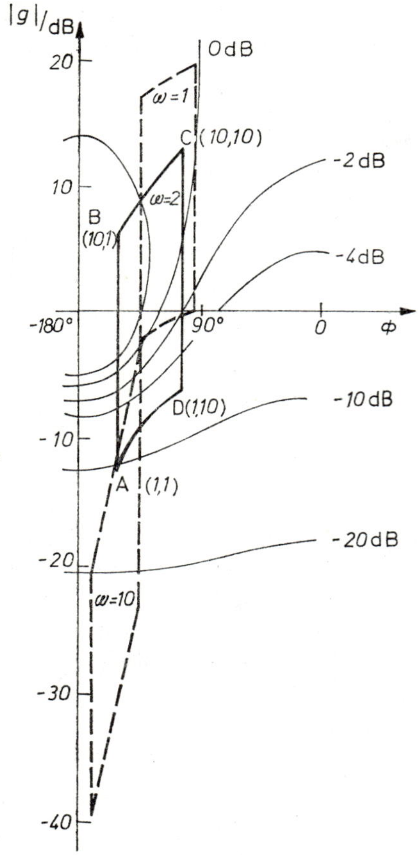

Fig. 8.2 Regions of $g(j\omega, \mathbf{a})$, $\underline{\mathbf{a}} \leq \mathbf{a} \leq \bar{\mathbf{a}}$

regions are obtained. These regions altogether represent the uncertain plant (7).

2. *Determination of bounds of* $l(j\omega, \mathbf{a})$ so as to satisfy conditions (1) and (3). As the controller $k(j\omega)$ transforms all points of the region ABCD in like manner, the distance arg $k(j\omega)$ to the right and $|k(j\omega)|$ to the top, the whole region is shifted while retaining its form and not being rotated (Fig. 8.3(a)). This should be exploited to satisfy the condition (3), which, for our example, reads as

$$\Delta\, |\tilde{t}(j2)|_{dB} = 6.5\,.$$

The tolerance $\Delta\, |\tilde{t}(j2)|_{dB}$ can be read off the Nichols chart as difference of the parameters belonging to the extreme curves through the region of $l(j2, \mathbf{a})$. For instance, suppose we have shifted the region ABCD of $g(j2, \mathbf{a})$ into the region A′B′C′D′ of $l(j2, \mathbf{a})$. Then the magnitude of $\tilde{t}(j2)$ varies between the values −0.49 dB and −5.7 dB that correspond to the parameters of the contours passing the points C′ and A′. Hence, we have $\Delta\, |\tilde{t}(j2)|_{dB} = 5.2$. As 6.5 dB is the allowable tolerance, the region can be shifted down to A″B″C″D″ where the uncertainty is 6.5 dB the extreme value of $|\tilde{t}(j2)|_{dB}$ being −0.7 dB at C″ and −7.2 dB at A″. Note that due to the form of the contours in the Nichols chart the tolerance $\Delta\, |\tilde{t}(j\omega)|_{dB}$ decreases if the region is shifted upward.

If this manipulation of the region is repeated along other vertical lines corresponding to other open-loop phases, new points A″ are received. They

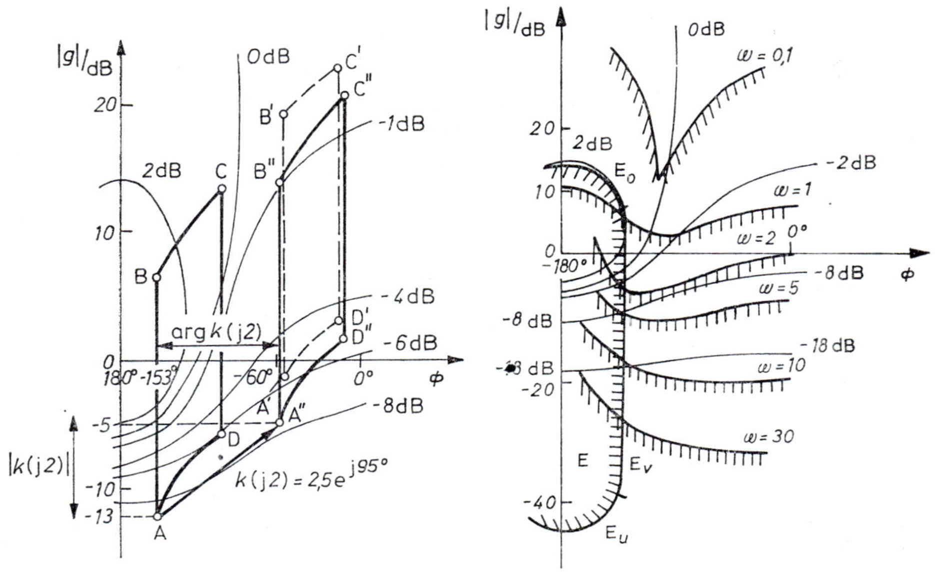

Fig. 8.3 Determination of bounds of $l(j\omega, \mathbf{a})$

a) Move of the region ABCD to the position A″B″C″D″

b) Bounds of $l(j\omega, \hat{\mathbf{a}})$ for different frequencies altogether bound the permissible region of the Nichols chart in which the region ABCD has to be transferred in order to satisfy the condition (3). Such bounds for different frequencies are drawn in Fig. 8.3(b).

A further bound on $l(j\omega, \mathbf{a})$ is given by the design requirement (1) with $d = 2$. Since the whole region ABCD for all frequencies ω must not lie in regions of the Nichols chart where contours with parameters greater than 2 dB exist, this condition leads to the boundary E in Fig. 8.3(b). The parts E_o, E_v and E_u of E are determined so that the point A, the line AB or the point C, respectively, of the region $l(j\omega, \mathbf{a})$ lie on the 2 dB-contour line.

As the result of this design step, the problem of finding an appropriate feedback compensator $k(j\omega)$ for the given set of plants $g(j\omega, \mathbf{a})$ has been resolved to the problem of determining a function $l(j\omega, \hat{\mathbf{a}})$ that observes the bounds drawn in Fig. 8.3(b) for $\mathbf{a} = \hat{\mathbf{a}}$. In what follows, the plant can be considered to have the nominal parameters $\hat{\mathbf{a}} = \begin{pmatrix} a_1 \\ a_2 \end{pmatrix} = \begin{pmatrix} 1 \\ 1 \end{pmatrix}$ corresponding to point A of the region and we need not trouble about the possible effects of plant uncertainties or disturbances.

3. *Determination of* $l(j\omega, \mathbf{a})$ *and* $k(j\omega)$. For a given $l(j\omega, \hat{\mathbf{a}})$ the controller frequency response $k(j\omega) = |k(j\omega)|\, e^{j\phi_k(\omega)}$ can be read off the Nichols chart. $|k(j\omega)|_{dB}$ and ϕ_k are the vertical and horizontal distances between the points corresponding to $g(j\omega, \hat{\mathbf{a}}) = 1/j\omega(1 + j\omega)$ and to $l(j\omega, \hat{\mathbf{a}})$. For instance, if $l(j2, \hat{\mathbf{a}})$ is chosen to be point A″ in Fig. 8.3(a) then $|k(j2)|_{dB} = 8$ and $\phi_k(2) = 95°$. If A‴ in Fig. 8.4(a) is used, $|k(j2)|_{dB} = 10$ and $\phi_k(2) = 30°$ result. The values of $k(j\omega_i)$ received in this way for different frequencies ω_i can be approximated by a rational function $k(j\omega)$ in order to get an explicit feedback law.

The function $l(j\omega, \hat{\mathbf{a}})$ should be chosen so as to produce a controller of the least possible complexity (least number of poles and zeros). To date, no systematic way is known, so that this step requires intuition and experience. Some general guidelines for shaping $l(j\omega, \hat{\mathbf{a}})$ and the problems faced in this task will be discussed now for our examples.

In order to get a controller with the lowest gain and phase lead, we aim points $l(j\omega, \hat{\mathbf{a}})$ that are close to the boundaries. This is restricted by the correspondences between the gain and the phase of low-order controllers. For example, a slope $d\,|l(j\omega, \hat{\mathbf{a}})|/d\omega$ of -20 dB is accompanied by a phase lag of about $-90°$. For this reason, $l(j\omega, \hat{\mathbf{a}})$ may not follow the magnitude jump along the boundary E_v from $|l(j1, \hat{\mathbf{a}})| = 10$ dB to $|l(j50, \hat{\mathbf{a}})| = -40$ dB without phase variation (Fig. 8.3b).

Two tentative functions $l(j\omega, \hat{\mathbf{a}})$ are shown in Fig. 8.4(a). They start in the same initial point whose phase lag of $-90°$ is prescribed by the plant pole located in the origin. The middle part has a reasonable magnitude slope of $|l(j10, \mathbf{a})| - |l(j1, \mathbf{a})| = -20$ dB per decade and 30° phase lag. Using the dashed curve, which is completely outside the bounds, the controller has to bring about alternating lag corner and lead corner frequencies (dashed line in Fig. 8.4(b)) in order to ensure the gain increase of about 30 dB for

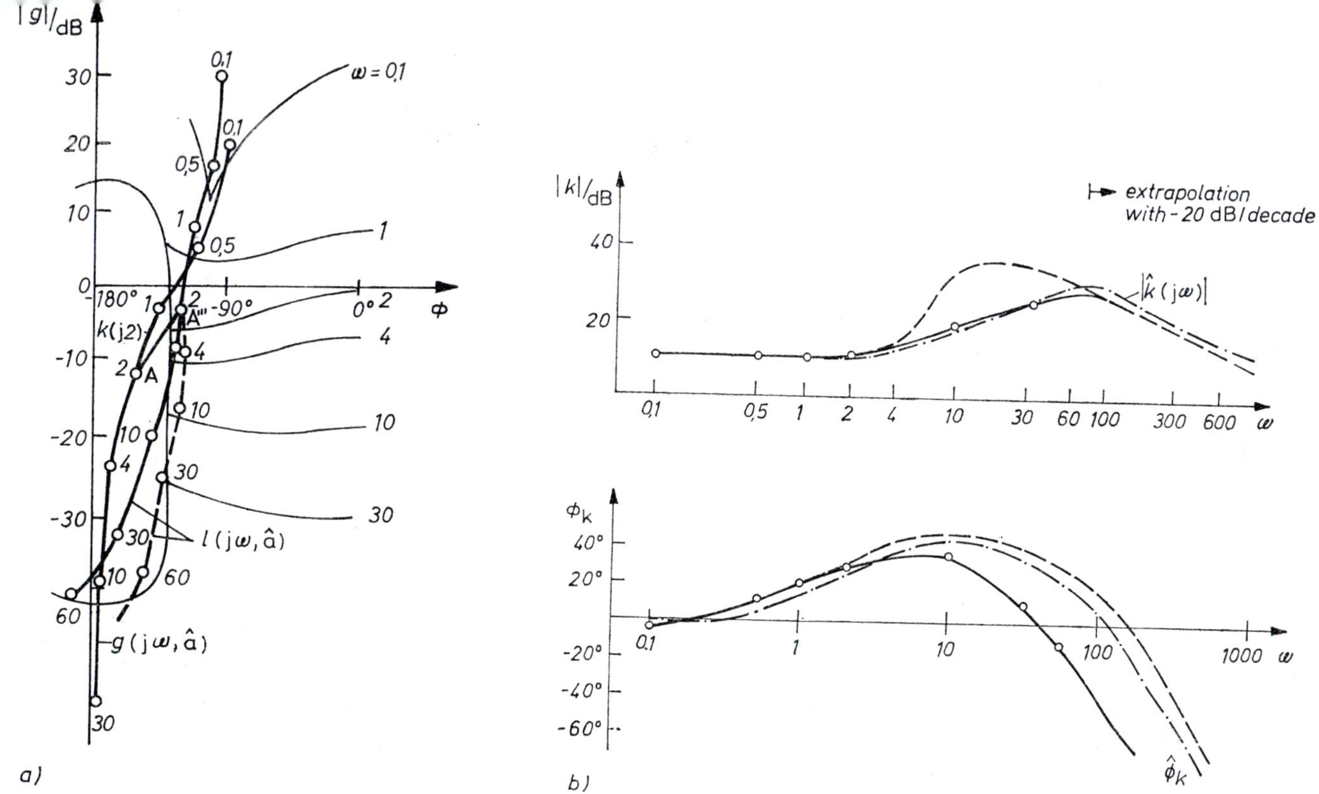

Fig. 8.4 Choice of $l(j\omega, \mathbf{a})$ and $k(j\omega)$

(a) Tentative functions $l(j\omega, \mathbf{a})$ in the Nichols chart

(b) Bode plot of the controller $k(j\omega)$ corresponding to the loop transfer function $l(j\omega, \mathbf{a})$ of (a)

$\omega = 2 \dots 20$ with an accompanying phase lead of less than 40°. However, if we use the solid curve of $l(j\omega, \hat{a})$ a low-order controller is received (solid line in Fig. 8.4(b)) but for the price that the boundary E is violated. This control law can be approximated by

$$k(j\omega) = 3.2 \frac{1 + 0.25j\omega}{(1 + 0.02j\omega)(1 + 0.005j\omega)} \tag{10}$$

(—·— line). The example demonstrates that the problem of choosing a reasonable function $l(j\omega, \hat{a})$ can only be solved by trial-and-error and necessitates a great amount of practise and experience.

4. *Determination of the prefilter* $f(j\omega)$. The prefilter has to move the graph of the command response $t(j\omega, \hat{a})$ into the lower bound of the tolerance band shown in Fig. 8.1. Then $|t(j\omega, a)|$ will be greater than $|t(j\omega, \hat{a})|$ for all other parameters (of Fig. 8.2) but will remain inside the band. To accomplish this task read $|\tilde{t}(j\omega, \hat{a})|_{dB}$ off the Nichols chart as parameters of the contour lines crossing the chosen function $l(j\omega, \hat{a})$ (these contours are omitted in Fig. 8.4(a) but are the same as in Fig. 8.3). $|t(j\omega, \hat{a})|_{dB}$ is given by the lower bound shown in Fig. 8.1. Then

$$|f(j\omega)|_{dB} = |t(j\omega, \hat{a})|_{dB} - |\tilde{t}(j\omega, \hat{a})|_{dB}.$$

For the points indicated at the solid line in Fig. 8.4(a) the prefilter values are drawn in Fig. 8.5. They nearly coincide with the corresponding values of the rational function (dashed line)

$$f(j\omega) = \frac{1}{(1 + 0.25j\omega)(1 + 0.1j\omega)}. \tag{11}$$

Result: As illustrated in Fig. 8.6 the closed-loop system with feedback compensator (10) and prefilter (11) satisfies the objectives given in Fig. 8.1 and eqn (1). Because of the violation of the bound E in Fig. 8.3(b) and the approximations made to get rational transfer functions $k(p)$ and $f(p)$, some curves go slightly beyond the tolerance band.

Remarks. The determination of the region ABCD of $g(j\omega, a)$ for our example has been very simple, because the hull of this region is given by $g(j\omega, a)$ for the extreme parameter values. For other uncertainties, however, this task may become more involved as it is in general unknown for which $a \in \mathcal{A}$ the function $g(j\omega, a)$ lies at the boundary of the corresponding region (think eg of a plant with three uncertain parameters).

For large frequencies the region ABCD degenerates into a vertical line whose length corresponds to the uncertainty $\bar{k}_s$ of the static reinforcement k_s since $g(j\omega, a) \xrightarrow[\omega \to \infty]{} k_s(j\omega)^{\bar{n}-n}$ where $\bar{n} - n$ is the pole-zero excess of the plant. As for high frequencies

$$\Delta\,|t(j\omega)|_{dB} = \Delta\,|\tilde{t}(j\omega)|_{dB} \approx \Delta\,|g(j\omega)|_{dB} = |\bar{k}_s|_{dB}$$

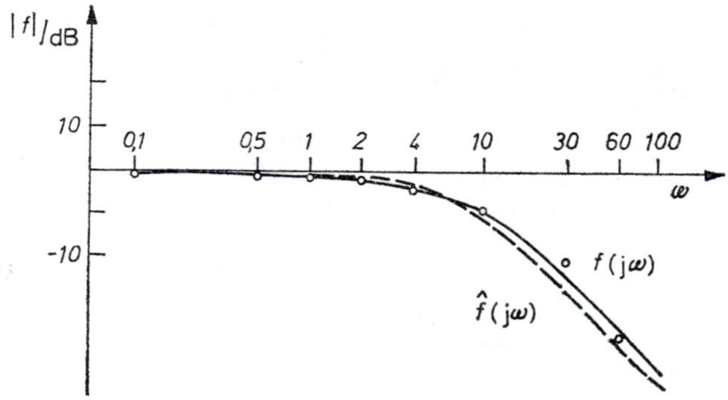

Fig. 8.5 Bode plot of the prefilter $f(j\omega)$

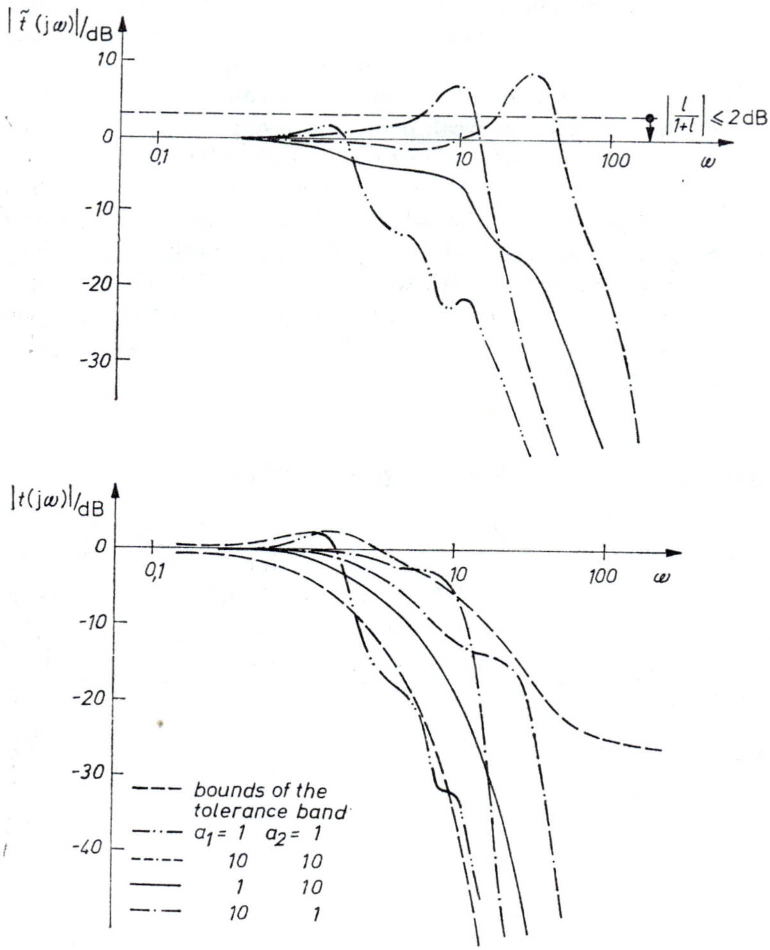

Fig. 8.6 Bode plot of $\tilde{t}(j\omega, \mathbf{a})$ and $t(j\omega, \mathbf{a})$

holds the allowable tolerances of $|t(j\omega, \mathbf{a})|$ in Fig. 8.1 must have a width of at least $\bar{k}_s$ for $\omega > \omega_h$ with some ω_h. Since the plant was assumed to be minimum-phase, this is the only condition for the existence of a solution to our design problem. The designer can be sure to find a robust controller even if the tolerance band is arbitrarily small for $\omega < \omega_h$ because this requirement can be met by means of high controller gains in the low frequency range. This, however, may cause the side effects of high bandwidth of $l(j\omega, \mathbf{a})$, high noise amplification, and high actuator signals as discussed in Section 6.4.

Discussion of the method

Horowitz's method represents a true robustness design method because the amount of feedback is directly adapted to the aim to guarantee the desired tolerance of a given range of plant uncertainties. Because of the transparency of the design process, a suitable compromise between the tolerances of the closed-loop behaviour, the disturbance attenuation, and the compensator complexity (number of poles and zeros, bandwidth, gain) can be found.

As seen from the example, the weak point of this approach is the lack of systematic methods for choosing the functions $l(j\omega, \hat{\mathbf{a}})$ or $k(j\omega)$, respectively. The controller found by intuition may have a larger complexity than necessary. For example, the question whether the problem might be solved by a static feedback remains open. These shortcomings and the impossibility to handle frequency response matrices by means of the Nichols chart restrict the application of the quantitative design method to single-loop controllers or the sequential design of the different control loops in a multivariable control system.

8.2 Description of the closed-loop original system

To prepare the robustness analysis of multi-input multi-output feedback systems subject to unknown-but-bounded uncertainties, the model of the closed-loop system will be derived from the model (5.3-3), (5.3-5) of the plant and the two-degrees-of-freedom controller shown in Fig. 2.5(a). The resulting equations are generalisations of those presented in Section 1.4.

Consider the feedback system shown in Fig. 8.7. The plant is described by

$$\mathbf{y}(p) = \mathbf{G}(p)\,\mathbf{u}(p) + \mathbf{d}(p) \tag{1}$$

where

$$\mathbf{G}(p) = \hat{\mathbf{G}}(p) + \delta\mathbf{G}(p) \tag{2}$$

and

$$\delta\mathbf{G}(p) = \mathbf{G}_{ys}(p)\,\mathbf{G}_2(p)\,\big(\mathbf{I} - \mathbf{G}_{zs}(p)\,\mathbf{G}_2(p)\big)^{-1}\,\mathbf{G}_{zu}(p) \tag{3}$$

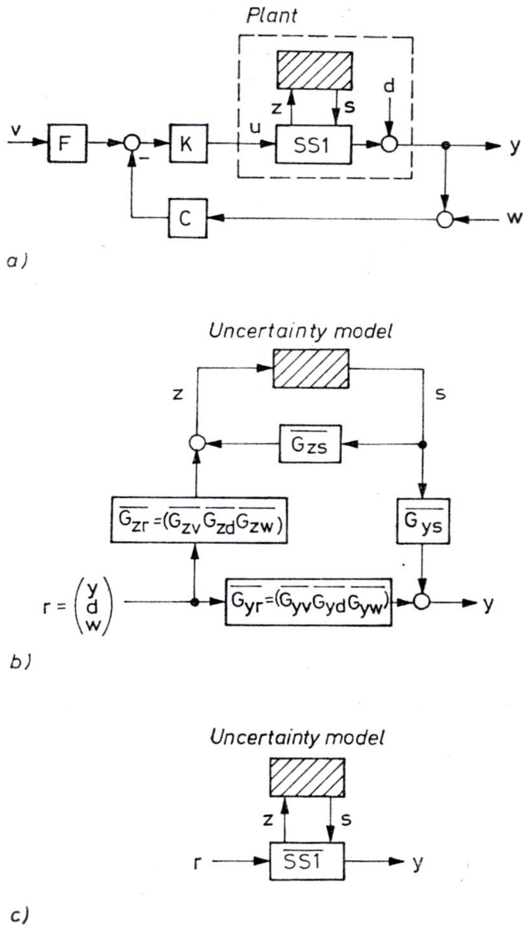

Fig. 8.7 Closed-loop original system
(a) General control configuration
(b) Detailed representation
(c) Decomposition into two parts

(cf eqn (5.3-7)). The control law has the form

$$\mathbf{u}(p) = \mathbf{K}(p)\,\mathbf{F}(p)\,\mathbf{v}(p) - \mathbf{K}(p)\,\mathbf{C}(p)\,\mathbf{w}(p) - \mathbf{K}(p)\,\mathbf{C}(p)\,\mathbf{y}(p)\,. \tag{4}$$

Hence, the closed-loop original system has the description

$$\begin{aligned}\mathbf{y}(p) &= (\mathbf{I} + \mathbf{GKC})^{-1}\,\mathbf{GKFv} - (\mathbf{I} + \mathbf{GKC})^{-1}\,\mathbf{GKCw} - (\mathbf{I} + \mathbf{GKC})^{-1}\,\mathbf{d} \\ &= \mathbf{G}_{\mathbf{yv}}\mathbf{v} + \mathbf{G}_{\mathbf{yw}}\mathbf{w} + \mathbf{G}_{\mathbf{yd}}\mathbf{d}\end{aligned} \tag{5}$$

with $\mathbf{G}(p)$ from eqns (2), (3), where $\mathbf{G}_2$ is only known to belong to some set $\mathscr{G}_2$ of possible candidates.

An alternative way of deriving this model should provide a decomposition of eqn (5) into a completely known and an uncertain part. It starts with lumping the completely known parts of the system. These are the controller (4) and the 'subsystem 1' of the plant model

$$\mathbf{y}(p) = \hat{\mathbf{G}}(p)\,\mathbf{u}(p) + \mathbf{G}_{ys}(p)\,\mathbf{s}(p) + \mathbf{d}(p)$$

$$\mathbf{z}(p) = \mathbf{G}_{zu}(p)\,\mathbf{u}(p) + \mathbf{G}_{zs}(p)\,\mathbf{s}(p)$$

which, in comparison with eqn (5.3-3), has been extended to include the plant disturbance $\mathbf{d}(p)$. Then, eqn (4) yields

$$\begin{aligned}\mathbf{y} &= (\mathbf{I} + \hat{\mathbf{G}}\mathbf{K}\mathbf{C})^{-1}\,\hat{\mathbf{G}}\mathbf{K}\mathbf{F}\mathbf{v} + (\mathbf{I} + \hat{\mathbf{G}}\mathbf{K}\mathbf{C})^{-1}\,\mathbf{d} - (\mathbf{I} + \hat{\mathbf{G}}\mathbf{K}\mathbf{C})^{-1}\,\hat{\mathbf{G}}\mathbf{K}\mathbf{C}\mathbf{w} \\ &\quad + (\mathbf{I} + \hat{\mathbf{G}}\mathbf{K}\mathbf{C})^{-1}\,\mathbf{G}_{ys}\mathbf{s} \\ &= \bar{\mathbf{G}}_{yv}\mathbf{v} + \bar{\mathbf{G}}_{yd}\mathbf{d} + \bar{\mathbf{G}}_{yw}\mathbf{w} + \bar{\mathbf{G}}_{ys}\mathbf{s}, \end{aligned} \tag{6}$$

$$\begin{aligned}\mathbf{u} &= (\mathbf{I} + \mathbf{K}\mathbf{C}\hat{\mathbf{G}})^{-1}\,\mathbf{K}\mathbf{F}\mathbf{v} - \mathbf{K}\mathbf{C}(\mathbf{I} + \hat{\mathbf{G}}\mathbf{K}\mathbf{C})^{-1}\,\mathbf{d} - \mathbf{K}\mathbf{C}(\mathbf{I} + \hat{\mathbf{G}}\mathbf{K}\mathbf{C})^{-1}\,\mathbf{w} \\ &\quad - \mathbf{K}\mathbf{C}(\mathbf{I} + \hat{\mathbf{G}}\mathbf{K}\mathbf{C})^{-1}\,\mathbf{G}_{ys}\mathbf{s} \\ &= \bar{\mathbf{G}}_{uv}\mathbf{v} + \bar{\mathbf{G}}_{ud}\mathbf{d} + \bar{\mathbf{G}}_{uw}\mathbf{w} + \bar{\mathbf{G}}_{us}\mathbf{s} \end{aligned} \tag{7}$$

and

$$\begin{aligned}\mathbf{z} &= \mathbf{G}_{zu}(\mathbf{I} + \mathbf{K}\mathbf{C}\hat{\mathbf{G}})^{-1}\,\mathbf{K}\mathbf{F}\mathbf{v} - \mathbf{G}_{zu}\mathbf{K}\mathbf{C}(\mathbf{I} + \hat{\mathbf{G}}\mathbf{K}\mathbf{C})^{-1}\,\mathbf{d} - \mathbf{G}_{zu}\mathbf{K}\mathbf{C}(\mathbf{I} + \hat{\mathbf{G}}\mathbf{K}\mathbf{C})^{-1}\,\mathbf{w} \\ &\quad + \left(\mathbf{G}_{zs} - \mathbf{G}_{zu}\mathbf{K}\mathbf{C}(\mathbf{I} + \hat{\mathbf{G}}\mathbf{K}\mathbf{C})^{-1}\,\mathbf{G}_{ys}\right)\mathbf{s} \\ &= \bar{\mathbf{G}}_{zv}\mathbf{v} + \bar{\mathbf{G}}_{zd}\mathbf{d} + \bar{\mathbf{G}}_{zw}\mathbf{w} + \bar{\mathbf{G}}_{zs}\mathbf{s}, \end{aligned} \tag{8}$$

where the meaning of the abbreviations becomes clear by a comparison of both lines of each equation. Eqns (6)—(8) describe the closed-loop approximate model, which is denoted as 'SS 1' in Fig. 8.7(c) and can be written in the compact form

$$\begin{aligned}\mathbf{y} &= (\bar{\mathbf{G}}_{yv}\bar{\mathbf{G}}_{yd}\bar{\mathbf{G}}_{yw})\begin{pmatrix}\mathbf{v}\\ \mathbf{d}\\ \mathbf{w}\end{pmatrix} + \bar{\mathbf{G}}_{ys}\mathbf{s} \\ \mathbf{z} &= (\bar{\mathbf{G}}_{zv}\bar{\mathbf{G}}_{zd}\bar{\mathbf{G}}_{zw})\begin{pmatrix}\mathbf{v}\\ \mathbf{d}\\ \mathbf{w}\end{pmatrix} + \bar{\mathbf{G}}_{zs}\mathbf{s}. \end{aligned} \tag{9}$$

Using the uncertainty model (5.3-5)

$$\mathbf{s}(p) = \mathbf{G}_2(p)\,\mathbf{z}(p) \tag{10}$$

we get

$$\begin{aligned}\mathbf{z} &= \bar{\mathbf{G}}_{zv}\mathbf{v} + \bar{\mathbf{G}}_{zd}\mathbf{d} + \bar{\mathbf{G}}_{zw}\mathbf{w} + \bar{\mathbf{G}}_{zs}\mathbf{G}_2\mathbf{z} \\ &= (\mathbf{I} - \bar{\mathbf{G}}_{zs}\mathbf{G}_2)^{-1}\,\bar{\mathbf{G}}_{zv}\mathbf{v} + (\mathbf{I} - \bar{\mathbf{G}}_{zs}\mathbf{G}_2)^{-1}\,\bar{\mathbf{G}}_{zd}\mathbf{d} + (\mathbf{I} - \bar{\mathbf{G}}_{zs}\mathbf{G}_2)^{-1}\,\bar{\mathbf{G}}_{zw}\mathbf{w}\end{aligned}$$

and

$$\begin{aligned}\mathbf{y} &= \left(\bar{\mathbf{G}}_{yv} + \bar{\mathbf{G}}_{ys}\mathbf{G}_2(\mathbf{I} - \bar{\mathbf{G}}_{zs}\mathbf{G}_2)^{-1}\bar{\mathbf{G}}_{zv}\right)\mathbf{v} + \left(\bar{\mathbf{G}}_{yd} + \bar{\mathbf{G}}_{ys}\mathbf{G}_2(\mathbf{I} - \bar{\mathbf{G}}_{zs}\mathbf{G}_2)^{-1}\bar{\mathbf{G}}_{zd}\right)\mathbf{d}\\ &\quad + \left(\bar{\mathbf{G}}_{yw} + \bar{\mathbf{G}}_{ys}\mathbf{G}_2(\mathbf{I} - \bar{\mathbf{G}}_{zs}\mathbf{G}_2)^{-1}\bar{\mathbf{G}}_{zw}\right)\mathbf{w}\\ &= (\hat{\mathbf{G}}_{yv} + \delta\mathbf{G}_{yv})\,\mathbf{v} + (\hat{\mathbf{G}}_{yd} + \delta\mathbf{G}_{yd})\,\mathbf{d} + (\hat{\mathbf{G}}_{yw} + \delta\mathbf{G}_{yw})\,\mathbf{w}. \end{aligned} \tag{11}$$

This equation provides a description of the original system (1), (4) where, as opposed to eqn (5), the terms depending on the model uncertainties are separated from those independent of $\mathbf{G}_2$

$$\begin{aligned} &\hat{\mathbf{G}}_{yv}(p) = \bar{\mathbf{G}}_{yv}(p), \qquad \hat{\mathbf{G}}_{yd}(p) = \bar{\mathbf{G}}_{yd}(p), \qquad \hat{\mathbf{G}}_{yw}(p) = \bar{\mathbf{G}}_{yw}(p)\\ &\delta\mathbf{G}_{yv}(p) = \bar{\mathbf{G}}_{ys}\mathbf{G}_2(\mathbf{I} - \bar{\mathbf{G}}_{zs}\mathbf{G}_2)^{-1}\bar{\mathbf{G}}_{zv}\\ &\delta\mathbf{G}_{yd}(p) = \bar{\mathbf{G}}_{ys}\mathbf{G}_2(\mathbf{I} - \bar{\mathbf{G}}_{zs}\mathbf{G}_2)^{-1}\bar{\mathbf{G}}_{zd}\\ &\delta\mathbf{G}_{yw}(p) = \bar{\mathbf{G}}_{ys}\mathbf{G}_2(\mathbf{I} - \bar{\mathbf{G}}_{zs}\mathbf{G}_2)^{-1}\bar{\mathbf{G}}_{zw}. \end{aligned} \tag{12}$$

If no model uncertainties occur or those encountered are ignored ($\mathbf{G}_2 = \mathbf{0}$) then

$$\hat{\mathbf{y}} = \bar{\mathbf{G}}_{yv}\mathbf{v} + \bar{\mathbf{G}}_{yd}\mathbf{d} + \bar{\mathbf{G}}_{yw}\mathbf{w} \tag{13}$$

is received as an approximate model of the closed-loop original system. Eqn (13) coincides with the closed-loop approximate model (6) for $\mathbf{s} = \mathbf{0}$.

In analogy to the frequency domain development, time domain considerations start from the state space models

$$\begin{aligned}\dot{\mathbf{x}} &= \mathbf{A}\mathbf{x} + \mathbf{B}\mathbf{u}\\ \mathbf{y} &= \mathbf{C}\mathbf{x} + \mathbf{D}\mathbf{u} + \mathbf{d}\end{aligned} \tag{14}$$

and

$$\begin{aligned}\dot{\mathbf{x}}_r &= \mathbf{A}_r\mathbf{x}_r + \mathbf{B}_y(\mathbf{y} + \mathbf{w}) + \mathbf{B}_v\mathbf{v}\\ \mathbf{u} &= \mathbf{C}_r\mathbf{x}_r - \mathbf{K}_y(\mathbf{y} + \mathbf{w}) + \mathbf{K}_v\mathbf{v}\end{aligned} \tag{15}$$

of the original plant and the controller, respectively, where all parts of the controller are described by a common model. Both equations together yield the relations of the closed-loop original system

$$\begin{aligned}\frac{d}{dt}\begin{pmatrix}\mathbf{x}\\ \mathbf{x}_r\end{pmatrix} &= \begin{pmatrix}\mathbf{A} - \mathbf{B}\tilde{\mathbf{K}}_y\mathbf{C} & \mathbf{B}\tilde{\mathbf{C}}_r\\ \mathbf{B}_y\tilde{\mathbf{M}}\mathbf{C} & \mathbf{A}_r + \mathbf{B}_y\mathbf{D}\tilde{\mathbf{C}}_r\end{pmatrix}\begin{pmatrix}\mathbf{x}\\ \mathbf{x}_r\end{pmatrix} + \begin{pmatrix}\mathbf{B}\tilde{\mathbf{K}}_v\\ \mathbf{B}_v - \mathbf{B}_y\mathbf{D}\tilde{\mathbf{K}}_v\end{pmatrix}\mathbf{v}\\ &\quad + \begin{pmatrix}-\mathbf{B}\tilde{\mathbf{K}}_y\\ \mathbf{B}_y\tilde{\mathbf{M}}\end{pmatrix}\mathbf{d} + \begin{pmatrix}-\mathbf{B}\tilde{\mathbf{K}}_y\\ \mathbf{B}_y\tilde{\mathbf{M}}\end{pmatrix}\mathbf{w}\\ \mathbf{y} &= (\tilde{\mathbf{M}}\mathbf{C} \quad \mathbf{D}\tilde{\mathbf{C}}_r)\begin{pmatrix}\mathbf{x}\\ \mathbf{x}_r\end{pmatrix} + \mathbf{D}\tilde{\mathbf{K}}_v\mathbf{v} + \tilde{\mathbf{M}}\mathbf{d} - \mathbf{D}\tilde{\mathbf{K}}_y\mathbf{w}\end{aligned} \tag{16}$$

with

$$\begin{aligned}&\tilde{\mathbf{M}} = (\mathbf{I} + \mathbf{D}\mathbf{K}_y)^{-1}\\ &\tilde{\mathbf{C}}_r = (\mathbf{I} + \mathbf{K}_y\mathbf{D})^{-1}\mathbf{C}_r, \quad \tilde{\mathbf{K}}_y = (\mathbf{I} + \mathbf{K}_y\mathbf{D})^{-1}\mathbf{K}_y, \quad \tilde{\mathbf{K}}_v = (\mathbf{I} + \mathbf{K}_y\mathbf{D})^{-1}\mathbf{K}_v.\end{aligned} \tag{17}$$

Eqn (16) can be abbreviated by

$$\begin{aligned}\dot{\tilde{\mathbf{x}}} &= \tilde{\mathbf{A}}\tilde{\mathbf{x}} + \tilde{\mathbf{B}}\mathbf{v} + \tilde{\mathbf{E}}_d\mathbf{d} + \tilde{\mathbf{E}}_w\mathbf{w} \\ \mathbf{y} &= \tilde{\mathbf{C}}\tilde{\mathbf{x}} + \tilde{\mathbf{D}}\mathbf{v} + \tilde{\mathbf{F}}_d\mathbf{d} + \tilde{\mathbf{F}}_w\mathbf{w}\end{aligned} \tag{18}$$

and reformulated as

$$\mathbf{y} = \mathbf{G}_{yv} * \mathbf{v} + \mathbf{G}_{yd} * \mathbf{d} + \mathbf{G}_{yw} * \mathbf{w} \tag{19}$$

with

$$\mathbf{G}_{yv}(t) = \tilde{\mathbf{D}}\delta(t) + \tilde{\mathbf{C}} \exp(\tilde{\mathbf{A}}t)\, \tilde{\mathbf{B}} \tag{20}$$

etc (compare eqns (5) and (19)). Instead of eqn (14) the approximate model

$$\begin{aligned}\dot{\hat{\mathbf{x}}} &= \hat{\mathbf{A}}\hat{\mathbf{x}} + \hat{\mathbf{B}}\mathbf{u} + \hat{\mathbf{E}}\mathbf{s} \\ \mathbf{y} &= \hat{\mathbf{C}}\hat{\mathbf{x}} + \hat{\mathbf{D}}\mathbf{u} + \hat{\mathbf{F}}\mathbf{s} + \mathbf{d} \\ \mathbf{z} &= \hat{\mathbf{C}}_z\hat{\mathbf{x}} + \hat{\mathbf{D}}_z\mathbf{u} + \hat{\mathbf{F}}_z\mathbf{s}\end{aligned} \tag{21}$$

and some bounds on the uncertainty model

$$\mathbf{s}(t) = \mathbf{G}_2(t) * \mathbf{z}(t) \tag{22}$$

are available. Then eqns (15) and (21) yield

$$\begin{aligned}\frac{d}{dt}\begin{pmatrix}\hat{\mathbf{x}}\\ \mathbf{x}_r\end{pmatrix} &= \begin{pmatrix}\hat{\mathbf{A}} - \hat{\mathbf{B}}\overline{\mathbf{K}}_y\hat{\mathbf{C}} & \hat{\mathbf{B}}\overline{\mathbf{C}}_r \\ \mathbf{B}_y\mathbf{M}\hat{\mathbf{C}} & \mathbf{A}_r + \mathbf{B}_y\hat{\mathbf{D}}\overline{\mathbf{C}}_r\end{pmatrix}\begin{pmatrix}\hat{\mathbf{x}}\\ \mathbf{x}_r\end{pmatrix} + \begin{pmatrix}\hat{\mathbf{B}}\overline{\mathbf{K}}_v \\ \mathbf{B}_v - \mathbf{B}_y\hat{\mathbf{D}}\overline{\mathbf{K}}_y\end{pmatrix}\mathbf{v} \\ &\quad + \begin{pmatrix}-\hat{\mathbf{B}}\overline{\mathbf{K}}_y \\ \mathbf{B}_y\mathbf{M}\end{pmatrix}\mathbf{d} + \begin{pmatrix}-\hat{\mathbf{B}}\overline{\mathbf{K}}_y \\ \mathbf{B}_y\mathbf{M}\end{pmatrix}\mathbf{w} + \begin{pmatrix}\hat{\mathbf{E}} - \hat{\mathbf{B}}\overline{\mathbf{K}}_y\hat{\mathbf{F}} \\ \mathbf{B}_y\mathbf{M}\hat{\mathbf{F}}\end{pmatrix}\mathbf{s} \\ \mathbf{y} &= (\mathbf{M}\hat{\mathbf{C}} \quad \hat{\mathbf{D}}\overline{\mathbf{C}}_r)\begin{pmatrix}\hat{\mathbf{x}}\\ \mathbf{x}_r\end{pmatrix} + \hat{\mathbf{D}}\overline{\mathbf{K}}_v\mathbf{v} + \mathbf{M}\mathbf{d} - \hat{\mathbf{D}}\overline{\mathbf{K}}_y\mathbf{w} + \mathbf{M}\hat{\mathbf{F}}\mathbf{s} \\ \mathbf{z} &= (\hat{\mathbf{C}}_z - \hat{\mathbf{D}}_z\overline{\mathbf{K}}_y\hat{\mathbf{C}} \quad \hat{\mathbf{D}}_z\overline{\mathbf{C}}_r)\begin{pmatrix}\hat{\mathbf{x}}\\ \mathbf{x}_r\end{pmatrix} + \hat{\mathbf{D}}_z\overline{\mathbf{K}}_v\mathbf{v} - \hat{\mathbf{D}}_z\overline{\mathbf{K}}_y\mathbf{d} - \hat{\mathbf{D}}_z\overline{\mathbf{K}}_y\mathbf{w} + (\hat{\mathbf{F}}_z - \hat{\mathbf{D}}_z\overline{\mathbf{K}}_y\hat{\mathbf{F}})\,\mathbf{s}\end{aligned} \tag{23}$$

with

$$\begin{aligned}\mathbf{M} &= (\mathbf{I} + \hat{\mathbf{D}}\mathbf{K}_y)^{-1} \\ \overline{\mathbf{C}}_r &= (\mathbf{I} + \mathbf{K}_y\hat{\mathbf{D}})^{-1}\,\mathbf{C}_r, \quad \overline{\mathbf{K}}_y = (\mathbf{I} + \mathbf{K}_y\hat{\mathbf{D}})^{-1}\,\mathbf{K}_y, \quad \overline{\mathbf{K}}_v = (\mathbf{I} + \mathbf{K}_y\hat{\mathbf{D}})^{-1}\,\mathbf{K}_v.\end{aligned} \tag{24}$$

Eqn (23) will be abbreviated by

$$\begin{aligned}\dot{\overline{\mathbf{x}}} &= \overline{\mathbf{A}}\overline{\mathbf{x}} + \overline{\mathbf{B}}\mathbf{v} + \overline{\mathbf{E}}_d\mathbf{d} + \overline{\mathbf{E}}_w\mathbf{w} + \overline{\mathbf{E}}\mathbf{s} \\ \mathbf{y} &= \overline{\mathbf{C}}\overline{\mathbf{x}} + \overline{\mathbf{D}}\mathbf{v} + \overline{\mathbf{F}}_d\mathbf{d} + \overline{\mathbf{F}}_w\mathbf{w} + \overline{\mathbf{F}}\mathbf{s} \\ \mathbf{z} &= \overline{\mathbf{C}}_z\overline{\mathbf{x}} + \overline{\mathbf{D}}_z\mathbf{v} + \overline{\mathbf{G}}_d\mathbf{d} + \overline{\mathbf{G}}_w\mathbf{w} + \overline{\mathbf{G}}\mathbf{s}.\end{aligned} \tag{25}$$

It can be converted into the I/O-description

$$\begin{aligned}\mathbf{y} &= \overline{\mathbf{G}}_{yv} * \mathbf{v} + \overline{\mathbf{G}}_{yd} * \mathbf{d} + \overline{\mathbf{G}}_{yw} * \mathbf{w} + \overline{\mathbf{G}}_{ys} * \mathbf{s} \\ \mathbf{z} &= \overline{\mathbf{G}}_{zv} * \mathbf{v} + \overline{\mathbf{G}}_{zd} * \mathbf{d} + \overline{\mathbf{G}}_{zw} * \mathbf{w} + \overline{\mathbf{G}}_{zs} * \mathbf{s}\end{aligned} \tag{26}$$

with

$$\bar{\mathbf{G}}_{yv}(t) = \bar{\mathbf{D}}\delta(t) + \bar{\mathbf{C}} \exp(\bar{\mathbf{A}}t)\, \bar{\mathbf{B}}$$
$$\bar{\mathbf{G}}_{zs}(t) = \bar{\mathbf{G}}\delta(t) + \bar{\mathbf{C}}_z \exp(\bar{\mathbf{A}}t)\, \bar{\mathbf{E}} \qquad \text{etc} \tag{27}$$

or, in abbreviated form,

$$\mathbf{y} = (\bar{\mathbf{G}}_{yv}\; \bar{\mathbf{G}}_{yd}\; \bar{\mathbf{G}}_{yw}) * \begin{pmatrix} \mathbf{v} \\ \mathbf{d} \\ \mathbf{w} \end{pmatrix} + \bar{\mathbf{G}}_{ys} * \mathbf{s}$$
$$\mathbf{z} = (\bar{\mathbf{G}}_{zv}\; \bar{\mathbf{G}}_{zd}\; \bar{\mathbf{G}}_{zw}) * \begin{pmatrix} \mathbf{v} \\ \mathbf{d} \\ \mathbf{w} \end{pmatrix} + \bar{\mathbf{G}}_{zs} * \mathbf{s} \tag{28}$$

in analogy to eqn (9). Combining eqns (22) and (28) yields

$$\mathbf{z} = \tilde{\mathbf{G}} * \bar{\mathbf{G}}_{zv} * \mathbf{v} + \tilde{\mathbf{G}} * \bar{\mathbf{G}}_{zd} * \mathbf{d} + \tilde{\mathbf{G}} * \bar{\mathbf{G}}_{zw} * \mathbf{w} \tag{29}$$

with

$$\tilde{\mathbf{G}} = \delta(t)\, \mathbf{I} + \bar{\mathbf{G}}_{zs} * \mathbf{G}_2 * \tilde{\mathbf{G}} \tag{30}$$

and, finally,

$$\begin{aligned}\mathbf{y} &= (\bar{\mathbf{G}}_{yv} + \bar{\mathbf{G}}_{ys} * \mathbf{G}_2 * \tilde{\mathbf{G}} * \bar{\mathbf{G}}_{zv}) * \mathbf{v} + (\bar{\mathbf{G}}_{yd} + \bar{\mathbf{G}}_{ys} * \mathbf{G}_2 * \tilde{\mathbf{G}} * \bar{\mathbf{G}}_{zd}) * \mathbf{d} \\ &\quad + (\bar{\mathbf{G}}_{yw} + \bar{\mathbf{G}}_{ys} * \mathbf{G}_2 * \tilde{\mathbf{G}} * \bar{\mathbf{G}}_{zw}) * \mathbf{w} \\ &= (\hat{\mathbf{G}}_{yv} + \delta\mathbf{G}_{yv}) * \mathbf{v} + (\hat{\mathbf{G}}_{yd} + \delta\mathbf{G}_{yd}) * \mathbf{d} + (\hat{\mathbf{G}}_{yw} + \delta\mathbf{G}_{yw}) * \mathbf{w}.\end{aligned} \tag{31}$$

Ignoring the model uncertainties gives the approximate closed-loop system

$$\hat{\mathbf{y}} = \bar{\mathbf{G}}_{yv} * \mathbf{v} + \bar{\mathbf{G}}_{yd} * \mathbf{d} + \bar{\mathbf{G}}_{yw} * \mathbf{w}. \tag{32}$$

The model errors are

$$\delta\mathbf{G}_{yv}(t) = \bar{\mathbf{G}}_{ys}(t) * \mathbf{G}_2(t) * \tilde{\mathbf{G}}(t) * \bar{\mathbf{G}}_{zv}(t)$$
$$\delta\mathbf{G}_{yd}(t) = \bar{\mathbf{G}}_{ys}(t) * \mathbf{G}_2(t) * \tilde{\mathbf{G}}(t) * \mathbf{G}^{zd}(t) \tag{33}$$
$$\delta\mathbf{G}_{yw}(t) = \bar{\mathbf{G}}_{ys}(t) * \mathbf{G}_2(t) * \tilde{\mathbf{G}}(t) * \mathbf{G}_{zw}(t).$$

The presentation of the closed-loop original system (5) or (19) as a composition of the completely known $\overline{\text{SS1}}$ (9) or (28), respectively, and the uncertainty model (10) or (22), respectively, shows that the problem of analysing the original system raises the basic question how to analyse a system of the structure depicted in Fig. 8.7(c). In the frequency domain the closed-loop original system is described by some completely known model of the form

$$\mathbf{y}(p) = \bar{\mathbf{G}}_{yr}(p)\, \mathbf{r}(p) + \bar{\mathbf{G}}_{ys}(p)\, \mathbf{s}(p)$$
$$\mathbf{z}(p) = \bar{\mathbf{G}}_{zr}(p)\, \mathbf{r}(p) + \bar{\mathbf{G}}_{zs}(p)\, \mathbf{s}(p) \tag{34}$$

in combination with the uncertainty model

$$\mathbf{s}(p) = \mathbf{G}_2(p)\,\mathbf{z}(p)\,. \tag{35}$$

The overall system (34), (35) is represented by

$$\mathbf{y}(p) = \mathbf{G}_{\mathbf{yr}}(p)\,\mathbf{r}(p) \tag{36}$$

with

$$\mathbf{G}_{\mathbf{yr}}(p) = \overline{\mathbf{G}}_{\mathbf{yr}}(p) + \overline{\mathbf{G}}_{\mathbf{ys}}(p)\,\mathbf{G}_2(p)\left(\mathbf{I} - \overline{\mathbf{G}}_{\mathbf{zs}}(p)\,\mathbf{G}_2(p)\right)^{-1}\overline{\mathbf{G}}_{\mathbf{zr}}(p)\,, \tag{37}$$

which is actually an abbreviation of eqn (11). The time domain alternatives of eqns (34)—(37) are

$$\begin{aligned} \mathbf{y}(t) &= \overline{\mathbf{G}}_{\mathbf{yr}}(t) * \mathbf{r}(t) + \overline{\mathbf{G}}_{\mathbf{ys}}(t) * \mathbf{s}(t) \\ \mathbf{z}(t) &= \overline{\mathbf{G}}_{\mathbf{zr}}(t) * \mathbf{r}(t) + \overline{\mathbf{G}}_{\mathbf{zs}}(t) * \mathbf{s}(t) \end{aligned} \tag{38}$$

with uncertainty model

$$\mathbf{s}(t) = \mathbf{G}_2(t) * \mathbf{z}(t) \tag{39}$$

and

$$\mathbf{y}(t) = \mathbf{G}_{\mathbf{yr}}(t) * \mathbf{r}(t) \tag{40}$$

with

$$\mathbf{G}_{\mathbf{yr}}(t) = \overline{\mathbf{G}}_{\mathbf{yr}}(t) + \overline{\mathbf{G}}_{\mathbf{ys}}(t) * \mathbf{G}_2(t) * \tilde{\mathbf{G}}(t) * \overline{\mathbf{G}}_{\mathbf{zr}}(t) \tag{41}$$

$$\tilde{\mathbf{G}}(t) = \delta(t)\,\mathbf{I} + \overline{\mathbf{G}}_{\mathbf{zs}}(t) * \mathbf{G}_2(t) * \tilde{\mathbf{G}}(t)\,. \tag{42}$$

The following sections present methods for analysing such a kind of systems on the basis of norm or element-by-element bounds on the uncertainty $\mathbf{G}_2$. Because of the simple correspondences between the eqns (34) and (9) or (38) and (28), respectively, these results can be easily applied to the original feedback system.

As we aim at finding controllers that produce a robustly stable closed-loop system and as the approximate plant model describes a possible representative of the original plant, it is reasonable to assume that the controller has been designed so as to ensure the stability of the closed-loop approximate model (9) or (28) with regard to all inputs and outputs. That is, $\overline{\text{SS1}}$ in Fig. 8.7(c) and the corresponding blocks in Fig. 8.7(b) are stable.

8.3 Frequency domain stability criteria

At first, we consider the stability of the system (8.2-34), (8.2-35). The system is called *robustly stable* if it remains stable for all uncertainties $\mathbf{G}_2$ of a given set $\mathcal{S}_2$ of the form (5.3-12) or (5.3-18). Before specifying the set $\mathcal{S}_2$ we present two general stability criteria.

General stability conditions

To present the first criterion we make the assumption that all elements of the set $\mathcal{S}_2$ describe stable systems (8.2-35). According to Theorem 2.1 the system (36), (37) is stable, if the matrix norm of $\mathbf{G}_{yr}(p)$ is bounded for all p with $\mathrm{Re}\,[p] \geqq 0$. Since

$$\|\mathbf{G}_{yr}\| \leqq \|\bar{\mathbf{G}}_{yr}\| + \|\bar{\mathbf{G}}_{ys}\| \, \|\mathbf{G}_2\| \, \|(\mathbf{I} - \bar{\mathbf{G}}_{zs}\mathbf{G}_2)^{-1}\| \, \|\bar{\mathbf{G}}_{zr}\|$$

holds and $\|\mathbf{G}_2\|$ is bounded due to our assumption, the inequality

$$\|(\mathbf{I} - \bar{\mathbf{G}}_{zs}\mathbf{G}_2)^{-1}\| < \infty \quad \text{for all } \mathbf{G}_2 \in \mathcal{S}_2 \quad \text{for all } p \text{ with } \mathrm{Re}\,[p] \geqq 0 \tag{1}$$

provides a sufficient condition for robust stability. That is, the closed-loop on the top of Fig. 8.7(b) with open-loop transfer function matrix $\bar{\mathbf{G}}_{zs}\mathbf{G}_2$ must be stable.

Eqn (1) is also necessary for robust stability since $\mathbf{G}_2(p)$ can be continuously varied within the set $\mathcal{S}_2$. Therefore, there exist some $\mathbf{G}_2 \in \mathcal{S}_2$ such that $\|\bar{\mathbf{G}}_{ys}\mathbf{G}_2(\mathbf{I} - \bar{\mathbf{G}}_{zs}\mathbf{G}_2)^{-1}\,\bar{\mathbf{G}}_{zr}\|$ is infinite whenever eqn (1) is not satisfied. With other words, the influence of unstable modes occuring in the $\bar{\mathbf{G}}_{zs}\mathbf{G}_2$-loop may not be 'blocked' for all $\mathbf{G}_2 \in \mathcal{S}_2$ by $\bar{\mathbf{G}}_{ys}$ and $\bar{\mathbf{G}}_{zr}$ with appropriate transmission zeros because these zeros are fixed whereas the frequencies of the unstable modes vary with $\mathbf{G}_2$ (see the chain connection including $\bar{\mathbf{G}}_{zr}$, the loop, and $\bar{\mathbf{G}}_{ys}$ in Fig. 8.7(b)).

Lemma 8.1 Assume that the system (34) is stable with respect to all its inputs and outputs and that all elements $\mathbf{G}_2(p) \in \mathcal{S}_2$ are stable. Then the original system (8.2-34), (8.2-35) is robustly stable if and only if, eqn (1) is satisfied or, equivalently, the image of the Nyquist contour $\mathcal{D}$ under $\det\,(\mathbf{I} - \bar{\mathbf{G}}_{zs}\mathbf{G}_2)$ does not encircle the origin of the complex plane.

Due to the direct correspondences between (8.2-34) and the closed-loop approximate model (8.2-9) we can give the following stability condition.

Theorem 8.1 Assume that the closed-loop approximate model (8.2-9) is stable and that all elements of $\mathcal{S}_2$ are stable. Then the closed-loop system (8.2-5) is robustly stable if and only if for all $\mathbf{G}_2 \in \mathcal{S}_2$ the image of the Nyquist contour $\mathcal{D}$ under $\det\big(\mathbf{I} - \bar{\mathbf{G}}_{zs}(p)\,\mathbf{G}_2(p)\big)$ does not encircle the origin of the complex plane.

Since $\mathbf{G}_2 \in \mathcal{S}_2$ implies $\varepsilon\mathbf{G}_2 \in \mathcal{S}_2$ for $0 \leqq \varepsilon \leqq 1$, this stability condition is equivalent to

$$\sup_{\mathbf{G}_2 \in \mathcal{S}_2} \varrho[\bar{\mathbf{G}}_{zs}(p)\,\mathbf{G}_2(p)] < 1\,, \qquad p \in \mathcal{D} \tag{2}$$

where $\varrho[.]$ denotes the spectral radius of the given matrix.

The assumptions of Theorem 8.1 are restrictive insofar as they claim the stability of the uncertainty model. For additive and multiplicative perturbations (cf Fig. 5.2(a) and (c)) this assumption demands that the approximate model $\hat{\mathbf{G}}(p)$ has the same right-half plane poles as the original system $\mathbf{G}(p)$, ie the unstable part of the plant has to be exactly known. This restriction can be removed by applying the Nyquist stability criterion directly to the closed-loop original system (8.2-1), (8.2-4), while breaking the system at the input channel, (Fig. 8.7(a)). We assume that the approximate model and the original system share the same number of right-half plane poles but permit that these poles have different locations. Then the images of the Nyquist contour under $\det\left(\mathbf{I} + \mathbf{K}(p)\,\mathbf{C}(p)\,\hat{\mathbf{G}}(p)\right)$ and $\det\left(\mathbf{I} + \mathbf{K}(p)\,\mathbf{C}(p)\,\mathbf{G}(p)\right)$ must have the same number of encirclements of the origin of the complex plane.

To exploit this idea we develop the following relation

$$\begin{aligned}\det(\mathbf{I} + \mathbf{KCG}) &= \det(\mathbf{I} + \mathbf{KC}\hat{\mathbf{G}} + \mathbf{KC}\delta\mathbf{G})\\ &= \det(\mathbf{I} + \mathbf{KC}\hat{\mathbf{G}})\det\left(\mathbf{I} + (\mathbf{I} + \mathbf{KC}\hat{\mathbf{G}})^{-1}\,\mathbf{KC}\delta\mathbf{G}\right).\end{aligned} \tag{3}$$

By using eqn (8.2-3) and the identity $\det(\mathbf{I} + \mathbf{AB}) = \det(\mathbf{I} + \mathbf{BA})$, the second term in the equation above yields

$$\begin{aligned}&\det\left(\mathbf{I} + (\mathbf{I}+\mathbf{KC}\hat{\mathbf{G}})^{-1}\,\mathbf{KC}\delta\mathbf{G}\right) = \det\left(\mathbf{I} + (\mathbf{I}+\mathbf{KC}\hat{\mathbf{G}})^{-1}\,\mathbf{KCG}_{ys}\mathbf{G}_2(\mathbf{I} - \mathbf{G}_{zs}\mathbf{G}_2)^{-1}\,\mathbf{G}_{zu}\right)\\ &= \det\left(\mathbf{I} + \mathbf{G}_2(\mathbf{I} - \mathbf{G}_{zs}\mathbf{G}_2)^{-1}\,\mathbf{G}_{zu}(\mathbf{I} + \mathbf{KC}\hat{\mathbf{G}})^{-1}\,\mathbf{KCG}_{ys}\right)\\ &= \det\left(\mathbf{I} + (\mathbf{I} - \mathbf{G}_2\mathbf{G}_{zs})^{-1}\,\mathbf{G}_2\mathbf{G}_{zu}\mathbf{KC}(\mathbf{I} + \hat{\mathbf{G}}\mathbf{KC})^{-1}\,\mathbf{G}_{ys}\right)\\ &= \det(\mathbf{I} - \mathbf{G}_2\mathbf{G}_{zs})^{-1}\det\left(\mathbf{I} - \mathbf{G}_2(\mathbf{G}_{zs} - \mathbf{G}_{zu}\mathbf{KC}(\mathbf{I} + \hat{\mathbf{G}}\mathbf{KC})^{-1}\,\mathbf{G}_{ys}\right)\\ &= \det(\mathbf{I} - \mathbf{G}_2\mathbf{G}_{zs})^{-1}\det(\mathbf{I} - \mathbf{G}_2\overline{\mathbf{G}}_{zs})\end{aligned} \tag{4}$$

with $\overline{\mathbf{G}}_{zs}$ from eqn (8.2-8). As $\det(\mathbf{I} + \mathbf{KC}\hat{\mathbf{G}})$ and $\det(\mathbf{I} + \mathbf{KCG})$ have to encircle the critical point the same number of times, $\det(\mathbf{I} - \mathbf{G}_2\mathbf{G}_{zs})^{-1} \times \det(\mathbf{I} - \mathbf{G}_2\overline{\mathbf{G}}_{zs})$ must not encircle this point.

Theorem 8.2 Assume that $\hat{\mathbf{G}}(p)$ and $\mathbf{G}(p)$ share the same number of right-half plane poles and that the closed-loop approximate model is stable. Then, the closed-loop system is robustly stable if and only if for all $\mathbf{G}_2 \in \mathcal{G}_2$

$$n_1 = n_2 \tag{5}$$

holds where n_1 and n_2 are the numbers of encirclement of the origin by the images of the Nyquist contour under $\det(\mathbf{I} - \mathbf{G}_2\mathbf{G}_{zs})$ and $\det(\mathbf{I} - \mathbf{G}_2\overline{\mathbf{G}}_{zs})$.

Norm-bounded uncertainties

Norm-bounded perturbations are described by eqn (5.2-12)

$$\|\mathbf{G}_2(p)\|_M \leqq v_2(p). \tag{6}$$

By using properties of the matrix norm, the stability condition (2) can be transformed via

$$\sup_{\mathbf{G}_2} \varrho[\overline{\mathbf{G}}_{zs}(p)\,\mathbf{G}_2(p)] \leqq \sup_{\mathbf{G}_2} \|\overline{\mathbf{G}}_{zs}\mathbf{G}_2\|_M \leqq \sup_{\mathbf{G}_2} \|\overline{\mathbf{G}}_{zs}\|_M\,\|\mathbf{G}_2\|_M \leqq \|\overline{\mathbf{G}}_{zs}\|_M\,v_2\,(p)$$

into

$$\|\bar{\mathbf{G}}_{zs}(p)\|_M < \frac{1}{v_2(p)}, \qquad p \in \mathcal{D} \tag{7}$$

which is sufficient for robust stability. After writing the complete expression of $\bar{\mathbf{G}}_{zs}(p)$, eqn (7) reads as

$$\left\|\mathbf{G}_{zs}(p) - \mathbf{G}_{zu}(p)\,\mathbf{K}(p)\,\mathbf{C}(p)\left(\mathbf{I} + \hat{\mathbf{G}}(p)\,\mathbf{K}(p)\,\mathbf{C}(p)\right)^{-1}\mathbf{G}_{ys}(p)\right\|_M < \frac{1}{v_2(p)}, \qquad p \in \mathcal{D}. \tag{8}$$

Lemma 8.2 Consider a plant whose uncertainties are norm-bounded according to eqn (6). Assume that the closed-loop approximate model (8.2-6) is stable and that the model uncertainties $\mathbf{G}_2$ are known to be stable. Then the closed-loop system (8.2-5) is robustly stable if eqn (8) is satisfied.

This stability criterion can be applied for an arbitrary matrix norm $\|.\|_M$. If it is used for the spectral norm (5.3-16), eqn (7) changes to

$$\sigma_{\max}[\bar{\mathbf{G}}_{zs}(p)] < \frac{1}{v_2(p)} \tag{9}$$

where $v_2(p)$ represents an upper bound of the maximum singular value of $\mathbf{G}_2(p)$

$$\sigma_{\max}[\mathbf{G}_2(p)] \leqq v_2(p)\,. \tag{10}$$

Then the necessity of the condition (9) can be proved by contradiction. Assuming that eqn (9) does not hold at the frequency point ω_1, ie

$$\sigma_{\max}[\bar{\mathbf{G}}_{zs}(\omega_1)] \geqq 1/v_2(\omega_1)\,,$$

an uncertainty model $\tilde{\mathbf{G}}_2(p)$ can be constructed in such a way that $\bar{\mathbf{G}}_{zs}(\omega_1)\,\tilde{\mathbf{G}}_2(\omega_1)$ has some eigenvalue at -1 and, thus, violates eqn (2). Based on the singular value decomposition $\bar{\mathbf{G}}_{zs}(\omega_1) = \mathbf{U}\operatorname{diag}\sigma_i\mathbf{V}^*$ (cf Appendix 2), $\tilde{\mathbf{G}}_2(p)$ is chosen so as to be strictly proper, stable, satisfying eqn (10) and yielding

$$\mathbf{G}_2(\omega_1) = \begin{pmatrix} \mathbf{v}_{11} \\ \vdots \\ \mathbf{v}_{n1} \end{pmatrix} - \frac{1}{\sigma_1}\left(u_{11}^* \dots u_{n1}^*\right) = \mathbf{V}\begin{pmatrix} \sigma_1 & & & \\ & 0 & & \\ & & \ddots & \\ & & & 0 \end{pmatrix}\mathbf{U}^*$$

for $p = \mathrm{j}\omega_1$. Hence, $\bar{\mathbf{G}}_{zs}(\omega_2)\,\tilde{\mathbf{G}}_2(\omega_1) = \mathbf{U}\begin{pmatrix} -1 & & & \\ & 0 & & \\ & & \ddots & \\ & & & 0 \end{pmatrix}\mathbf{U}^*$ has an eigenvalue at -1.

Theorem 8.3 (*Stability criterion of the singular value approach*). Consider a plant model where the model uncertainty is bounded by eqn (10). Assume that the closed-loop approximate model (8.2-6) is stable and that the model uncertainties $\mathbf{G}_2$ are known to be stable. Then the closed-loop system (8.2-5) is robustly stable if and only if eqn (9) is satisfied.

Note that this condition is *necessary and sufficient* for robust stability. As the counterexample above demonstrates, if eqn (9) is violated then there exists some transfer function matrix $\tilde{\mathbf{G}}_2$ which cannot be excluded as representation of the uncertainty model (8.2-10) on the basis of the information available about the plant and which causes instability of the closed-loop system. As opposed to Theorem 8.3, for many problems we are only able to state sufficient *or* necessary conditions. Eqn (9) can be reformulated as a condition on $\sigma_{\min}[\bar{\mathbf{G}}_{zs}]$, which can be interpreted as a distance of $\bar{\mathbf{G}}_{zs}$ from singularity

$$\sigma_{\min}[\bar{\mathbf{G}}_{zs}(p)] = \sigma_{\min}[\mathbf{G}_{zs} - \mathbf{G}_{zu}\mathbf{K}\mathbf{C}(\mathbf{I} + \hat{\mathbf{G}}\mathbf{K}\mathbf{C})^{-1}\,\mathbf{G}_{ys}] > v_2(p)\,, \qquad p \in \mathcal{D}\,.$$

Specific results for the model structures shown in Fig. 5.2 are summarised in Fig. 8.8. To provide a better view of the differences between these inequalities, the bounds on the singular values of the approximate closed-loop system are given for unity-feedback systems too. It is an important observation that the restrictions that must be imposed on the closed-loop approximate model in order to ensure robust stability vary in dependence upon the entry point of the uncertainty model. Whereas for the model structures (a) and (d) the minimum singular value of the return-difference matrix $(\mathbf{I} + \hat{\mathbf{G}})$ or, equivalently, the maximum singular value of the sensitivity matrix $\mathbf{S} = (\mathbf{I} + \hat{\mathbf{G}})^{-1}$ of the approximate model must be bounded

$$\sigma_{\max}[(\mathbf{I} + \hat{\mathbf{G}})^{-1}] < \frac{1}{v_2} \tag{11}$$

uncertainties located as in the models (b) and (c) lead to bounds on the maximum singular value of the complementary sensitivity matrix $\mathbf{T} = (\mathbf{I} + \hat{\mathbf{G}})^{-1}\,\hat{\mathbf{G}}$

$$\sigma_{\max}[(\mathbf{I} + \hat{\mathbf{G}})^{-1}\,\hat{\mathbf{G}}] < \frac{1}{v_2}. \tag{12}$$

Hence, decreasing the sensitivity (as described by $\mathbf{S}$) is not appropriate for improving the robustness in any case. What has to be decreased is the transmission $\bar{\mathbf{G}}_{zs}$ between the signals $\mathbf{s}$ and $\mathbf{z}$. Disturbance attenuation with regard to this particular ‘disturbance’ $\mathbf{s}$ leads to stability robustness. This contrasts with the sensitivity approach, which aims at attenuating the plant disturbance $\mathbf{d}$ independent of the kind of parameter deviations considered (cf Chapter 3).

By the way, eqn (12) is equivalent to

$$\sigma_{\max}[\mathbf{I} + \hat{\mathbf{G}}^{-1}] > \|\hat{\mathbf{G}}^{-1} - \mathbf{G}^{-1}\|$$

which refers to the inverse system descriptions (cf Fig. 6.4 with $\mathbf{K} = \mathbf{I}$).

Example 8.1 Let us consider the system described in Example 5.2 under unity feedback. The uncertainty model is stable. If we use

$$\hat{g}(p) = \frac{\hat{\lambda}_1\hat{\lambda}_2}{(p - \hat{\lambda}_1)\,(p - \hat{\lambda}_2)}$$

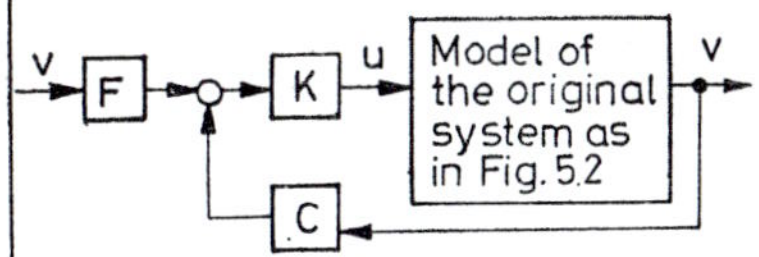	$p \in \mathfrak{D}$ Norm-bounded model uncertainties general results: $\lVert G_2 \rVert_M \le v_2$.	specific results: $C = K = I$ $\delta_{max}(G_2) \le v_2$	Element-by-element-bounded model uncertainties $\lvert G_2 \rvert \le v_2$
General model structure	$\lVert G_{zs} - G_{zu} KC(I+\hat{G}KC)^{-1} G_{ys} \rVert < \frac{1}{v_2}$	$\delta_{min}(G_{zs} - G_{zu}(I+\hat{G})^{-1} G_{ys})^{-1} > v_2$	$\lambda_p[\lvert G_{zs} - G_{zu} KC(I+\hat{G}K)^{-1} G_{ys} \rvert V_2] < 1$
Model structure (a)	$\lVert KC(I+\hat{G}KC)^{-1} \rVert < \frac{1}{v_2}$	$\delta_{min}(I+\hat{G}) > v_2$	$\lambda_p[\lvert KC(I+\hat{G}K)^{-1} \rvert V_2] < 1$
Model structure (b)	$\lVert (I+\hat{G}KC)^{-1}\hat{G} \rVert \le \frac{1}{v_2}$	$\delta_{min}(I+\hat{G}^{-1}) > v_2$	$\lambda_p[\lvert (I+\hat{G}KC)^{-1}\hat{G} \rvert V_2] < 1$
Model structure (c)	$\lVert KC(I+\hat{G}KC)^{-1}\hat{G} \rVert \le \frac{1}{v_2}$	$\delta_{min}(I+\hat{G}^{-1}) > v_2$	$\lambda_p[\lvert KC(I+\hat{G}KC)^{-1}\hat{G} \rvert V_2] < 1$
Model structure (d)	$\lVert (I+KC\hat{G})^{-1} \rVert < \frac{1}{v_2}$	$\delta_{min}(I+\hat{G}) > v_2$	$\lambda_p[\lvert (I+KC\hat{G})^{-1} \rvert V_2] < 1$
Model structure (e)	$\lVert (I+KC\hat{G})^{-1}(I-KC\hat{G}) \rVert < \frac{1}{v_2}$	$\delta_{min}[(I-\hat{G})^{-1}(I+\hat{G})] > v_2$	$\lambda_p[\lvert (I+KC\hat{G})^{-1}(I-KC\hat{G}) \rvert V_2] < 1$

Fig. 8.8 Frequency domain conditions for robust stability

with $\hat{\lambda}_1, \hat{\lambda}_2 < 0$ as approximate model then the closed-loop approximate model is stable as well. Hence, we may apply Theorem 8.3. Since model structure (d) and unity feedback are used, the closed-loop original system is robustly stable if and only if

$$|1 + g(j\omega)| = \left|\frac{(j\omega)^2 - j(\hat{\lambda}_1 + \hat{\lambda}_2) + 2\hat{\lambda}_1\hat{\lambda}_2}{(j\omega - \hat{\lambda}_1)(j\omega - \hat{\lambda}_2)}\right| > v_2(j\omega) = \left|\frac{k_0\lambda_u}{(j\omega - \lambda_u)}\right|.$$

As Fig. 8.9 shows this condition is satisfied for $k_o = 0.8$, $\lambda_u = -2.5$, $\hat{\lambda}_1 = -1$, $\hat{\lambda}_2 = -5$.

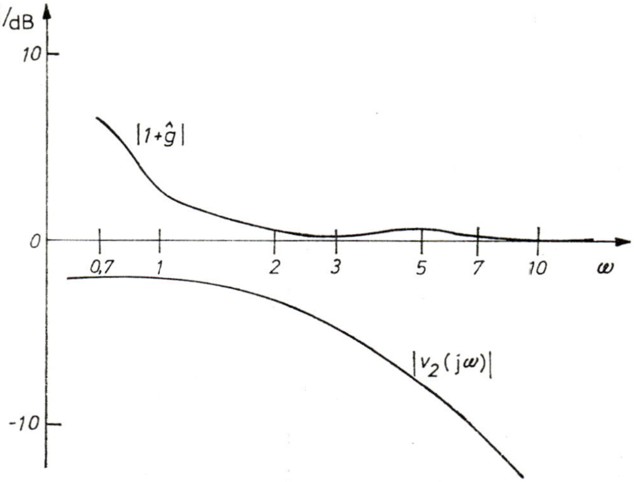

Fig. 8.9 Bode plot of $|1 + g|$ and v_2 showing that the stability condition $|1 + g| \geqq v_2$ is satisfied

Note that the necessity of the condition above refers to the stability under all possible perturbations $g_2(p)$ that satisfy the relation $|g_2(j\omega)| \leqq v_2(j\omega)$. The set of these functions $g_2(p)$ is larger than the set of functions $g_2(p) = -k\lambda/(p - \lambda)$, which had been taken as a basis for modelling the original system in Example 5.2. Besides this, it is worth mentioning that the stability of the closed-loop original system has been proved although the original system and the approximate model have a different number of unstable poles and may have a different number of right-half plane zeros. □

Element-by-element-bounded uncertainties

Now let us consider uncertainties bounded by

$$|\mathbf{G}_2(p)| \leqq \mathbf{V}_2(p). \tag{13}$$

For these uncertainties a simple sufficient condition can be derived by means of the equation

$$\varrho[\mathbf{AB}] \leqq \lambda_p[|\mathbf{AB}|] \leqq \lambda_p[|\mathbf{A}| \cdot |\mathbf{B}|]$$

which relates the spectral radius of the product **AB** to the spectral radius of $|\mathbf{A}| \cdot |\mathbf{B}|$, which, in turn, coincides with the Perron root (maximum eigenvalue) of the given nonnegative matrix (cf Appendix 1). Hence, eqn (2) is satisfied if

$$\lambda_p[|\bar{\mathbf{G}}_{zs}(p)|\,\mathbf{V}_2(p)] < 1, \qquad p \in \mathcal{D} \tag{14}$$

since then

$$\sup_{\mathbf{G}_2} \varrho[\bar{\mathbf{G}}_{zs}\mathbf{G}_2] \leqq \sup_{\mathbf{G}_2} \lambda_p[|\bar{\mathbf{G}}_{zs}\mathbf{G}_2|] \leqq \sup_{\mathbf{G}_2} \lambda_p[|\bar{\mathbf{G}}_{zs}|\,\mathbf{G}_2] = \lambda_p[|\bar{\mathbf{G}}_{zs}|\,\mathbf{V}_2] < 1\,.$$

Theorem 8.4 Consider a plant model where the model uncertainty is bounded by eqn (13). Assume that the closed-loop approximate model (8.2-6) is stable and that the model uncertainties $\mathbf{G}_2$ are known to be stable. Then the closed-loop system (8.2-5) is robustly stable if eqn (14) is satisfied.

The determination of the Perron root is very simple indeed and does not require the solution of the eigenvalue problem (cf Appendix 1).

The stability condition of Theorem 8.4 is weaker (better) than that from Theorem 8.3 because it exploits more information about the original system. However, the condition (14) is not necessary for robust stability because there may not exist any $\mathbf{G}_2(p)$ satisfying eqn (13) which causes closed-loop instability even if eqn (14) is violated. Therefore, Theorem 8.4 is said to be *conservative.*

Example 8.2 The following simple example illustrates that Theorem 8.4 yields a better robustness assessment than Theorem 8.3. Consider the model with parallel structure shown in Fig. 5.2 (a) with

$$\hat{\mathbf{G}}(p) = \begin{pmatrix} \hat{g}_1(p) & \\ & \hat{g}_2(p) \end{pmatrix}, \qquad \delta\mathbf{G}(p) = \begin{pmatrix} \delta g_1(p) & \\ & \delta g_2(p) \end{pmatrix}$$

under feedback

$$\mathbf{u}(p) = \mathbf{K}(p)\,\big(\mathbf{v}(p) - \mathbf{y}(p)\big) \qquad \text{with} \qquad \mathbf{K}(p) = \begin{pmatrix} k_1(p) & \\ & k_2(p) \end{pmatrix}.$$

Then both the closed-loop approximate model and the closed-loop original system consist of two independent loops. Obviously, they are stable if and only if both loops are stable.

Using the uncertainty bound (13)

$$|\delta\mathbf{G}(p)| = \begin{pmatrix} |g_1(p)| & \\ & |g_2(p)| \end{pmatrix} \leqq \begin{pmatrix} v_{11}(p) & \\ & v_{22}(p) \end{pmatrix} = \mathbf{V}_2(p)$$

this conclusion is actually reflected by the corresponding stability condition of Theorem 8.4 since $\lambda_p[|\mathbf{K}(\mathbf{I} + \mathbf{GK})^{-1}|\,\mathbf{V}_2] < 1$ for all $p \in \mathcal{D}$ can be decomposed into two independent inequalities

$$\left|\frac{k_i(p)}{1+\hat{g}_i(p)\,k_i(p)}\right| < \frac{1}{v_{ii}(p)} \qquad (i = 1, 2).$$

For arbitrary uncertainty bounds $v_{ii}(p)$, these conditions are necessary and sufficient for the robust stability of the closed-loop system. Hence, the stability condition of Theorem 8.4 for the system under consideration is as weak as possible.
For the singular value approach the uncertainty bound has to be brought into the form

$$\sigma_{\max}[\delta\mathbf{G}(p)] \leqq v_2(p) = \max\,[v_{11}(p), v_{22}(p)].$$

Then the stability condition (9) leads to the inequality

$$\max\left[\left|\frac{k_1}{1+g_1k_1}\right|, \left|\frac{k_1}{1+g_2k_2}\right|\right] \leqq \frac{1}{\max\,[v_{11}, v_{22}]}$$

which obviously is stronger than the condition above.

As a second example, consider the given parallel model with

$$|\delta\mathbf{G}(p)| \leqq \begin{pmatrix} 0 & v_{12}(p) \\ 0 & 0 \end{pmatrix} = \mathbf{V}_2(p).$$

Then the closed-loop original system consists of two loops in a chain connection and is stable if both loops are stable. Using Theorem 8.4 we get

$$\lambda_p\left[\begin{pmatrix} 0 & \left|\dfrac{k_1(p)\,v_{12}(p)}{1+\hat{g}_1(p)\,k_1(p)}\right| \\ 0 & 0 \end{pmatrix}\right] < 1, \qquad p \in \mathcal{D}$$

which is fulfilled for arbitrary functions $v_{12}(p)$. Again, the condition (14) is not at all conservative. The singular value approach yields

$$\max\left[\left|\frac{k_1(p)}{1+\hat{g}_1(p)\,k_1(p)}\right|, \left|\frac{k_2(p)}{1+\hat{g}_2(p)\,k_2(p)}\right|\right] < \frac{1}{v_{12}(p)}, \qquad p \in \mathcal{D}$$

which can be satisfied only for sufficiently small $v_{12}(p)$.

These examples provoke two observations. First, although we began with the same information about the original system, the results received by means of the element-by-element uncertainty bounds are better than the results of the singular value approach and, for these examples, represent the best possible results for robust stability. Hence, the available information can be better exploited if we use the element-by-element bounds.

Secondly, having our examples in mind it seems paradoxical that Theorem 8.3 states a necessary and sufficient stability condition whereas the criterion of Theorem 8.4, which yielded the better result, is said to be conservative. The explanation for this is given by the fact that the conservatism of the stability criteria is usually assessed with regard to the set $\mathcal{G}$ of plant representative rather than in comparison with the primarily available information about the plant. The conditions derived above by means of the singular value approach are necessary and sufficient for the preservation of the stability under all perturbations that are bounded by $\sigma_{\max}[\delta \mathbf{G}(p)] \leqq \max\,[v_{11}(p), v_{22}(p)]$ or $\sigma_{\max}[\delta \mathbf{G}(p)] \leqq v_{12}(p)$, respectively. Some of these perturbations have substantial non-diagonal elements, which can be tolerated only if the strong conditions given above are satisfied. As such perturbations are excluded by element-by-element bounds the conditions received for these uncertainty bounds are weaker. □

As the condition (14) is not necessary for robust stability there exists some weaker condition that may prove the stability of systems for which condition (14) fails.

The endeavour to find this condition turns out to be very complicated. To start with, remember that the matrix norm provides an upper bound of the spectral radius

$$\varrho[\mathbf{AB}] \leqq \sigma_{\max}[\mathbf{AB}] \leqq \sigma_{\max}[\mathbf{A}]\,\sigma_{\max}[\mathbf{B}]. \tag{15}$$

The inequality sign may cause rather conservative relations. However, by means of scaling techniques the norms may be changed while the eigenvalues and the spectral radius of **AB** remain unaltered. Then

$$\begin{aligned}\varrho[\mathbf{AB}] &= \varrho[\mathbf{R}^{-1}\mathbf{A}\mathbf{L}^{-1}\mathbf{L}\mathbf{B}\mathbf{R}] \leqq \sigma_{\max}[\mathbf{R}^{-1}\mathbf{A}\mathbf{L}^{-1}\mathbf{L}\mathbf{B}\mathbf{R}] \\ &\leqq \sigma_{\max}[\mathbf{R}^{-1}\mathbf{A}\mathbf{L}^{-1}]\,\sigma_{\max}[\mathbf{L}\mathbf{B}\mathbf{R}]\end{aligned} \tag{16}$$

holds. We use this relation for our purpose in the form

$$\begin{aligned}\sup_{\mathbf{G}_2} \varrho[\bar{\mathbf{G}}_{zs}\mathbf{G}_2] &= \sup_{\mathbf{G}_2} \varrho[\mathbf{R}^{-1}\bar{\mathbf{G}}_{zs}\mathbf{L}^{-1}\mathbf{L}\mathbf{G}_2\mathbf{R}] \leqq \sup_{\mathbf{G}_2} \sigma_{\max}[\mathbf{R}^{-1}\bar{\mathbf{G}}_{zs}\mathbf{L}^{-1}\mathbf{L}\mathbf{G}_2\mathbf{R}] \\ &\leqq \sup_{\mathbf{G}_2} \sigma_{\max}[\mathbf{R}^{-1}\bar{\mathbf{G}}_{zs}\mathbf{L}^{-1}]\,\sigma_{\max}[\mathbf{L}\mathbf{G}_2\mathbf{R}] = \sigma_{\max}[\mathbf{R}^{-1}\bar{\mathbf{G}}_{zs}\mathbf{L}^{-1}]\,\sigma_{\max}[\mathbf{L}\mathbf{V}_2\mathbf{R}],\end{aligned} \tag{17}$$

where the scaling matrices are restricted to be real nonnegative diagonal matrices $\mathbf{R} = \operatorname{diag} r_i$ and $\mathbf{L} = \operatorname{diag} l_i$.

An improvement of eqn (14) can be reached by *Perron scaling*

$$\mathbf{R}_p = \operatorname{diag} \sqrt{\frac{u_i}{v_i}}, \quad \mathbf{L}_p = \operatorname{diag} \sqrt{\frac{y_i}{z_i}} \tag{18}$$

where **u**, **v** and **y**, **z** are the right and left eigenvectors of the Perron root λ_p of $|\bar{\mathbf{G}}_{zs}|\,\mathbf{V}_2$ or $\mathbf{V}_2\,|\bar{\mathbf{G}}_{zs}|$, respectively (cf Appendix 1). Then the inequality

$$\sup_{\mathbf{G}_2} \varrho[\bar{\mathbf{G}}_{zs}\mathbf{G}_2] \leqq \sigma_{\max}[\mathbf{R}_p^{-1}\bar{\mathbf{G}}_{zs}\mathbf{L}_p^{-1}]\,\sigma_{\max}[\mathbf{L}_p\mathbf{V}_2\mathbf{R}_p] \leqq \lambda_p[|\bar{\mathbf{G}}_{zs}|\,\mathbf{V}_2] \tag{19}$$

holds. Consequently,

$$\sigma_{\max}[\mathbf{R}_p^{-1}\bar{\mathbf{G}}_{zs}\mathbf{L}_p^{-1}]\,\sigma_{\max}[\mathbf{L}_p\mathbf{V}_2\mathbf{R}_p] < 1, \quad p \in \mathcal{D} \tag{20}$$

represents a sufficient condition for robust stability, which is weaker than eqn (14). The application of this condition necessitates the determination of the Perron root and the corresponding eigenvectors of $|\bar{\mathbf{G}}_{zs}|\,\mathbf{V}_2$ and $\mathbf{V}_2\,|\bar{\mathbf{G}}_{zs}|$ for all $p \in \mathcal{D}$.

A further improvement of the stability condition can be made if the right-hand side of eqn (17) is minimised with respect to the scaling matrices aiming at bounds that are as tight as possible. A preliminary result is given by the following lemma which is an extension of Lemma A2.1.

Lemma 8.3 Equation (15) holds with equality signs if and only if the input and output major principal directions of $\mathbf{A}$ are aligned with the output and input major principal directions of $\mathbf{B}$, i.e.

$$\begin{aligned} \mathbf{v_A} &= e^{j\phi_1}\mathbf{u_B} \\ \mathbf{u_A} &= e^{j\phi_2}\mathbf{v_B}, \end{aligned} \tag{21}$$

where $\mathbf{u_A}$, $\mathbf{v_A}$ and $\mathbf{u_B}$, $\mathbf{v_B}$ are the eigenvectors of $\mathbf{AA}^*$, $\mathbf{A}^*\mathbf{A}$, $\mathbf{BB}^*$ and $\mathbf{B}^*\mathbf{B}$, respectively, belonging to the maximum eigenvalues of these matrices.

For the definition of the principal directions see Appendix 2. We do not want to prove this result but give a plausible explanation. Eqns (A2—9) and (21) yield

$$\begin{aligned} \mathbf{ABv_B} &= \sigma_{\max}[\mathbf{B}]\,\mathbf{Au_B} \\ &= \sigma_{\max}[\mathbf{B}]\,e^{-j\phi_1}\mathbf{Av_A} \\ &= \sigma_{\max}[\mathbf{A}]\,\sigma_{\max}[\mathbf{B}]\,e^{-j\phi_1}\mathbf{u_A} \\ &= \sigma_{\max}[\mathbf{A}]\,\sigma_{\max}[\mathbf{B}]\,e^{j(\phi_2-\phi_1)}\mathbf{v_B}. \end{aligned}$$

Hence, $\sigma_{\max}[\mathbf{A}]\,\sigma_{\max}[\mathbf{B}]\,e^{j(\phi_2-\phi_1)}$ is an eigenvalue of $\mathbf{AB}$. Since its modulus equals $\sigma_{\max}[\mathbf{A}]\,\sigma_{\max}[\mathbf{B}]$ eqn (16) holds with equality signs. Eqn (21) is referred to as the *major principal direction alignment property*.

Due to this lemma, we try to find scaling matrices $\mathbf{R}$ and $\mathbf{L}$ so that the principal directions of $\mathbf{R}^{-1}\bar{\mathbf{G}}_{ZS}\mathbf{L}^{-1}$ and $\mathbf{L}\tilde{\mathbf{G}}_2\mathbf{R}$ can be aligned by appropriately choosing $\tilde{\mathbf{G}}_2$ and, thus, eqn (17) holds with the equality signs. The following lemma shows that diagonal scaling matrices $\mathbf{R}$ and $\mathbf{L}$ are sufficiently general to produce this property.

Lemma 8.4 For optimal scaling eqn (17) holds with equality signs

$$\sup_{\mathbf{G}_2} \varrho[\bar{\mathbf{G}}_{ZS}\mathbf{G}_2] = \inf_{\mathbf{R},\mathbf{L}} \sigma_{\max}[\mathbf{R}^{-1}\bar{\mathbf{G}}_{ZS}\mathbf{L}^{-1}]\,\sigma_{\max}[\mathbf{L}\mathbf{V}_2\mathbf{R}]. \tag{22}$$

Proof. The lemma can be proved in two steps. To outline the first step consider the optimisation problem

$$\mathbf{I} = \sigma_{\max}[\mathbf{R}^{-1}\bar{\mathbf{G}}_{ZS}\mathbf{L}^{-1}]\,\sigma_{\max}[\mathbf{L}\mathbf{V}_2\mathbf{R}] = \frac{\sigma_{\max}[\mathbf{L}\mathbf{V}_2\mathbf{R}]}{\sigma_{\min}[\mathbf{L}\bar{\mathbf{G}}_{ZS}\mathbf{R}]} \underset{l_i, r_i}{\to} \min. \tag{23}$$

It can be shown that the conditions

$$|\mathbf{u}_{\mathbf{LV}_2\mathbf{R}}| = \left|\mathbf{v}_{\mathbf{R}^{-1}\bar{\mathbf{G}}_{ZS}\mathbf{L}^{-1}}\right|, \quad |\mathbf{v}_{\mathbf{LV}_2\mathbf{R}}| = \left|\mathbf{u}_{\mathbf{R}^{-1}\bar{\mathbf{G}}_{ZS}\mathbf{L}^{-1}}\right| \tag{24}$$

are necessary for attaining the optimum. That is, the moduli of the corresponding elements of the major principal directions of $\mathbf{L}\mathbf{V}_2\mathbf{R}$ and $\mathbf{R}^{-1}\bar{\mathbf{G}}_{ZS}\mathbf{L}^{-1}$ are equalised by optimal scaling.

As second step we will prove that under optimal scaling there exists some $\tilde{\mathbf{G}}_2$ that produces the major principal alignment property of $\mathbf{L}^{-1}\bar{\mathbf{G}}_{ZS}\mathbf{R}^{-1}$ and $\mathbf{L}\tilde{\mathbf{G}}_2\mathbf{R}$ and, thus, necessitates the equality sign in eqn (17). Choose

$$\tilde{\mathbf{G}}_2 = \operatorname{diag} e^{j\phi_i}\mathbf{V}_2 \operatorname{diag} e^{-j\psi_i}$$

where ϕ_i and ψ_i are the phases of the i-th elements of the major principal directions $\mathbf{v}_{\mathbf{R}^{-1}\overline{\mathbf{G}}_{\mathrm{ZS}}\mathbf{L}^{-1}}$ and $\mathbf{u}_{\mathbf{R}^{-1}\overline{\mathbf{G}}_{\mathrm{ZS}}\mathbf{L}^{-1}}$ of the matrix $\mathbf{R}^{-1}\overline{\mathbf{G}}_{\mathrm{ZS}}\mathbf{L}^{-1}$ for optimal scaling. Obviously, $\tilde{\mathbf{G}}_2$ satisfies eqn (13). As the diagonal matrices are unitary, $\sigma_{\max}[\mathbf{L}\tilde{\mathbf{G}}_2\mathbf{R}] = \sigma_{\max}[\mathbf{L}\mathbf{V}_2\mathbf{R}]$ holds. Because of

$$\mathbf{L}\tilde{\mathbf{G}}_2\mathbf{R}\ \mathrm{diag}\ \mathrm{e}^{\mathrm{j}\psi_i}\mathbf{v}_{\mathbf{L}\mathbf{V}_2\mathbf{R}} = \mathrm{diag}\ \mathrm{e}^{\mathrm{j}\phi_i}\mathbf{L}\mathbf{V}_2\mathbf{R}\mathbf{v}_{\mathbf{L}\mathbf{V}_2\mathbf{R}} = \sigma_{\max}[\mathbf{L}\mathbf{V}_2\mathbf{R}]\ \mathrm{diag}\ \mathrm{e}^{\mathrm{j}\phi_i}\mathbf{v}_{\mathbf{L}\mathbf{V}_2\mathbf{R}}$$

we have for the major principal directions of $\mathbf{L}\tilde{\mathbf{G}}_2\mathbf{R}$

$$\mathbf{v}_{\mathbf{L}\tilde{\mathbf{G}}_2\mathbf{R}} = \mathrm{e}^{\mathrm{j}\psi_i}\mathbf{v}_{\mathbf{L}\mathbf{V}_2\mathbf{R}}, \quad \mathbf{u}_{\mathbf{L}\tilde{\mathbf{G}}_2\mathbf{R}} = \mathrm{e}^{\mathrm{j}\phi_i}\mathbf{v}_{\mathbf{L}\mathbf{V}_2\mathbf{R}}. \tag{25}$$

This and eqn (24) yield

$$|\mathbf{v}_{\mathbf{L}\tilde{\mathbf{G}}_2\mathbf{R}}| = |\mathbf{v}_{\mathbf{L}\mathbf{V}_2\mathbf{R}}| = \left|\mathbf{v}_{\mathbf{R}^{-1}\overline{\mathbf{G}}_{\mathrm{ZS}}\mathbf{L}^{-1}}\right|, \quad |\mathbf{u}_{\mathbf{L}\tilde{\mathbf{G}}_2\mathbf{R}}| = |\mathbf{u}_{\mathbf{L}\mathbf{V}_2\mathbf{R}}| = \left|\mathbf{v}_{\mathbf{R}^{-1}\overline{\mathbf{G}}_{\mathrm{ZS}}\mathbf{L}^{-1}}\right|. \tag{26}$$

Since $\mathbf{L}\mathbf{V}_2\mathbf{R}$ represents a nonnegative matrix, its major principal directions can be chosen so as to have zero phase. Then eqns (25) and (26) prove the major principal direction alignment property of $\mathbf{L}^{-1}\overline{\mathbf{G}}_{\mathrm{ZS}}\mathbf{R}^{-1}$ and $\mathbf{L}\tilde{\mathbf{G}}_2\mathbf{R}$. Hence, the spectral radius $\varrho[\overline{\mathbf{G}}_{\mathrm{ZS}}\mathbf{G}_2]$ assumes its supremum for $\mathbf{G}_2 = \tilde{\mathbf{G}}_2$. □

Theorem 8.5 (*Necessary and sufficient stability condition for elementwise bounded uncertainties*) Consider a plant model where the model uncertainty is bounded by eqn (13). Assume that the closed-loop approximate model (8.2-6) is stable and that the model uncertainties $\mathbf{G}_2$ are known to be stable. Then the closed-loop system (8.2-5) is robustly stable if and only if

$$\inf_{\mathbf{L},\mathbf{R}} \sigma_{\max}[\mathbf{R}^{-1}\overline{\mathbf{G}}_{\mathrm{ZS}}\mathbf{L}^{-1}]\ \sigma_{\max}[\mathbf{L}\mathbf{V}_2\mathbf{R}] < 1\,, \qquad p \in \mathscr{D}\,. \tag{27}$$

As this theorem is based on the exact determination of $\sup_{\mathbf{G}_2} \varrho[\overline{\mathbf{G}}_{\mathrm{ZS}}\mathbf{G}_2]$ it fully exploits the information (13) about the model uncertainties. The optimisation problem (27) or, equivalently, eqn (23) can be solved by means of standard procedures. Eqn (24) may be used to check the termination of the algorithm. The use of Theorem 8.5 provides a significant computational burden because eqn (27) must be applied for all, or at least all 'critical' frequencies p on the Nyquist contour $\mathscr{D}$. It is, therefore, advisable to try to prove the robust stability by means of Theorems 8.3 and 8.4 first and to use eqn (27) only if these tests fail.

Extensions to unstable perturbations

Until now we have presupposed the stability of the uncertainty model. This might be too restrictive as it virtually necessitates the correct presentation of the unstable parts by the approximate model. Much effort was and is being made to remove this assumption and to derive more general stability conditions. We will present some of the results here in order to show how they work and that they necessitate additional assumptions concerning the model uncertainties.

Assuming that the original plant and the approximate model share the same number of right-half plane poles, we will use Theorem 8.2 here for additive and multiplicative perturbations. As $\mathbf{G}_{zs} = 0$ holds (cf Fig. 5.2) $\det(\mathbf{I} - \mathbf{G}_2\overline{\mathbf{G}}_{zs})$ must not encircle the origin of the complex plane for $p \in \mathcal{D}$. This condition coincides with eqn (2), which has been used to derive the stability conditions of Theorems 8.3 through 8.5. Therefore, the range of validity of these conditions can be extended.

Theorem 8.6 (*Extension of Theorems 8.3, 8.4 and 8.5*). The results of Theorems 8.3—8.5 hold for unstable additive or multiplicative perturbations provided that the original system $\mathbf{G}(p)$ and the approximate model $\hat{\mathbf{G}}(p)$ share the same number of right-half plane poles.

Notice that the inequalities (9), (14), and (27) are not changed. But we must be sure that the model uncertainties do not bring about additional right-half plane poles. This must be proved by additional a-priori information about the plant since this assumption may not be checked by means of the upper uncertainty bounds $v_2(p)$ and $\mathbf{V}_2(p)$ for $p \in \mathcal{D}$. Hence, Theorem 8.6 does not prove the closed-loop stability for any plant of the set $\mathcal{G}$ defined in (5.3-19) or (5.3-20), respectively, but only for those that observe the pole condition.

Example 8.3 Consider the robust stabilisation of the plant described in Example 5.2. The system is known to possess exactly one pole in the right-half complex plane. Obviously, the apprximate model ($\hat{\lambda} = 0.5$) is stabilised by unity feedback. According to Theorem 8.6 the closed-loop original system is robustly stable under all perturbation $\delta g(p)$ that satisfy $|\delta g(p)| \leqq v_2(p)$ and do not bring about additional right-half plane poles if and only if

$$|1 + \hat{g}(p)| = \left|\frac{p - \hat{\lambda} + 1}{p - \hat{\lambda}}\right| = \left|\frac{p + 0.5}{p - 0.5}\right| \geqq v_2(p), \qquad p \in \mathcal{D}$$

is satisfied. Since the original plant is known to be strictly proper we have only to consider $p = \mathrm{j}\omega$. The stability condition is satisfied because $v_2(\mathrm{j}\omega) < 1$ whereas $|1 + \hat{g}(\mathrm{j}\omega)| = 1$ (see Fig. 5.5). □

A similar extension can be made for the model structures (b) and (d) in Fig. 5.2 on the basis of the inverse Nyquist stability criterion. In analogy to eqns (3) and (4) we get for the structure (b)

$$\begin{aligned}\det\left(\mathbf{I} + (\mathbf{KCG})^{-1}\right) &= \det\left(\mathbf{I} + (\mathbf{KC}\hat{\mathbf{G}}(\mathbf{I} + \mathbf{G}_2\hat{\mathbf{G}})^{-1})^{-1}\right)\\ &= \det\left(\mathbf{I} + (\mathbf{I} + \mathbf{G}_2\hat{\mathbf{G}})\,(\mathbf{KC}\hat{\mathbf{G}})^{-1}\right)\\ &= \det\left(\mathbf{I} + (\mathbf{KC}\hat{\mathbf{G}})^{-1}\right)\det\left(\mathbf{I} + \mathbf{G}_2\hat{\mathbf{G}}(\mathbf{I} + \mathbf{KC}\hat{\mathbf{G}})^{-1}\right). \qquad (28)\end{aligned}$$

Furthermore, for this model structure

$$\overline{\mathbf{G}}_{zs} = -\hat{\mathbf{G}}(\mathbf{I} + \mathbf{KC}\hat{\mathbf{G}})^{-1} \qquad (29)$$

can be derived. Hence, we have

$$\det\left(\mathbf{I} + (\mathbf{KCG})^{-1}\right) = \det\left(\mathbf{I} + (\mathbf{KC}\hat{\mathbf{G}})^{-1}\right)\det(\mathbf{I} - \mathbf{G}_2\overline{\mathbf{G}}_{zs}). \qquad (30)$$

If the original system and the approximate model have the same number of right-half plane zeros then the image of the Nyquist contour under det $\left(\mathbf{I} + (\mathbf{KCG})^{-1}\right)$ and det $\left(\mathbf{I} + (\mathbf{KC\hat{G}})^{-1}\right)$ must encircle the origin the same number of times. Due to eqn (30) we have to check again the condition (2). The same happens if the model structure of Fig. 5.2 (d) is used.

Theorem 8.7 (*Extension of Theorems 8.3, 8.4, and 8.5*). The results of Theorems 8.3—8.5 hold for unstable perturbations occuring in the plant model of structures (b) and (d) in Fig. 5.2 provided that the original system $\mathbf{G}(p)$ and the approximate model $\hat{\mathbf{G}}(p)$ share the same number of right-half plane zeros.

Further extensions are possible by applying the results of Theorems 8.6 and 8.7 for complementary frequency intervals. Then pole and zero movements across the imaginary axis of the complex plane can be permitted.

The claim that the original system and the approximate model must have identical numbers of right-half plane poles or zeros is, in principle, an artificial restriction that arises from the use of the Nyquist criterion as basis of the stability investigations. That it may be dropped is demonstrated by Example 8.1 or Lemma 6.2, which can be applied to systems with different numbers of right-half plane poles and zeros.

Extension to nonlinear perturbations

To outline the possibility of extending our results to nonlinear plants we use a static nonlinear uncertainty model

$$s = n(z) \tag{31}$$

with scalar signals s and z instead of the linear model (8.2-10) or (8.2-35), respectively. The nonlinearity is assumed to lie inside the sector

$$f_{\mathrm{u}} \leqq \frac{n(z)}{z} \leqq f_{\mathrm{o}} \,. \tag{32}$$

Then the same argumentation which led to Lemma 8.1 can be applied to the nonlinear system (8.2-34), (31) saying that the stability of the loop consisting of the linear element with transfer function $\bar{g}_{\mathrm{zs}}(p)$ and the nonlinear element (31) is sufficient for the stability of the whole system (cf Fig. 8.7(b)). This condition can be proved by means of the generalised circle criterion (Theorem 2.4). Applying it to the closed-loop system (8.2-4), (8.2-6), (31) we get the following result.

Theorem 8.8 (*Robust stability of nonlinear original systems*). Consider the nonlinear plant described by a linear approximate model (8.2-6) and a single-variable static nonlinear uncertainty model (31) which is known to lie inside the sector (32). Assume that the closed-loop approximate model (8.2-4), (8.2-6) is stable and that $\bar{g}_{\mathrm{zs}}(p)$ is strictly proper. Then the closed-loop original system (8.2-4), (8.2-6), (31) is robustly stable if the Nyquist diagram of $-\bar{g}_{zs}(p)$ does not encircle or cross the area defined by the circle through the points $-\frac{1}{f_u}$ and $-\frac{1}{f_o}$ with centre at the real axis.

The negative sign of $-\bar{g}_{zs}$ results from the absence of such a sign within the loop on the top of Fig. 8.7 (b).

Discussion

The conditions presented in this section concern the preservation of closed-loop stability under a quantitatively specified class $\mathscr{S}_2$ of perturbations. In comparison with the sensitivity analysis they have two remarkable properties. First, they give *quantitative* bounds on the model uncertainties and, hence, on the neighbourhood of $\hat{\mathbf{G}}(p)$ for which the stability of the closed-loop system is ensured by the stability of the closed-loop approximate model and the robustness of feedback systems. As the examples confirmed, the tolerable uncertainties are by no means restricted to be differentially small.

Second, the indicator of robustness varies with the entry points of the uncertainty model. The stability robustness may be described by both the sensitivity matrix and the complementary sensitivity matrix (cf eqns (11), (12)). Contrary to this, sensitivity considerations uniformly led to the advice to strive at the smallest possible sensitivity matrix.

For stability considerations it suffices to know the uncertainty bounds $v_2(p)$ and $\mathbf{V}_2(p)$ for p on the Nyquist contour $\mathscr{D}$ or, if $\bar{\mathbf{G}}_{zs}\mathbf{G}_2$ is known to be strictly proper, for p on the imaginary axis part of $\mathscr{D}$. Modelling problems will only occur in the surrounding of imaginary axis poles of $\bar{\mathbf{G}}_{zs}\mathbf{G}_2(p)$ where the Nyquist contour has indentations to the left-half complex plane. Further qualitative properties of $\mathbf{G}_2$ as the stability or the number of right-half plane poles or zeros can be checked without additional quantitative uncertainty evaluations.

8.4 Time domain criteria for I/O-stability

This section is devoted to time domain criteria for robust I/O-stability. If the closed-loop original system (8.2-18) is known to be completely controllable and observable, I/O-stability is equivalent to internal stability. Some results that are directly concerned with internal stability will be presented in Section 8.7.

General I/O-stability condition

The stability of the system (8.2-38), (8.2-39) can be analysed in the same way as its frequency domain conterpart (8.2-34), (8.2-35). According to Theorem 2.1 the system $\mathbf{y}(t) = \mathbf{G}_{yr}(t) * \mathbf{r}(t)$ is stable if and only if there exists some finite scalar $m_{\mathbf{yr}}$ such that

$$\int_0^\infty \|\mathbf{G}_{\mathbf{yr}}(t)\| \, \mathrm{d}t \leqq m_{\mathbf{yr}} < \infty \tag{1}$$

holds. By assumption, the system (8.2-38) is stable with regard to all its inputs and outputs. We further assume that all elements $\mathbf{G}_2(t)$ of the set $\mathscr{G}_2$ are stable, ie

$$\int_0^\infty \|\mathbf{G}_2(t)\| \, \mathrm{d}t \leqq \overline{m}_2 < \infty \tag{2}$$

holds for all $\mathbf{G}_2(t) \in \mathscr{G}_2$ with the same finite scalar $\overline{m}_2$. Then eqn (8.2-41) leads to

$$\int_0^\infty \|\mathbf{G}_{\mathbf{yr}}(t)\| \, \mathrm{d}t \leqq \int_0^\infty \|\overline{\mathbf{G}}_{\mathbf{yr}}\| \, \mathrm{d}t + \int_0^\infty \|\overline{\mathbf{G}}_{\mathbf{ys}}\| \, \mathrm{d}t \int_0^\infty \|\mathbf{G}_2\| \, \mathrm{d}t \int_0^\infty \|\tilde{\mathbf{G}}\| \, \mathrm{d}t \int_0^\infty \|\overline{\mathbf{G}}_{\mathbf{zr}}\| \, \mathrm{d}t$$

$$\leqq m_{\mathbf{yr}} + m_{\mathbf{ys}}\overline{m}_2 \int_0^\infty \|\tilde{\mathbf{G}}(t)\| \, \mathrm{d}t \; m_{\mathbf{zr}}$$

where $m_{\mathbf{yr}}$, $m_{\mathbf{ys}}$, and $m_{\mathbf{zr}}$ satisfy relations similar to eqn (2) for $\overline{\mathbf{G}}_{\mathbf{yr}}(t)$, $\overline{\mathbf{G}}_{\mathbf{ys}}(t)$, and $\overline{\mathbf{G}}_{\mathbf{zr}}(t)$. If $\tilde{\mathbf{G}}(t)$ from eqn (8.2-42)

$$\tilde{\mathbf{G}}(t) = \delta(t)\,\mathbf{I} + \overline{\mathbf{G}}_{\mathbf{zs}}(t) * \mathbf{G}_2(t) * \tilde{\mathbf{G}}(t) \tag{3}$$

represents a stable system, then eqn (1) is satisfied. The stability of $\tilde{\mathbf{G}}(t)$ is necessary as well if rank $\overline{\mathbf{G}}_{\mathbf{ys}}(t) = \min\,[\dim \mathbf{y}, \dim \mathbf{s}]$ and rank $\overline{\mathbf{G}}_{\mathbf{zr}}(t) = \min\,[\dim \mathbf{z}, \dim \mathbf{r}]$ for at least one instant of time, which is true for models with the minimum dimensional signals $\mathbf{z}$ and $\mathbf{s}$.

Lemma 8.5 Assume that the completely known part (8.2-38) of the original system is stable with respect to all its inputs and outputs and that all elements $\mathbf{G}_2(t) \in \mathscr{G}_2$ are stable. Then the original system (8.2-38), (8.2-39) is robustly stable if and only if the condition

$$\int_0^\infty \|\tilde{\mathbf{G}}(t)\| \, \mathrm{d}t \leqq m < \infty \tag{4}$$

with $\tilde{\mathbf{G}}(t)$ from eqn (3) is satisfied for all $\mathbf{G}_2(t) \in \mathscr{G}_2$.

The correspondences between (8.2-38) and the closed-loop approximate model (8.2-28) leads to the following stability criterion.

Theorem 8.9 Assume that the closed-loop approximate model (8.2-28) is I/O-stable and that all elements of $\mathscr{G}_2$ are stable. Then the closed-loop original system (8.2-31) is robustly stable if and only if condition (4) is satisfied for all $\mathbf{G}_2(t) \in \mathscr{G}_2$.

Norm-bounded uncertainties

If $\mathbf{G}_2(t)$ is norm-bounded as in eqn (5.3-32)

$$\|\mathbf{G}_2(t)\| \leqq v_2(t) \tag{5}$$

for an arbitrary matrix norm $\|.\|$, eqn (3) leads to

$$\|\tilde{\mathbf{G}}(t)\| \leqq \delta(t) + \|\overline{\mathbf{G}}_{zs}\| * \|\mathbf{G}_2\| * \|\tilde{\mathbf{G}}\| \leqq \delta(t) + \|\overline{\mathbf{G}}_{zs}\| * v_2 * \|\tilde{\mathbf{G}}\|. \tag{6}$$

Hence, we have

$$\int_0^\infty \|\tilde{\mathbf{G}}(t)\|\, dt \leqq 1 + \int_0^\infty \|\overline{\mathbf{G}}_{zs}(t)\|\, dt \int_0^\infty v_2(t)\, dt \int_0^\infty \|\tilde{\mathbf{G}}(t)\|\, dt$$

and

$$\int_0^\infty \|\tilde{\mathbf{G}}(t)\|\, dt \leqq \frac{1}{1 - \int_0^\infty \|\overline{\mathbf{G}}_{zs}(t)\|\, dt \int_0^\infty v_2(t)\, dt} < \infty$$

provided that

$$\int_0^\infty \|\overline{\mathbf{G}}_{zs}(t)\|\, dt \int_0^\infty v_2(t)\, dt < 1. \tag{7}$$

Equation (7) proves the robust stability of the original system.

Theorem 8.10 Consider a plant model where the model uncertainties are bounded by eqn (5). Assume that the closed-loop approximate model (8.2-26) is stable and that the stability of $\mathbf{G}_2(t) \in \mathcal{G}_2$ can be proved by

$$m_2 = \int_0^\infty v_2(t)\, dt < \infty. \tag{8}$$

Then the closed-loop system (8.2-31) is robustly stable if

$$\int_0^\infty \|\overline{\mathbf{G}}_{zs}(t)\|\, dt < \frac{1}{m_2}. \tag{9}$$

As opposed to the frequency domain results presented in Theorem 8.3, this stability condition is merely sufficient and in general not necessary. The stability of $\mathbf{G}_2(t)$ can be easily proved by means of the given function $v_2(t)$.

Element-by-element-bounded uncertainties

If the set $\mathcal{G}_2$ is described by some matrix function $\mathbf{V}_2(t)$ bounding $|\mathbf{G}_2(t)|$

$$|\mathbf{G}_2(t)| \leqq \mathbf{V}_2(t) \tag{10}$$

an analog line of reasoning leads to

$$\int_0^\infty |\tilde{\mathbf{G}}(t)|\, dt \leq \mathbf{I} + \int_0^\infty |\bar{\mathbf{G}}_{zs}(t)|\, dt \int_0^\infty \mathbf{V}_2(t)\, dt \int_0^\infty |\tilde{\mathbf{G}}(t)|\, dt$$

and

$$\left(\mathbf{I} - \int_0^\infty |\bar{\mathbf{G}}_{zs}(t)|\, dt \int_0^\infty \mathbf{V}_2(t)\, dt\right) \int_0^\infty |\tilde{\mathbf{G}}(t)|\, dt \leq \mathbf{I}. \tag{11}$$

This inequality can be resolved for $\int_0^\infty |\tilde{\mathbf{G}}(t)|\, dt$ only if the inverse matrix of

$$\mathbf{V} = \mathbf{I} - \int_0^\infty |\bar{\mathbf{G}}_{zs}(t)|\, dt \int_0^\infty \mathbf{V}_2(t)\, dt \tag{12}$$

is nonnegative. This property is ensured if

$$\lambda_p \left[\int_0^\infty |\bar{\mathbf{G}}_{zs}(t)|\, dt \int_0^\infty \mathbf{V}_2(t)\, dt\right] < 1 \tag{13}$$

holds, where λ_p is the Perron root (maximum eigenvalue) of the given nonnegative matrix. Under the condition (13) the matrix $\mathbf{V}$ in eqn (12) represents an M-matrix and, thus, has a nonnegative inverse (for details of M-matrices see Appendix 1). Then $\int_0^\infty |\tilde{\mathbf{G}}(t)|\, dt \leq \mathbf{V}^{-1} < \infty$ holds which proves the stability of $\tilde{\mathbf{G}}(t)$.

Theorem 8.11 Consider a plant model where the model uncertainties are bounded by eqn (10). Assume that the closed-loop approximate model (8.2-26) is stable and that the stability of $\mathbf{G}_2(t) \in \mathcal{G}_2$ can be proved by

$$\mathbf{M}_2 = \int_0^\infty \mathbf{V}_2(t)\, dt < \infty. \tag{14}$$

Then the closed-loop system (8.2-31) is robustly stable if eqn (13) is satisfied.

Note the analogy of this result and that of Theorem 8.4.

Theorem 8.11 has a nice graphical interpretation. In the condition (13) the model uncertainty is merely described by the matrix $\mathbf{M}_2$ from eqn (14). In general, $\int |\mathbf{G}_2|\, dt \geqq |\mathbf{H}_2(\infty)|$ holds where $\mathbf{H}_2(t)$ denotes the step response of the uncertainty model. If we have a *monotone modelling error*, ie all elements of $\mathbf{H}_2(t)$ are either nonnegative or nonpositive for all t (sign-definite), then the elements of $\mathbf{M}_2$ are the largest static reinforcements of all $\mathbf{G}_2(t) \in \mathcal{G}_2$.

More generally, $\int |\mathbf{G}_2|\, dt$ represents the *absolute matrix variation* of $\mathbf{H}(t)$. All its elements can be determined by adding the ups and downs of the corresponding step responses. For the approximation error encountered in Example 1.1 the absolute variation $\int |g_2| \, dt$ amounts to 0.18 (Fig. 8.10). Using the first-order approximate model with $T_1 = 1$, the closed-loop stability can be proved by condition (13) for all P-controllers with $-0.9 \leq k \leq 3.5$.

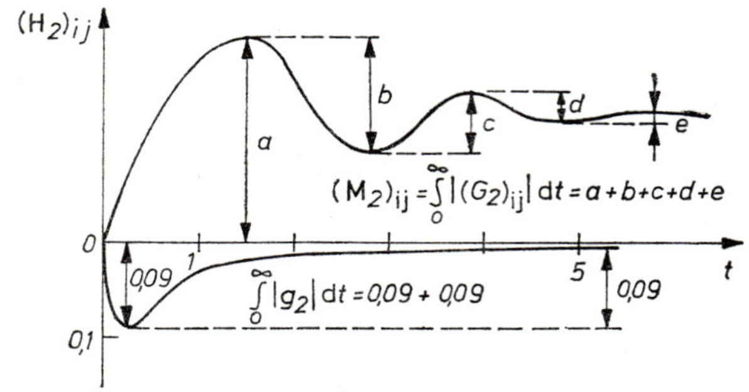

Fig. 8.10 Total variation $\mathbf{M}_2$ of $\mathbf{H}_2(t)$
Above: The way of determining $\mathbf{M}_2$
Below: Total variation of the approximation error of Example 1.1

8.5 Analysis of the I/O-performance

After we have succeeded in proving the robust stability of the closed-loop original system we turn our attention to a more detailed investigation of the system performance. Again we consider the system (8.2-34), (8.2-35) or (8.2-38), (8.2-39) first and apply the results to the feedback system then.

Frequency domain analysis

In what follows we deal with the system frequency response ($p = j\omega$). Ignoring the model uncertainties (8.2-35), eqn (8.2-34) gives the appropriate model

$$\hat{\mathbf{y}}(j\omega) = \overline{\mathbf{G}}_{yr}(j\omega)\, \mathbf{r}(j\omega) \tag{1}$$

of the whole system (8.2-34), (8.2-35). Robustness considerations of the I/O-performance necessitates an answer to the question how far the original system output $\mathbf{y}$ may deviate from the approximate $\hat{\mathbf{y}}$. Subtracting (1) from (8.2-34) yields

$$\mathbf{y} - \hat{\mathbf{y}} = \overline{\mathbf{G}}_{ys}\mathbf{s} = \overline{\mathbf{G}}_{ys}\mathbf{G}_2\mathbf{z}\,. \tag{2}$$

$\mathbf{z}$ is given by eqns (8.2-34) and (8.2-35)

$$\mathbf{z} = \overline{\mathbf{G}}_{zr}\mathbf{r} + \overline{\mathbf{G}}_{zs}\mathbf{G}_2\mathbf{z}\,. \tag{3}$$

The further way of the analysis will be explained for *norm-bounded* uncertainties

$$\|\mathbf{G}_2\| \leqq v_2 \tag{4}$$

first. Using norm inequalities eqn (3) yields

$$\|\mathbf{z}\| \leqq \|\bar{\mathbf{G}}_{zr}\| \, \|\mathbf{r}\| + \|\bar{\mathbf{G}}_{zs}\| \, \|\mathbf{G}_2\| \, \|\mathbf{z}\| \leqq \|\bar{\mathbf{G}}_{zr}\| \, \|\mathbf{r}\| + \|\bar{\mathbf{G}}_{zs}\| \, v_2 \, \|\mathbf{z}\| .$$

If

$$\|\bar{\mathbf{G}}_{zs}(j\omega)\| \, v_2(j\omega) < 1 \tag{5}$$

holds, then the equation above can be transformed into

$$\|\mathbf{z}\| \leqq \frac{\|\bar{\mathbf{G}}_{zr}\|}{1 - \|\bar{\mathbf{G}}_{zs}\| \, v_2} \|\mathbf{r}\| .$$

With eqn (2) we have

$$\|\mathbf{y} - \hat{\mathbf{y}}\| \leqq \frac{\|\bar{\mathbf{G}}_{ys}\| \, v_2 \, \|\bar{\mathbf{G}}_{zr}\|}{1 - \|\bar{\mathbf{G}}_{zs}\| \, v_2} \|\mathbf{r}\| . \tag{6}$$

Lemma 8.6 Eqn (1) gives an approximate model of the system (8.2-34), (8.2-35). An upper bound of the approximation error is given by eqn (6) for norm-bounded uncertainties provided that the condition (5) is satisfied.

Applying this result to our feedback system (8.2-9), (8.2-10) we use

$$\hat{\mathbf{y}}(j\omega) = \bar{\mathbf{G}}_{yv}(j\omega) \, \mathbf{v}(j\omega) + \bar{\mathbf{G}}_{yd}(j\omega) \, \mathbf{d}(j\omega) + \bar{\mathbf{G}}_{yw}(j\omega) \, \mathbf{w}(j\omega) \tag{7}$$

as approximate model. Eqn (6) leads to

$$\|\mathbf{y} - \hat{\mathbf{y}}\| \leqq \frac{\|\bar{\mathbf{G}}_{ys}\| \, v_2 \, \|\bar{\mathbf{G}}_{zv}\|}{1 - \|\bar{\mathbf{G}}_{zs}\| \, v_2} \|\mathbf{v}\| + \frac{\|\bar{\mathbf{G}}_{ys}\| \, v_2 \, \|\bar{\mathbf{G}}_{zd}\|}{1 - \|\bar{\mathbf{G}}_{zs}\| \, v_2} \|\mathbf{d}\| + \frac{\|\bar{\mathbf{G}}_{ys}\| \, v_2 \, \|\bar{\mathbf{G}}_{zw}\|}{1 - \|\bar{\mathbf{G}}_{zs}\| \, v_2} \|\mathbf{w}\| . \tag{8}$$

In view of eqns (8.2-11) and (8.2-12) we have

$$\begin{aligned}
\|\delta\mathbf{G}_{yv}(j\omega)\| &\leqq \frac{\|\bar{\mathbf{G}}_{ys}(j\omega)\| \, v_2(j\omega) \, \|\bar{\mathbf{G}}_{zv}(j\omega)\|}{1 - \|\bar{\mathbf{G}}_{zs}(j\omega)\| \, v_2(j\omega)} = v_{yv}(j\omega) \\
\|\delta\mathbf{G}_{yd}(j\omega)\| &\leqq \frac{\|\bar{\mathbf{G}}_{ys}(j\omega)\| \, v_2(j\omega) \, \|\bar{\mathbf{G}}_{zv}(j\omega)\|}{1 - \|\bar{\mathbf{G}}_{zs}(j\omega)\| \, v_2(j\omega)} = v_{yd}(j\omega) \\
\|\delta\mathbf{G}_{yw}(j\omega)\| &\leqq \frac{\|\bar{\mathbf{G}}_{ys}(j\omega)\| \, v_2(j\omega)\| \, \|\bar{\mathbf{G}}_{zw}(j\omega)\|}{1 - \|\bar{\mathbf{G}}_{zs}(j\omega)\| \, v_2(j\omega)} = v_{yw}(j\omega) .
\end{aligned} \tag{9}$$

Theorem 8.13 Eqn (7) gives an approximate model of the I/O-behaviour of the closed-loop original system (8.2-5). For norm-bounded uncertainties the approximation errors are bounded by eqn (9) provided that the condition (5) is satisfied.

Note that eqn (5) holds if the robust stability has been proved by Lemma 8.2. The error bound can be illustrated in the Nyquist diagram of the closed-loop system as circles around $\bar{\mathbf{G}}_{yv}(j\omega)$, $\bar{\mathbf{G}}_{yd}(j\omega)$, or $\bar{\mathbf{G}}_{yw}(j\omega)$ with radius $v_{yv}(j\omega)$ etc (Fig. 8.11). In the Bode plot the amplitude response of the original system (8.2-5) is enclosed in the tolerance band given by

$$\left| \|\bar{\mathbf{G}}_{yv}(j\omega)\| - v_{yv}(j\omega) \right|_{dB} \leqq \left| \|\mathbf{G}_{yv}(j\omega)\| \right|_{dB} \leqq \left| \|\bar{\mathbf{G}}_{yv}(j\omega)\| + v_{yv}(j\omega) \right|_{dB} \qquad (10)$$

etc. The evaluation of $\mathbf{G}_{yv}(j\omega)$ according to the given design requirements has to consider all functions $\mathbf{G}_{yv}(j\omega)$ that lie within the tolerance band. If the band is sufficiently small the Nyquist or Bode plots give a suitable impression on whether the closed loop will satisfy the design objectives.

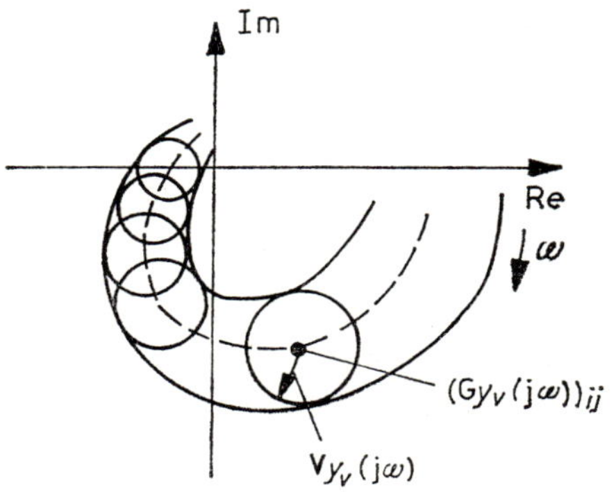

Fig. 8.11 Tolerance band of the frequency response of the closed-loop original system (element ij)

For single-input single-output systems the bounds $v_{yv}(j\omega)$, $v_{yd}(j\omega)$, and $v_{yw}(j\omega)$ are tight, ie eqn (8) holds with equality sign. For multi-input multi-output systems the norm inequalities are conservative. There is in general no original system for which $\|\mathbf{G}_{yv}(j\omega)\|$ touches the bounds $\|\bar{\mathbf{G}}_{yv}(j\omega)\| \pm v_{yv}(j\omega)$ etc. In other words, the bounds given in eqn (8) are broader than they must be in order to enclose all closed-loop systems whose uncertainties are limited by eqn (4). A further conservatism results from lumping together the whole information about the multivariable system into scalar uncertainty measures v_{yv} etc, which do not differentiate between the uncertainties of the different signal paths through the closed-loop system.

If *element-by-element bounds* are used

$$|\mathbf{G}_2(j\omega)| \leqq \mathbf{V}_2(j\omega) \qquad (11)$$

eqn (3) yields

$$|\mathbf{z}| \leqq |\bar{\mathbf{G}}_{zr}|\,|\mathbf{r}| + |\bar{\mathbf{G}}_{zs}|\,\mathbf{V}_2\,|\mathbf{z}|\,.$$

If the relation

$$\lambda_p[|\bar{\mathbf{G}}_{zs}(j\omega)|\,\mathbf{V}_2(j\omega)] < 1 \qquad (12)$$

holds and, thus, $(\mathbf{I} - |\bar{\mathbf{G}}_{zs}|\,\mathbf{V}_2)$ represents an M-matrix,

$$|\mathbf{z}| \leqq (\mathbf{I} - |\bar{\mathbf{G}}_{zs}|\,\mathbf{V}_2)^{-1}\,|\bar{\mathbf{G}}_{zr}|\,|\mathbf{r}|$$

is received. With eqn (2) we get

$$|\mathbf{y} - \hat{\mathbf{y}}| \leqq |\bar{\mathbf{G}}_{ys}|\,\mathbf{V}_2(\mathbf{I} - |\bar{\mathbf{G}}_{zs}|\,\mathbf{V}_2)^{-1}\,|\mathbf{G}_{zr}|\,|\mathbf{r}|\,. \tag{13}$$

Lemma 8.7 Eqn (13) gives an upper bound of the approximation error for element-by-element-bounded model uncertainties provided that the condition (12) is satisfied.

For the feedback system (8.2-9), (8.2-10) we get

$$|\mathbf{y} - \hat{\mathbf{y}}| \leqq \mathbf{V}_{yv}\,|\mathbf{v}| + \mathbf{V}_{yd}\,|\mathbf{d}| + \mathbf{V}_{yw}\,|\mathbf{w}| \tag{14}$$

with

$$\begin{aligned}
\mathbf{V}_{yv}(j\omega) &= |\bar{\mathbf{G}}_{ys}(j\omega)|\,\mathbf{V}_2(j\omega)\,\big(\mathbf{I} - |\bar{\mathbf{G}}_{zs}(j\omega)|\,\mathbf{V}_2(j\omega)\big)^{-1}\,|\bar{\mathbf{G}}_{zv}(j\omega)| \\
\mathbf{V}_{yd}(j\omega) &= |\bar{\mathbf{G}}_{ys}(j\omega)|\,\mathbf{V}_2(j\omega)\,\big(\mathbf{I} - |\bar{\mathbf{G}}_{zs}(j\omega)|\,\mathbf{V}_2(j\omega)\big)^{-1}\,|\bar{\mathbf{G}}_{zd}(j\omega)| \qquad (15) \\
\mathbf{V}_{yw}(j\omega) &= |\bar{\mathbf{G}}_{ys}(j\omega)|\,\mathbf{V}_2(j\omega)\,\big(\mathbf{I} - |\bar{\mathbf{G}}_{zs}(j\omega)|\,\mathbf{V}_2(j\omega)\big)^{-1}\,|\bar{\mathbf{G}}_{zw}(j\omega)|\,.
\end{aligned}$$

Theorem 8.14 Eqn (7) gives an approximate model of the I/O-behaviour of the closed-loop original system (8.2-5). For element-by-element-bounded uncertainties the approximation errors are bounded by eqn (14) provided that the condition (12) is satisfied.

Eqn (14) provides better results than eqn (8) since the individual relations between $y_i - \hat{y}_i$ and v_j, d_j, or w_j are evaluated separately. In particular, the strength of the interactions between different control loops can be distinguished from the main control loops. This will be used in Section 11.2 for assessing the effects of ignored interactions on the closed-loop system behaviour. However, alike the bounds (9), eqn (15) describes conservative bounds, that, in general, are not reached by perturbations satisfying eqn (11).

The bounds given in eqn (9) and (15) are monotonically increasing functions of $v_2(j\omega)$ and $\mathbf{V}_2(j\omega)$. For eqn (15) this is due to monotonicity properties of M-matrices. Therefore, the product $\|\bar{\mathbf{G}}_{zs}\|\,v_2$ and the Perron root (12) are indicators not only of robust stability but also of the width of the tolerance bands. By experience, if their values exceed 0.5 the tolerances v_{yv} and $\mathbf{V}_{yv}$ etc of the closed-loop performance are so large that no reasonable design specifications (III) can be met. Then, we may expect a substantial stability degree, but a detailed analysis of the I/O-behaviour on the basis of the tolerance band is impossible.

Time domain analysis

In direct analogy to eqns (7), (8) and (14) the following time domain results can be derived.

Theorem 8.15 An approximate model of the I/O-behaviour of the closed-loop original system (8.2-19) is given by

$$\hat{\mathbf{y}}(t) = \bar{\mathbf{G}}_{\mathbf{yv}} * \mathbf{v} + \bar{\mathbf{G}}_{\mathbf{yd}} * \mathbf{d} + \bar{\mathbf{G}}_{\mathbf{yw}} * \mathbf{w}. \tag{16}$$

The approximation errors $\delta\mathbf{G}_{\mathbf{yv}}$, $\delta\mathbf{G}_{\mathbf{yd}}$, and $\delta\mathbf{G}_{\mathbf{yw}}$ are bounded by

$$\begin{aligned}
\|\delta\mathbf{G}_{\mathbf{yv}}(t)\| &\leqq \|\bar{\mathbf{G}}_{\mathbf{ys}}(t)\| * v_2(t) * \tilde{g}(t) * \|\bar{\mathbf{G}}_{\mathbf{zv}}(t)\| = v_{\mathbf{yv}}(t) \\
\|\delta\mathbf{G}_{\mathbf{yd}}(t)\| &\leqq \|\bar{\mathbf{G}}_{\mathbf{ys}}(t)\| * v_2(t) * \tilde{g}(t) * \|\bar{\mathbf{G}}_{\mathbf{zd}}(t)\| = v_{\mathbf{yd}}(t) \\
\|\delta\mathbf{G}_{\mathbf{yw}}(t)\| &\leqq \|\bar{\mathbf{G}}_{\mathbf{ys}}(t)\| * v_2(t) * \tilde{g}(t) * \|\bar{\mathbf{G}}_{\mathbf{zw}}(t)\| = v_{\mathbf{yw}}(t)
\end{aligned} \tag{17}$$

with

$$\tilde{g}(t) = \delta(t) + \|\bar{\mathbf{G}}_{\mathbf{zs}}(t)\| * v_2(t) * \tilde{g}(t) \tag{18}$$

for norm-bounded uncertainties

$$\|\mathbf{G}_2(t)\| \leqq v_2(t) \tag{19}$$

provided that the stability condition (8.4-9)

$$\int_0^\infty \|\bar{\mathbf{G}}_{\mathbf{zs}}(t)\|\,\mathrm{d}t \int_0^\infty v_2(t)\,\mathrm{d}t < 1 \tag{20}$$

is satisfied. For element-by-element-bounded uncertainties

$$|\mathbf{G}_2(t)| \leqq \mathbf{V}_2(t) \tag{21}$$

the error bounds are given by

$$\begin{aligned}
|\delta\mathbf{G}_{\mathbf{yv}}(t)| &\leqq |\bar{\mathbf{G}}_{\mathbf{ys}}(t)| * \mathbf{V}_2(t) * \tilde{\mathbf{G}}(t) * |\bar{\mathbf{G}}_{\mathbf{zv}}(t)| = \mathbf{V}_{\mathbf{yv}}(t) \\
|\delta\mathbf{G}_{\mathbf{yd}}(t)| &\leqq |\bar{\mathbf{G}}_{\mathbf{ys}}(t)| * \mathbf{V}_2(t) * \tilde{\mathbf{G}}(t) * |\bar{\mathbf{G}}_{\mathbf{zd}}(t)| = \mathbf{V}_{\mathbf{yd}}(t) \\
|\delta\mathbf{G}_{\mathbf{yw}}(t)| &\leqq |\bar{\mathbf{G}}_{\mathbf{ys}}(t)| * \mathbf{V}_2(t) * \tilde{\mathbf{G}}(t) * |\bar{\mathbf{G}}_{\mathbf{zw}}(t)| = \mathbf{V}_{\mathbf{yw}}(t)
\end{aligned} \tag{22}$$

with

$$\tilde{\mathbf{G}}(t) = \delta(t)\,\mathbf{I} + |\bar{\mathbf{G}}_{\mathbf{zs}}(t)| * \mathbf{V}_2(t) * \tilde{\mathbf{G}}(t) \tag{23}$$

provided that the stability condition (8.4-13)

$$\lambda_p\left[\int_0^\infty |\bar{\mathbf{G}}_{\mathbf{zs}}(t)|\,\mathrm{d}t \int_0^\infty \mathbf{V}_2(t)\,\mathrm{d}t\right] < 1 \tag{24}$$

is satisfied.

Example 8.4 Consider the design of a state-feedback controller $u(t) = -\mathbf{k}'x(t)$ for the vehicle of Example 5.3 so that initial deviations $x_2(0)$ of the distance between two vehicles are removed monotonically (without overshoot). The monotonicity requirement is imposed to avoid oscillations between the neighbouring vehicles within a long vehicle string. As an initial deviation $x_2(0)$ has the same effect as a disturbance $d(t) = x_2(0)\,\delta(t)$ this requirement claims that the response of the closed-loop system between $d(t)$ and $y = x_2(t)$ be nonnegative. The approximate model subject to state feedback is given by

$$\frac{d}{dt}\begin{pmatrix}x_1\\x_2\end{pmatrix} = \begin{pmatrix}-1-k_1 & -k_2\\ -1 & 0\end{pmatrix}\begin{pmatrix}x_1\\x_2\end{pmatrix} + \begin{pmatrix}0\\1\end{pmatrix} d + \begin{pmatrix}-1\\-1\end{pmatrix} s$$

$$y = (0 \quad 1)\,\mathbf{x}$$

$$z = (-k_1 \quad -k_2)\,\mathbf{x} + (-k_1)\,s$$

where $\mathbf{k}' = (k_1, k_2)$ has been used (cf Example 5.3). For the controller $\mathbf{k}' = (-0.22 \;\; 0.22)$ the stability is proved by eqn (20) where $\int |\bar{g}_{zs}|\,dt \int v_2\,dt = 0.23 < 1$ and the tolerance band of the impulse response between $d(t)$ and $y(t)$ described by $\bar{g}_{yd}(t)$ and $v_{yd}(t)$ is in fact (almost) nonnegative (Fig. 8.12). □

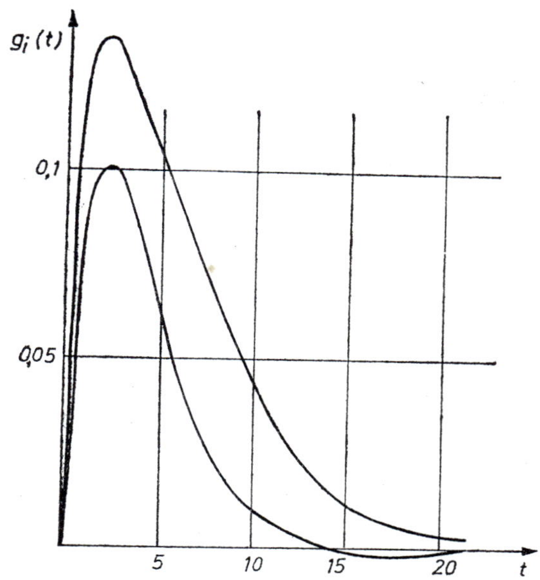

Fig. 8.12 Tolerance band of the impulse response of the closed-loop original system

Suitability and conservatism of the method

In summary, the original closed-loop system can be analysed in the following steps:

(1) Check the stability of the approximate closed-loop system (8.2-9) or (8.2-25).

(2) Check the robust stability of the closed-loop original system by Theorems 8.3—8.7 or 8.10—8.11.

(3) Determine bounds of the I/O-behaviour by Theorems 8.13—8.15.

Some general remarks are now in order concerning this way of analysis. For this, we consider again the aggregate model shown in Fig. 8.7 (c). Eqns (8.2-34), (8.2-35) describe the set

$$\mathscr{G}_c = \{\bar{\mathbf{G}}_{yr} + \delta\mathbf{G}_{yr} \text{ with } \delta\mathbf{G}_{yr} = \bar{\mathbf{G}}_{ys}\mathbf{G}_2(\mathbf{I} - \bar{\mathbf{G}}_{zs}\mathbf{G}_2)^{-1}\bar{\mathbf{G}}_{zu},\ |\mathbf{G}_2| \leqq \mathbf{V}_2\} \tag{25}$$

if element-by-element-bounded uncertainties are concerned. What has been done in this section is, in essence, the transformation of the assertion $\mathbf{G}_{yr} \in \mathscr{G}_c$ into the assertion $\mathbf{G}_{yr} \in \tilde{\mathscr{G}}_c$ where

$$\tilde{\mathscr{G}}_c = \{\bar{\mathbf{G}}_{yr} + \delta\mathbf{G}_{yr} \text{ with } |\delta\mathbf{G}_{yr}| \leqq \mathbf{V}_{yr}\} \tag{26}$$

with

$$\mathbf{V}_{yr} = |\bar{\mathbf{G}}_{ys}|\,\mathbf{V}_2(\mathbf{I} - |\bar{\mathbf{G}}_{zs}|\,\mathbf{V}_2)^{-1}\,|\bar{\mathbf{G}}_{zr}| \tag{27}$$

(cf eqns (1) and (13)). The merit of this transformation is that in eqn (26) the uncertainty $\delta\mathbf{G}_{yr}$ is *explicitly* characterised where $\mathbf{V}_{yr}$ can be easily determined for given $\mathbf{V}_2$.

The conservatism of this method is reflected by the difference between $\mathscr{G}_c$ and $\tilde{\mathscr{G}}_c$. As $\mathscr{G}_c$ includes the candidates of the closed-loop original system for the given characterisation $|\mathbf{G}_2| \leqq \mathbf{V}_2$ of the plant uncertainties, all elements of $\Delta = \tilde{\mathscr{G}}_c \setminus \mathscr{G}_c$ are known not to represent the original system. To assess the conservatism of the analysis method we have to 'count' or at least to characterise the elements of Δ. Obviously, these elements originate from two reasons. At first, as we use merely an upper bound $\mathbf{V}_{yr} \geqq |\delta\mathbf{G}_{yr}|$ to describe $\delta\mathbf{G}_{yr}$, the form of $\tilde{\mathscr{G}}_c$ is restricted. The second source is given by the way in which the bound $\mathbf{V}_{yr}$ is determined. Eqn (27) has only regard to the moduli $|\bar{\mathbf{G}}_{ys}|$, $|\bar{\mathbf{G}}_{zr}|$, and $|\bar{\mathbf{G}}_{zs}|$, although all these matrices are exactly known.

However, in concluding these remarks it should be emphasised that for many analysis problems the suitability of having an explicit description of $\mathbf{V}_{yr}$ is advantageous enough to take the conservatism of this method into the bargain. The determination of the best possible bound $\mathbf{V}_{yr}$ would necessitate the solution of the optimisation problem

$$\mathbf{V}_{yr} = \sup_{|\mathbf{G}_2| \leqq \mathbf{V}_2} |\bar{\mathbf{G}}_{ys}\mathbf{G}_2(\mathbf{I} - \bar{\mathbf{G}}_{zs}\mathbf{G}_2)^{-1}\bar{\mathbf{G}}_{zr}| \tag{28}$$

which needs unreasonable efforts unless very restrictive assumptions on the dimensions of $\mathbf{s}$ and $\mathbf{z}$ and the uncertainty model can be made.

8.6 Design of robust multivariable controllers in presence of bounded uncertainties

On the basis of the results presented in the previous sections we can design robust feedback controllers in the following way:

(1) Describe the uncertain plant by a model of the form (8.2-6) (8.2-10) or (8.2-21), (8.2-22), respectively, and give bounds for the uncertainty model $\mathbf{G}_2$.

(2) Find a controller (8.2-4) or (8.2-15) such that the closed-loop approximate model (8.2-9) or (8.2-26) is stable with regard to all its inputs and outputs and satisfies the design objectives (II) and (III) for $s = 0$.

(3) Check the robust stability. Determine tolerance bands for the I/O-behaviour of the closed-loop original system by means of the results presented in Section 8.3 and 8.4 and check the robustness concerning the dynamical requirements (III).

(For an explanation of the design requirements (I)...(III) see Section 1.1)

In this design strategy the intricate problem of choosing the controller with respect to a set $\mathcal{S}$ of plants is decomposed into two subproblems. First, the usual servomechanism problem (I)—(III) for the approximate model (8.2-6) or (8.2-21) (with $s = 0$) is to be solved by means of standard methods of multivariable linear control theory. Then, the robustness of the controller is to be investigated by criteria for robust stability and the envelopes of the system performance are:

- If one of the stability criteria is satisfied the closed-loop system satisfies the design requirement (I)
- If the controller includes an internal model of the command and disturbance signals specified in the requirement (II) and if it ensures robust stability then asymptotic regulation occurs in all linear closed-loop systems $\mathbf{G}_{yr} \in \tilde{\mathcal{S}}_c$ (cf (8.5-26)). For special controller structures such as PI-controllers this result holds true even for nonlinear sytems that can be modelled by (8.2-21), (8.2-22) (cf Section 10.2).
- If all trajectories possible within the tolerance bands calculated for specified input signals satisfy the requirement (III) then the closed-loop original system is robust according to these design specifications and the given class of plant perturbations.

While in Step 2 the controller is chosen so as to produce a well-suited 'nominal behaviour', Step 3 claims that the controller be chosen so as to satisfy the stability conditions and to yield a sufficiently small uncertainty bound. For great model errors these aims are contradictory and necessitate an iterative design procedure (Fig. 8.13). Hence, during the design process the question whether the quantitatively given model uncertainties are small enough to allow the satisfaction of all the design objectives can be answered. The design strategy is particularly useful if hard design specifications concern the nominal behaviour whereas the robustness requirements have regard to weaker specifications.

Conceptual similarities of time domain and frequency domain methods become obvious by the elaboration of the common design strategy above. Note, however, that these similarities do not imply that the results in both domains correspond via the Laplace transform. To the contrary, as uncertainty bounds may not be easily transformed from one domain to the other,

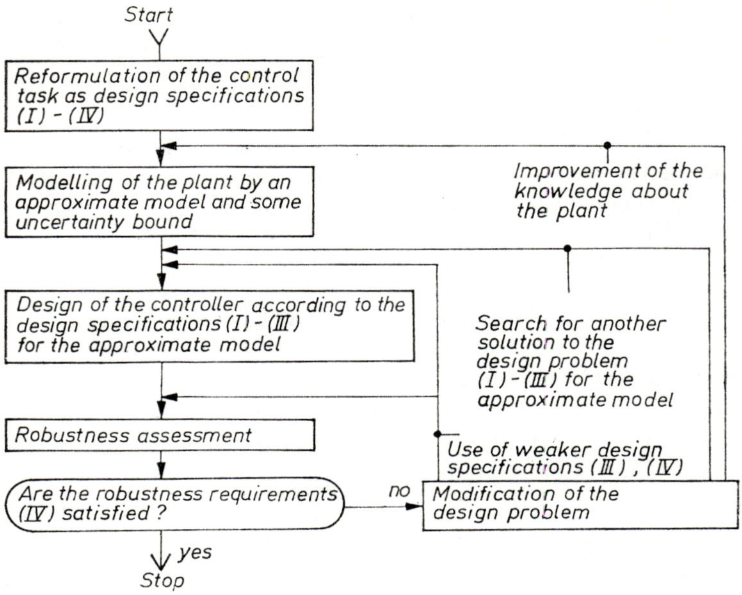

Fig. 8.13 Design of robust feedback controllers for unknown-but-bounded uncertainties

design methods in both domains do have their own right. The closed-loop uncertainties have to be assessed in the domain in which the plant uncertainties have been bounded.

8.7 Robustness analysis using state-space models

In this section we present results concerning the robust stability of linear systems that are described by state-space models with parameter uncertainties. As opposed to our considerations in the previous sections we do not assume that the model has been decomposed into a completely known and an uncertain subsystem but start directly from the model (5.3-1)

$$\begin{aligned} \dot{\mathbf{x}} &= \mathbf{A}(\mathbf{a})\,\mathbf{x} + \mathbf{B}(\mathbf{a})\,\mathbf{u}, \qquad \mathbf{x}(0) = \mathbf{x}_0 \\ \mathbf{y} &= \mathbf{C}(\mathbf{a})\,\mathbf{x} + \mathbf{D}(\mathbf{a})\,\mathbf{u}. \end{aligned} \tag{1}$$

None the less, we are concerned with unknown-but-bounded uncertainties because the parameter vector $\mathbf{a}$ is assumed to be an element of a bounded subset $\mathcal{A} \subset \mathbb{R}^q$. $\mathcal{A}$ be connected, ie $\mathbf{a}$ can be varied continuously within the whole set $\mathcal{A}$.

At first, we present structural constraints which, if imposed on the location of uncertainties within the matrices $\mathbf{A}(\mathbf{a})$ and $\mathbf{B}(\mathbf{a})$, guarantee the stabilisability of the original system (1) for arbitrarily large but finite untertainties. The second part is devoted to systems with element-by-element-bounded matrix $\mathbf{A}(\mathbf{a})$.

Stabilisability under matching conditions

We consider the stabilisability of the uncertain system (1) by means of constant-gain state feedback

$$\mathbf{u}(t) = -\mathbf{K}\mathbf{x}(t)\,. \tag{2}$$

The state equation is written in the form

$$\dot{\mathbf{x}} = \big(\hat{\mathbf{A}} + \delta\mathbf{A}(\mathbf{a})\big)\,\mathbf{x} + \big(\hat{\mathbf{B}} + \delta\mathbf{B}(\mathbf{a})\big)\,\mathbf{u}\,, \qquad \mathbf{x}(0) = \mathbf{x}_0\,. \tag{3}$$

The manner in which the uncertain parameter $\mathbf{a}$ enters into the equation is restricted by

$$\delta\mathbf{A}(\mathbf{a}) = \hat{\mathbf{B}}\mathbf{D}(\mathbf{a})\,, \qquad \delta\mathbf{B}(\mathbf{a}) = \hat{\mathbf{B}}\mathbf{E}(\mathbf{a}) \tag{4}$$

where $\mathbf{D}(\mathbf{a})$ and $\mathbf{E}(\mathbf{a})$ are matrices of appropriate dimensions which are continuous in the parameter vector $\mathbf{a}$. Equations (4) are called '*matching conditions*'. They are structural in nature because they only restrict the entry point but do not refer to the amount of uncertainties. The uncertainties are norm-bounded by some scalars d and e

$$\max_{\mathbf{a}\in\mathcal{A}} \|\mathbf{D}(\mathbf{a})\| \leqq d\,, \qquad \max_{\mathbf{a}\in\mathcal{A}} \|\mathbf{E}(\mathbf{a})\| \leqq e \tag{5}$$

where the spectral matrix norm $\|.\|$ is used. The approximate model is assumed to be stable. Then $w(\mathbf{x}) = \mathbf{x}'\mathbf{P}\mathbf{x}$ is a Lyapunov function of the approximate model where $\mathbf{P}$ is the positive definite solution of the Lyapunov equation

$$\hat{\mathbf{A}}'\mathbf{P} + \mathbf{P}\hat{\mathbf{A}} = -\mathbf{Q} \tag{7}$$

for an arbitrarily given positive definite matrix $\mathbf{Q}$. Now, we can prove the following results.

Theorem 8.16 Assume that the system (3) meets the matching conditions (4) and that the uncertainties are bounded by eqn (5) with some scalars d and e.

(i) If $e < 1$ then the system (3) can be robustly stabilised by some state feedback (2).

(ii) If $e < 1$ and the matrix $\hat{\mathbf{A}}$ is stable, then a controller that stabilises the original system is given by

$$\mathbf{K} = k\hat{\mathbf{B}}'\mathbf{P} \tag{8}$$

with

$$k \geqq \frac{d^2}{4(1-e)\,\lambda_{\min}[\mathbf{Q}]} \tag{9}$$

and $\mathbf{P}$ and $\mathbf{Q}$ from eqn (7).

The theorem states the interesting result that under matching assumptions arbitrarily large uncertainties within the system matrix **A** can be tolerated by a constant-gain state feedback controller (2). Using the feedback matrix (8) we only have to make the controller gain k sufficiently large. Restrictions must be imposed only on the amount of the uncertainties of the input matrix **B**.

Proof. The proof is based on the derivation of conditions under which the Lyapunov function (6) of the approximate model is also a Lyapunov function of the closed-loop original system (2), (3). The derivative of the Lyapunov function along the trajectory of this approximate model $\dot{\mathbf{x}} = \hat{\mathbf{A}}\mathbf{x}$ is given by

$$\dot{w}_{ol} = \left(\frac{\partial w}{\partial \mathbf{x}}\right)' \mathbf{x} = -\mathbf{x}'\mathbf{Q}\mathbf{x}. \tag{10}$$

For the closed-loop system (2), (3) we have

$$\dot{w} = \left(\frac{\partial w}{\partial \mathbf{x}}\right)' \left(\hat{\mathbf{A}}\mathbf{x} + \hat{\mathbf{B}}\mathbf{D}(\mathbf{a})\,\mathbf{x} + \hat{\mathbf{B}}\mathbf{u} + \hat{\mathbf{B}}\mathbf{E}(\mathbf{a})\,\mathbf{u}\right) = \dot{w}_{ol} + \mathbf{r}'\mathbf{D}(\mathbf{a})\,\mathbf{x} - k\,||\mathbf{r}||^2 - k\mathbf{r}'\mathbf{E}(\mathbf{a})\,\mathbf{r}$$

where the abbreviation $\mathbf{r}(t) = 2\hat{\mathbf{B}}'\mathbf{P}\mathbf{x}(t)$ and the Euclidean vector norm $||\mathbf{r}||^2 = \mathbf{r}'\mathbf{r}$ have been used. Since

$$\mathbf{r}'\mathbf{D}\mathbf{x} \leqq ||\mathbf{D}(\mathbf{a})||\,||\mathbf{r}||\,||\mathbf{x}|| \leqq d\,||\mathbf{r}||\,||\mathbf{x}||$$

$$-\mathbf{r}'\mathbf{E}\mathbf{r} \leqq ||\mathbf{E}(\mathbf{a})||\,||\mathbf{r}||^2 \leqq e\,||\mathbf{r}||^2$$

hold, we get

$$\begin{aligned}\dot{w} &\leqq \dot{w}_{ol} - k(1-e)\,||\mathbf{r}||^2 + d\,||\mathbf{r}||\,||\mathbf{x}|| \\ &= \dot{w}_{ol} - k(1-e)\left(||\mathbf{r}||^2 - \frac{d\,||\mathbf{x}||}{2k(1-e)}\right)^2 + \frac{d^2\,||\mathbf{x}||^2}{4k(1-e)} \leqq \dot{w}_{ol} + \frac{d^2\,||\mathbf{x}||^2}{4k\,(1-e)}\end{aligned}$$

where the condition $e < 1$ has been utilised. With

$$-\mathbf{x}'\mathbf{Q}\mathbf{x} \leqq -\lambda_{\min}[\mathbf{Q}]\,||\mathbf{x}||^2 \quad \text{we get} \quad \dot{w} \leqq \left(-\lambda_{\min}[\mathbf{Q}] + \frac{d^2}{4k(1-e)}\right) ||\mathbf{x}||$$

which is less than zero for all $\mathbf{x} \neq \mathbf{0}$ if k satisfies the condition (9). This proves part (ii) of the theorem.

To show the validity of part (i) we only have to demonstrate that the use of a state feedback controller

$$\mathbf{u}(t) = -\mathbf{K}\mathbf{x}(t) + \mathbf{u}_1(t) \tag{11}$$

which stabilises the approximate model does not affect the uncertainty bound e and the matching conditions. Eqns (3) and (11) yield

$$\dot{\mathbf{x}} = (\hat{\mathbf{A}} - \hat{\mathbf{B}}\mathbf{K} + \delta\mathbf{A} - \delta\mathbf{B}\mathbf{K})\,\mathbf{x} + (\hat{\mathbf{B}} + \delta\mathbf{B})\,\mathbf{u}_1 = (\hat{\mathbf{A}} + \widetilde{\delta\mathbf{A}})\,x + (\hat{\mathbf{B}} + \delta\mathbf{B})\,\mathbf{u}_1. \tag{12}$$

If the original system $(\hat{\mathbf{A}} + \delta\mathbf{A}, \hat{\mathbf{B}} + \delta\mathbf{B})$ satisfies the matching conditions (4) then the system (12) satisfies the matching condition (4) and eqn (5) with the same e because $\widetilde{\delta\mathbf{A}} = \delta\mathbf{A} - \delta\mathbf{B}\mathbf{K} = \hat{\mathbf{B}}\big(\mathbf{D}(\mathbf{a}) - \mathbf{K}\big)$ holds. □

The proof outlines a new methodology for constructing robust feedback controllers. Eqn (4) is only an example of many different 'matching conditions' that have been established recently in order to ensure robust stabilisation by the following design steps:

(1) Stabilise the approximate model and construct a Lyapunov function $w(\mathbf{x})$ for the closed loop.
(2) Design a controller such that $w(\mathbf{x})$ is a Lyapunov function of the closed-loop original system as well.

However, the most general matching conditions that necessarily have to be imposed to ensure stabilisability for arbitrarily large uncertainties remained to be elaborated.

Stability analysis in presence of unstructured uncertainties

We conclude this chapter with some remarks on the stability analysis of systems (1), for which we only know bounds on the perturbation $\delta\mathbf{A}$, ie

$$\mathbf{A}(\mathbf{a}) \in \mathcal{A} = \{\hat{\mathbf{A}} + \delta\mathbf{A}(\mathbf{a}) \text{ with } |\delta\mathbf{A}(\mathbf{a})| \leqq \bar{\mathbf{A}}\}. \tag{13}$$

$\mathbf{A}$ is an *interval matrix* since each element is described by

$$\hat{a}_{ij} - \bar{a}_{ij} \leqq a_{ij} \leqq \hat{a}_{ij} + \bar{a}_{ij}. \tag{14}$$

We aim at finding conditions under which all $\mathbf{A} \in \mathcal{A}$ are stable.

In principle, each matrix $\mathbf{A} \in \mathcal{A}$ has a characteristic equation

$$f\big(p, (c_n, c_{n-1}, \dots, c_1)\big) = \det(p\mathbf{I} + \mathbf{A}) = p^n + c_1 p^{n-1} + \cdots + c_n = 0 \tag{15}$$

and we can transform the intervals (14) into intervals

$$\bar{\bar{c}}_i \leqq c_i \leqq \bar{c}_i \tag{16}$$

for the polynomial coefficients. Then we can consider the set

$$\mathcal{F} = \{f(p, \mathbf{c}) \text{ with } \bar{\bar{c}}_i \leqq c_i \leqq \bar{c}_i\} \tag{17}$$

and ask the same question: Under what conditions have all polynomials $f \in \mathcal{F}$ exclusively zeros with negative real parts? To answer this question we define the set

$$\mathfrak{F} = \{f(p, \mathbf{c}) \text{ with } c_i = \bar{\bar{c}}_i \text{ or } c_i = \bar{c}_i\}, \tag{18}$$

which is formed from all possible combinations of the endpoints of the coefficient intervals. Then the following lemma holds.

Lemma 8.8 All polynomials of the set $\mathcal{F}$ have exclusively zeros with negative real parts if and only if all polynomials of the set $\mathfrak{F}$ possess this property.

Hence, we only have to consider the polynomials with extreme coefficient values. Although this is a considerable reduction there remain 2^n

polynomials to be checked. It is, therefore, a very important result that we have finally to consider merely four polynomials of $\mathfrak{F}$.

Theorem 8.17 All polynomials of the set $\mathcal{F}$ have exclusively zeros with negative real parts if and only if the following four polynomials possess this property

$$f(p, \mathbf{c}^1) = f\big(p, (\bar{c}_n, \bar{c}_{n-1}, \bar{\bar{c}}_{n-2}, \bar{\bar{c}}_{n-3}, \bar{c}_{n-4}, \ldots)\big)$$

$$f(p, \mathbf{c}^2) = f\big(p, (\bar{\bar{c}}_n, \bar{\bar{c}}_{n-1}, \bar{c}_{n-2}, \bar{c}_{n-3}, \bar{\bar{c}}_{n-4}, \ldots)\big)$$

$$f(p, \mathbf{c}^3) = f\big(p, (\bar{c}_n, \bar{\bar{c}}_{n-1}, \bar{c}_{n-2}, \bar{\bar{c}}_{n-3}, \bar{c}_{n-4}, \ldots)\big)$$

$$f(p, \mathbf{c}^4) = f\big(p, (\bar{\bar{c}}_n, \bar{c}_{n-1}, \bar{\bar{c}}_{n-2}, \bar{c}_{n-3}, \bar{\bar{c}}_{n-4}, \ldots)\big).$$

To come back to our primary problem, we have to extend these results to interval matrices. In analogy to $\mathfrak{F}$ we define

$$\mathcal{\dot{A}} = \{\hat{\mathbf{A}} + \delta\mathbf{A} \text{ with } \delta a_{ij} = \bar{a}_{ij} \text{ or } \delta a_{ij} = -\bar{a}_{ij}\}\,. \tag{19}$$

Unfortunately, for $\mathbf{A} \in \mathcal{\dot{A}}$ the polynomial coefficients c_i do not generally assume all extreme values of the corresponding intervals (16). Therefore, for $n > 2$ the stability of $\mathbf{A} \in \mathcal{\dot{A}}$ is not sufficient for the stability of all $\mathbf{A} \in \mathcal{A}$. This can be demonstrated by the *counterexample*

$$\mathbf{A}(\delta a_{11}) = \begin{pmatrix} -0.5 & 0 & 2 \\ 1/6 & 0 & 1 \\ 1 & -3 & -1 \end{pmatrix} + \begin{pmatrix} \delta a_{11} & 0 & 0 \\ 0 & 0 & 0 \\ 0 & 0 & 0 \end{pmatrix}$$

with $|\delta a_{11}| < 0.8$. The characteristic polynomial is given by

$$f(p, \mathbf{c}) = p^3 + (1.5 - \delta a_{11})\, p^2 + (1.5 - \delta a_{11})\, p + (2.5 - 3\delta a_{11}) = 0\,. \tag{20}$$

The Hurwitz criterion claims $0.5 \leqq \delta a_{11} \leqq 1.5$ or $-0.833 \leqq \delta a_{11} \leqq -0.5$. For $\mathbf{A} \in \mathcal{\dot{A}}$, $\delta a_{11} = 0.8$ or $\delta a_{11} = -0.8$ holds which satisfy the Hurwitz criterion. Hence, all $\mathbf{A} \in \mathcal{\dot{A}}$ are stable. However, for all $|\delta a_{11}| < 0.5$ the matrix $\mathbf{A}(\delta a_{11})$ is unstable. Obviously, $c_1 = c_2$ holds in eqn (20) and is the reason for this result. Not all combinations of the extreme values of the intervals of c_i $(0.7 \leqq c_1, c_2 \leqq 2.3,\ 0.1 \leqq c_3 \leqq 4.9)$ are assumed for $A \in \mathcal{\dot{A}}$. In particular, the set of characteristic polynomials of $\mathbf{A} \in \mathcal{\dot{A}}$ does not even include the four polynomials described in Theorem 8.17. □

Some ways out of this situation are provided by restrictions on $\mathbf{A}$ under which the stability of $\mathbf{A} \in \mathcal{\dot{A}}$ is in fact sufficient for the stability of $\mathbf{A} \in \mathcal{A}$, or by sufficient conditions for the stability of $\mathbf{A} \in \mathcal{A}$ that are generally not necessary. An example for the first way is given by the following lemma.

Lemma 8.9 Consider symmetric matrices $\mathbf{A}(\mathbf{a})$, ie restrict $\mathcal{A}$ and $\mathcal{\dot{A}}$ in eqns (13) and (19) to symmetric matrices. Then the stability of all $\mathbf{A} \in \mathcal{\dot{A}}$ is necessary and sufficient for the stability of all $\mathbf{A} \in \mathcal{A}$.

This lemma can be proved by induction. Examples for the second way are provided by the following lemmata, which circumvent even the laborious stability test of the 2^{n^2} matrices of $\tilde{\mathcal{A}}$.

Lemma 8.10 All $\mathbf{A} \in \mathcal{A}$ are stable if the condition

$$\hat{a}_{ii} + \bar{a}_{ii} + \sum_{\substack{j=1\\ j\neq i}}^{n} \max\left[|\hat{a}_{ij} + \bar{a}_{ij}|, |\hat{a}_{ij} - \bar{a}_{ij}|\right] < 0$$

is satisfied.

The proof is based on the Gershgorin-theorem, which states that the eigenvalues $\lambda_k[\mathbf{A}]$ lie in the union of the circular areas

$$|p - a_{ii}| \leqq \sum_{\substack{j=1\\ j\neq i}}^{n} |a_{ij}|.$$

Lemma 8.11 All $\mathbf{A} \in \mathcal{A}$ are stable if $\hat{\mathbf{A}}$ is stable and

$$\sigma_{\max}[\bar{\mathbf{A}}'\,|\mathbf{P}|] < 1 \tag{21}$$

holds where $\mathbf{P}$ denotes the positive definite solution of the Lyapunov equation $\hat{\mathbf{A}}'\mathbf{P} + \mathbf{P}\hat{\mathbf{A}} = -2\mathbf{I}$.

Proof: Eqn (21) proves $w(\mathbf{x}) = \mathbf{x}'\mathbf{P}\mathbf{x}$ is a Lyapunov function of all systems $\dot{\mathbf{x}} = \mathbf{A}\mathbf{x}$ with $\mathbf{A} \in \mathcal{A}$ since

$$\dot{w} = \dot{\mathbf{x}}'\mathbf{P}\mathbf{x} + \mathbf{x}'\mathbf{P}\dot{\mathbf{x}} = -2\mathbf{x}'\mathbf{x} + 2\mathbf{x}'\bar{\mathbf{A}}'\mathbf{P}\mathbf{x} \leqq (-2 + 2\sigma_{\max}[\bar{\mathbf{A}}'\,|\mathbf{P}|])\,\|\mathbf{x}\|^2 < 0. \;\square$$

Bibliographical notes

The quantitative design method has been developed by Horowitz and Sidi (1972), who also presented the design example used in Section 8.1. Extensions to nonlinear, time varying or non-minimum-phase plants are given in (Horowitz 1976, 1975; Horowitz and Sidi 1978). Multivariable control problems are tackled by Horowitz (1979, 1982), Shaked et al. (1976), Nwokah (1984), Walke et al. (1984), and Yaniv and Horowitz (1986), where the complete design problem is always decomposed into the problem of designing weakly coupled single-loop systems. A computer-aided design package has been implemented by Gräser et al. (1982). Industrial design examples can be found eg in the references by Walke et al. (1984), Tolle (1983), and Gräser and Tolle (1983). However, the theoretical foundation of this graphical method remains a matter of discussion, cf (Doyle, 1986).

The robustness of multivariable feedback systems in presence of unknown-but-bounded model uncertainties is tackled in a very large body of literature. Some of the first papers which dealt with norm-bounded uncertainties are those by Freeman (1973), Doyle and Stein (1981), and Carlucci and Vallauri (1981). Chen and Desoer (1982) proved the necessity of the conditions of Lemma 8.2 for systems subjected to additive or multiplicative perturbations. The extensions reported in Theorems 8.6 and 8.7 are due to

Postlethwaite and Foo (1985), while some weaker results have already been presented by Lehtomaki et al (1981b). While all these authors worked with frequency domain models, the alternative time domain approach has been proposed by El-Attar and Vidyasagar (1979), Aström (1980), Lunze (1981a, 1982c), and Owens and Chotai (1982).

The extension to element-by-element-bounded uncertainties has been initiated by Lunze (1982a, 1983c, 1984b), Kantor and Andres (1983), Owens and Chotai (1984). Theorems 8.10 and 8.11 are due to Lunze (1982c, 1984a). Many efforts have been made to improve the stability conditions and, eventually, come to necessary and sufficient conditions. The first of such conditions (Theorem 8.5) together with a constructive procedure for its practical implementation has been found by Kouvaritakis and Latchman (1985). Graphical robust-stability tests have been elaborated by Owens and Chotai (1983, 1986), who also introduced and exploited monotone modelling errors (1981).

The similarity of all these robustness considerations has been elaborated by Lunze (1984c), who presented a common systems theoretic framework whose specifications concerning the model structure and the uncertainty bounds yield many of the known results. In addition to that, this reference gives extensions of the robustness considerations from stability-oriented results to the investigation of the I/O-behvaiour, which has been approached only by very few researchers.

Very recent papers try to improve the stability conditions by exploiting additional information about the uncertainties. They are directed towards a reduction of the class of uncertainties $\mathcal{S}_2$ in such a way that the additional knowledge required in the modelling step can be fully exploited when checking the robust stability of the closed loop. For example, Lehtomaki et al. (1981a), Lee et al. (1982), and Yeh et al. (1985) derived stability conditions that exploit bounds on the maximum singular value and the second largest singular value of $\mathbf{G}_2(p)$. However, it remains questionable whether this additional information about the model uncertainties can be used to rule out directions of the perturbation that are relevant for physical reasons. In the present stage, these attempts are mainly influenced by the mathematical tools used to derive the robustness tests.

Other attempts to reduce the conservatism of the singular value approach restricted the transfer function matrix of the uncertainty model to diagonal matrices. Then, each diagonal element has to satisfy the same upper bound. The problem of determining the largest uncertainty bound which does not cause instability, the so-called *μ-problem*, has been posed by Doyle (1982) and Safonov (1982) and completely solved by Daniel et al. (1985b). An example from the field of filter design is given by Zafiriou and Morari (1986). The results turned out to be specifications of the stability conditions for element-by-element bounded uncertainties.

Whereas all the robustness investigations presented here are based on upper bounds of the modulus of the uncertainties, Postlethwaite et al. (1981)

and Hung and MacFarlane (1982) introduced the principal gains and principal phases which allow a characterisation of the phase of uncertain system parts and, in this way, to prove the stability for restricted phase but large and even inifinite gains.

The stability of interval matrices has attracted the interest of many researchers very recently. Initiated by Bialas (1983) whose necessary and sufficient conditions could be refuted by the counterexamples presented eg by Karl et al. (1984) or Barmish and Hollot (1984), many sufficient stability conditions (eg Lemmas 8.9—8.11) have been derived with different fields of application by Yeung (1983), Heinen (1984), Daoyi (1985), Zhou and Deng (1986), Shi and Gao (1986), and Yedavalli (1986). The rationale behind some of these criteria is presented in Section 8.7. Theorem 8.17 is due to Charitonov (1978). Because of the problems posed by the stability analysis of interval matrices and the conservatism of pertinent sufficient conditions, there are only a few papers in which these results are applied to the design of robust controllers, eg (Evans and Xianya 1985) and (Soh and Evans 1985).

The problem of deriving matching conditions on the model uncertainties under which the plant is robustly stabilisable for arbitrarily large but bounded uncertainties has been investigated in a much more general setting than that presented in Section 8.7. In general, nonlinear plants with nonlinear control laws are considered where the existence and the uniform boundedness of the solution pose severe problems. These results are outside the scope of robust linear control. Surveys can be found in (Gutman 1979) and (Leitmann 1983).

Specific investigations for linear original systems and linear controllers concern the question which kind of bounded but arbitrarily large uncertainties can be tolerated by some constant-gain state feedback. Results along this line have been presented by Thorp (1980), Barmish (1982), Willems and Willems (1983), and Barmish et al. (1983) who described the idea of the proof of Theorem 8.16. Leitmann et al. (1986) applied this methodology to consider the robustness against neglected actuator and sensor dynamics.

9 CONTROLLER DESIGN BY MEANS OF APPROXIMATE MODELS

The first part of this book was devoted to methodological aspects of robustness analysis and design in multivariable control. The present and the next two chapters deal with more practical aspects. They give examples and application studies which are solved by means of the methods presented so far and which should provide the reader with some experiences in robust control.

We begin with considering the design of controllers by means of simplified plant models. The widespread use of simplified models in single-loop process control is motivated by the observation that the whole design problem can be conceptually and computationally reduced if details of the plant dynamics are voluntarily neglected. This line of action is particularly successful if the plant has 'simple' dynamical properties which can be reasonably approximated by, for example, delay-lag models of first or second order, if only modest design objectives have to be satisfied, or if we aim at a quick preliminary insight into the design problem before a more detailed analysis is undertaken. In the following we investigate two design methods for multivariable controllers which are based on a rigorous simplification of the plant model since, in addition to the aspects mentioned above, the simplified model is used as part of the controller. Therefore, the applicability of the controller relies heavily on their robustness.

9.1 The design of a Smith predictor controller

We consider a plant with separable time delay. It can be described by

$$\mathbf{y}(p) = \mathbf{G}_d(p)\, \mathbf{G}_p(p)\, \mathbf{u}(p) \qquad (1)$$

where the (m, m)-matrices $\mathbf{G}_d$ and $\mathbf{G}_p$ denote the pure time delay and the delay-free strictly proper part of the plant. A particularly suitable controller structure is given by the Smith predictor (Fig. 9.1 (a)). It presents a special controller structure which, of course, can be transformed into the standard feedback configuration of Fig. 2.5 (c1) (cf Fig. 9.1 (b)). As the plant model $\hat{G}_d\hat{G}_p$ itself constitutes a part of the controller it should be as simple as possible. We want to derive conditions on the tolerable model reduction.

If the original plant in Fig. 9.1 (a) is substituted by the simplified model $\hat{G}_d\hat{G}_p$ used within the predictor controller and the corresponding uncertainty model we have

$$\begin{aligned} \mathbf{y}(p) &= \hat{\mathbf{G}}_d(p)\,\hat{\mathbf{G}}_p(p)\,\mathbf{u}(p) + \mathbf{s}(p) \\ \mathbf{z}(p) &= \mathbf{u}(p) \end{aligned} \tag{2}$$

and

$$\mathbf{s}(p) = \left(\mathbf{G}_d(p)\,\mathbf{G}_p(p) - \hat{\mathbf{G}}_d(p)\,\hat{\mathbf{G}}_p(p)\right)\mathbf{z}(p)\,. \tag{3}$$

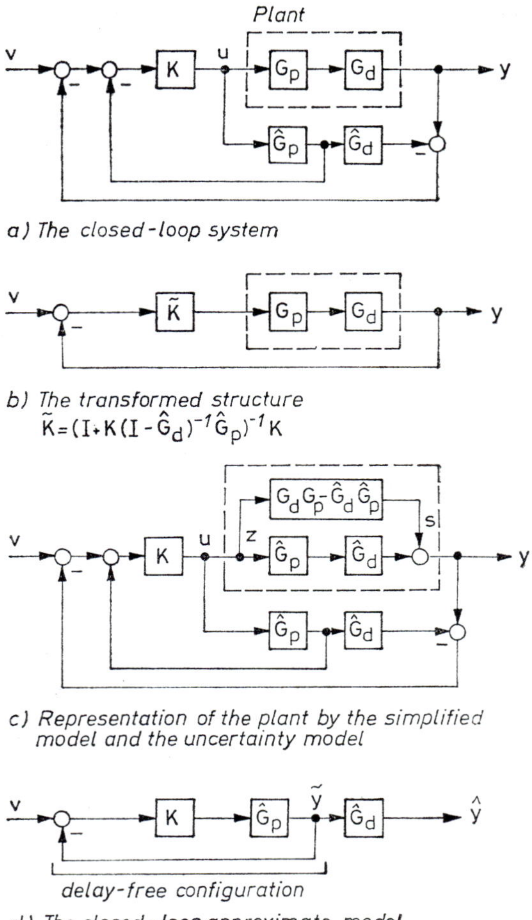

Fig. 9.1 Smith predictor controller configuration

(Fig. 9.1 (c)). The closed-loop approximate model is

$$\begin{aligned} \mathbf{y}(p) &= \hat{\mathbf{G}}_d \hat{\mathbf{G}}_p (\mathbf{I} + \mathbf{K}\hat{\mathbf{G}}_p)^{-1} \mathbf{K}\mathbf{v} + \big(\mathbf{I} + (\mathbf{I} - \hat{\mathbf{G}}_d)\,\hat{\mathbf{G}}_p \mathbf{K}\big)(\mathbf{I} + \hat{\mathbf{G}}_p \mathbf{K})^{-1}\,\mathbf{s} \\ \mathbf{u} &= (\mathbf{I} + \mathbf{K}\hat{\mathbf{G}}_p)^{-1} \mathbf{K}\mathbf{v} - (\mathbf{I} + \mathbf{K}\hat{\mathbf{G}}_p)^{-1} \mathbf{K}\mathbf{s}. \end{aligned} \tag{4}$$

If the correct plant model is used in the controller, ie $\hat{\mathbf{G}}_d \hat{\mathbf{G}}_p = \mathbf{G}_d \mathbf{G}_p$ and hence $\mathbf{s} = \mathbf{0}$, then eqn (4) yields

$$\mathbf{y} = \hat{\mathbf{G}}_d \hat{\mathbf{G}}_p (\mathbf{I} + \mathbf{K}\hat{\mathbf{G}}_p)^{-1} \mathbf{K}\mathbf{v}. \tag{5}$$

The closed-loop system behaves as the delay-free system

$$\tilde{\mathbf{y}} = \hat{\mathbf{G}}_p (\mathbf{I} + \mathbf{K}\hat{\mathbf{G}}_p)^{-1} \mathbf{K}\mathbf{v} \tag{6}$$

whose output is delayed according to $\mathbf{y} = \hat{\mathbf{G}}_d \tilde{\mathbf{y}}$ to give the original system output $\mathbf{y}$ (Fig. 9.1 (d)). If a simplified model is used, eqn (5) yields only an approximate $\hat{\mathbf{y}}$ of $\mathbf{y}$, but this approximate can be determined in the same simple way as before. Using an element-by-element uncertainty bound $\mathbf{V}_2$

$$\mathbf{V}_2(p) \geqq |\mathbf{G}_d(p)\,\mathbf{G}_p(p) - \hat{\mathbf{G}}_d(p)\,\hat{\mathbf{G}}_p(p)| \tag{7}$$

the closed-loop original system is proved to be robustly stable against the model errors brought about by the model simplification if

$$\lambda_P[|(\mathbf{I} + \mathbf{K}\hat{\mathbf{G}}_p)^{-1}\,\mathbf{K}|\,\mathbf{V}_2] < 1 \quad \text{for} \quad p \in \mathcal{D}. \tag{8}$$

Note that the left part of the matrix product does not include any time delay element whereas $\mathbf{V}_2(p)$ has regard to the time delays only with respect to the amplitude of the uncertainties to which they contribute. Eqn (8) describes a bound on the degree of simplification that is tolerated by the closed-loop system with regard to the stability property. The effect of the model error on the input-output behaviour can be determined by Theorem 8.14. This is left for the reader as an exercise.

9.2 Limiting the controller action to low-uncertainty frequencies

To deepen the insight into the robustness problem and, simultaneously, motivate a special controller structure, which will be described in this section, we investigate the robust stability of a one-degree-of-freedom control configuration subject to additive perturbations (cf Fig. 1.5). From eqn (8.2-8) and the equations of the model structure of Fig. 5.2(a) we get

$$\bar{\mathbf{G}}_{zs}(p) = -\mathbf{K}(\mathbf{I} + \hat{\mathbf{G}}\mathbf{K})^{-1} \tag{1}$$

and

$$\mathbf{G}_2 = \mathbf{G} - \hat{\mathbf{G}} = \delta\mathbf{G}, \tag{2}$$

where $\delta \mathbf{G}$ is described by some upper bound v_2 or $\mathbf{V}_2$ as in eqns (8.3-6) or (8.3-13). Assuming that the closed-loop approximate model and the uncertainty model are stable, a necessary and sufficient condition for robust stability is given by Theorem 8.1:

$$\sup_{\delta \mathbf{G} \in \mathcal{G}_2} \varrho[\overline{\mathbf{G}}_{zs}(p)\, \delta \mathbf{G}(p)] < 1\,, \qquad p \in \mathcal{D}. \tag{3}$$

As this relation is equivalent to

$$\det \big(\mathbf{I} - \overline{\mathbf{G}}_{zs}(p)\, \delta \mathbf{G}(p)\big) \neq 0 \quad \text{for all} \quad \delta \mathbf{G} \in \mathcal{G}_2,\, p \in \mathcal{D}$$

it can be transformed by means of eqn (1) and the identity

$$\begin{aligned} \det (\mathbf{I} - \overline{\mathbf{G}}_{zs}\delta \mathbf{G}) &= \det \big(\mathbf{I} + \mathbf{K}(\mathbf{I} + \hat{\mathbf{G}}\mathbf{K})^{-1}\, \delta \mathbf{G}\big) \\ &= \det (\mathbf{I} + \mathbf{K}\hat{\mathbf{G}})^{-1} \det \big(\mathbf{I} + \mathbf{K}(\hat{\mathbf{G}} + \delta \mathbf{G})\big) \end{aligned}$$

under the assumption the closed-loop approximate model be stable. Eqn (3) is equivalent to

$$\sup_{\delta \mathbf{G} \in \mathcal{G}_2} \varrho[\mathbf{K}(\hat{\mathbf{G}} + \delta \mathbf{G})] < 1\,, \qquad p \in \mathcal{D}. \tag{4}$$

Now, we restrict our considerations to norm-bound uncertainties

$$\|\delta \mathbf{G}\| < v_2 \tag{5}$$

although similar results can be received for other uncertainty bounds and, in fact, even for more general model structures. Using the admissible plant perturbation $\widetilde{\delta \mathbf{G}} = \hat{\mathbf{G}} \dfrac{v_2}{\|\hat{\mathbf{G}}\|}$ we have

$$1 > \sup_{\delta \mathbf{G} \in \mathcal{G}_2} \varrho[\mathbf{K}(\hat{\mathbf{G}} + \delta \mathbf{G})] > \varrho[\mathbf{K}(\hat{\mathbf{G}} + \widetilde{\delta \mathbf{G}})] = \varrho[\mathbf{K}\hat{\mathbf{G}}] \left(1 + \frac{v_2}{\|\hat{\mathbf{G}}\|}\right).$$

Therefore,

$$\varrho[\mathbf{K}\hat{\mathbf{G}}] < \frac{\|\hat{\mathbf{G}}\|}{\|\hat{\mathbf{G}}\| + v_2} \tag{6}$$

is a necessary condition for robust stability. That this condition limits the controller gain can be easily seen by decomposing the controller matrix according to $\mathbf{K} = k\hat{\mathbf{K}}$. The eqn (6) provides an upper bound on the gain factor k

$$k < \frac{\|\hat{\mathbf{G}}\|}{\|\hat{\mathbf{G}}\| + v_2} \frac{1}{\varrho[\hat{\mathbf{K}}\hat{\mathbf{G}}]} \tag{7}$$

which must be *necessarily* satisfied if the closed-loop system is to be robustly stable against all perturbations described by eqn (5). This demonstrates that the robustness requirement leads to an explicitly or implicitly given limitation of the loop gain and, in this way, restricts the range of validity of the general advice that high gain feedback makes the loop 'tight' and insensitive to model uncertainties (cf Section 3.4).

Having this result in mind, it is obvious that restrictions on the closed-loop uncertainty $\delta \mathbf{G}_{yv}$ lead to similar but stronger conditions on the loop gain. Therefore, we want to skip an investigation of this relation in favour of practical conclusions. Our results show that it is reasonable to limit the action of feedback controllers to those frequency ranges in which the original plant is well-known. In more detail, eqns (6) and (7) permit a high loop gain and, thus, very effective controller actions in frequency ranges where the uncertainty bound is low whereas they impose hard restrictions on the controller activity for frequencies where the uncertainties are large.

This provides the motivation of the controller structure depicted in Fig. 9.2(a). The loop including the controller **K** and the plant **G** is closed only for signals that pass the prefilter **P**. For the other frequencies the

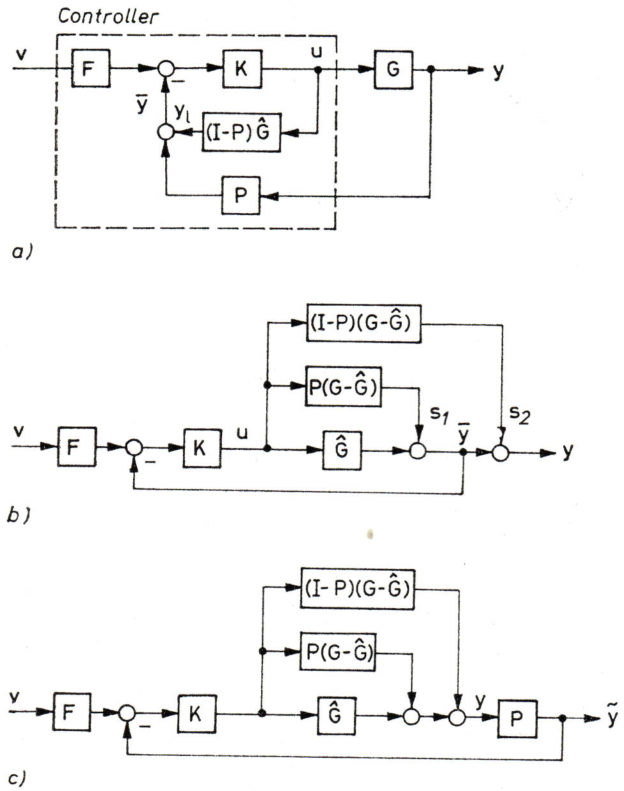

Fig. 9.2 Control configuration with limited-bandwidth controller
(a) The control structure
(b) Equivalent closed-loop system
(c) Equivalent closed-loop system for $\mathbf{y}_l = \mathbf{0}$

controller resorts to the information about the plant output which is given by the model $\hat{\mathbf{G}}$. To specify the prefilter, think of the situation that the low-frequency behaviour of the plant is well known, whereas the uncertainties grow significantly if the frequency exceeds a certain bound ω_0. Then it is reasonable to use a low-pass filter $\mathbf{P}$ with cut-off frequency ω_0, which limits the closed-loop activities of the controller to the low frequencies $\omega < \omega_0$ for which the plant behaviour is satisfactorily known.

To investigate the robustness we transform the control configuration of Fig. 9.2 (a) into our standard form. As

$$\bar{\mathbf{y}} = \mathbf{PGu} + (\mathbf{I} - \mathbf{P})\,\hat{\mathbf{G}}\mathbf{u} = \hat{\mathbf{G}}\mathbf{u} + \mathbf{s}_1 \tag{8}$$

with

$$\mathbf{s}_1 = \mathbf{P}(\mathbf{G} - \hat{\mathbf{G}})\,\mathbf{u} \tag{9}$$

and

$$\mathbf{y} = \hat{\mathbf{G}}\mathbf{u} + (\mathbf{G} - \hat{\mathbf{G}})\,\mathbf{u} = \hat{\mathbf{G}}\mathbf{u} + \mathbf{s}_1 + \mathbf{s}_2 \tag{10}$$

with

$$\mathbf{s}_2 = (\mathbf{I} - \mathbf{P})\,(\mathbf{G} - \hat{\mathbf{G}})\,\mathbf{u} \tag{11}$$

we get the configuration shown in Fig. 9.2 (b). The closed-loop approximate model is given by

$$\begin{aligned} \mathbf{y} &= (\mathbf{I} + \hat{\mathbf{G}}\mathbf{K})^{-1}\,\hat{\mathbf{G}}\mathbf{KFv} + \left((\mathbf{I} + \hat{\mathbf{G}}\mathbf{K})^{-1} \,\vdots\, \mathbf{I}\right) \begin{pmatrix} \mathbf{s}_1 \\ \mathbf{s}_2 \end{pmatrix} \\ \mathbf{z} &= (\mathbf{I} + \mathbf{K}\hat{\mathbf{G}})^{-1}\,\mathbf{KFv} + \left((\mathbf{I} + \mathbf{K}\hat{\mathbf{G}})^{-1}\mathbf{K} \,\vdots\, \mathbf{0}\right) \begin{pmatrix} \mathbf{s}_1 \\ \mathbf{s}_2 \end{pmatrix}. \end{aligned} \tag{12}$$

For $\mathbf{s} = \mathbf{0}$ we get

$$\hat{\mathbf{y}} = (\mathbf{I} + \hat{\mathbf{G}}\mathbf{K})^{-1}\,\hat{\mathbf{G}}\mathbf{KFv} \tag{13}$$

which is the same approximate model as we have got for the usual two-degrees-of-freedom structure of Fig. 2.5 (b1).

We assume that upper uncertainty bounds

$$\begin{aligned} \mathbf{V}_{11}(p) &\geqq \left|\mathbf{P}(p)\left(\mathbf{G}(p) - \hat{\mathbf{G}}(p)\right)\right| \\ \mathbf{V}_{21}(p) &\geqq \left|\left(\mathbf{I} - \mathbf{P}(p)\right)\left(\mathbf{G}(p) - \hat{\mathbf{G}}(p)\right)\right| \end{aligned} \tag{14}$$

are known. $\mathbf{V}_{11}(j\omega)$ describes the low-frequency uncertainties and vanishes for high ω whereas $\mathbf{V}_{21}$ provides a bound of the possibly large high-frequency uncertainties. The closed-loop system is robustly stable if

$$\lambda_P[|(\mathbf{I} + \mathbf{K}\hat{\mathbf{G}})^{-1}\,\mathbf{K}|\,\mathbf{V}_{11}] < 1, \qquad p \in \mathcal{D} \tag{15}$$

(cf eqn (8.3-14)). The advantage of our controller structure is that this condition refers merely to the low frequency uncertainty bound $\mathbf{V}_{11}$. That is, the loop gain is limited only for low frequencies and this limit depends merely upon the low-frequency uncertainties. Since $\mathbf{V}_{11}(j\omega)$ is usually small, the condition (15) is not very restrictive.

A bound for the uncertainties of the I/O-behaviour can be found by means of Theorem 8.14

$$|\mathbf{y}-\hat{\mathbf{y}}| \leqq \left(|(\mathbf{I}+\hat{\mathbf{G}}\mathbf{K})^{-1}|\,\mathbf{V}_{11}+\mathbf{V}_{21}\right)\left(\mathbf{I}+|(\mathbf{I}+\mathbf{K}\hat{\mathbf{G}})^{-1}\mathbf{K}|\,\mathbf{V}_{11}\right)^{-1} \times |(\mathbf{I}+\mathbf{K}\hat{\mathbf{G}})^{-1}\mathbf{K}\mathbf{F}|\,|\mathbf{v}.|$$

The high frequency uncertainties may cause a large deviation $\mathbf{y} - \hat{\mathbf{y}}$ for high ω. If, however, the command input excites only the low-frequency band or we restrict our consideration to the low-frequency behaviour $\tilde{\mathbf{y}}(p) = \mathbf{P}(p)\,\mathbf{y}(p)$ of the closed-loop original system, then the deviation of the original output $\tilde{\mathbf{y}}(p)$ from the approximate $\mathbf{P}(p)\,\hat{\mathbf{y}}(p)$ can be described by

$$|\tilde{\mathbf{y}}(p)-\mathbf{P}(p)\,\hat{\mathbf{y}}(p)| = |\mathbf{s}_1| \leqq \mathbf{V}_{11}\left(\mathbf{I}+|(\mathbf{I}+\mathbf{K}\hat{\mathbf{G}})^{-1}\mathbf{K}|\,\mathbf{V}_{11}\right)^{-1}|(\mathbf{I}+\mathbf{K}\hat{\mathbf{G}})^{-1}\mathbf{K}\mathbf{F}|\,|\mathbf{v}|.$$

This bound is independent of $\mathbf{V}_{21}$ and may be very small for small low-frequency uncertainties. Therefore, the low-frequency behaviour of the closed-loop original system exhibits a good tracking property although very large high-frequency model errors may occur.

There may arise the question why the controller possesses the internal loop including the model $(\mathbf{I} - \mathbf{P})\,\hat{\mathbf{G}}$. An answer can be easily received by investigating the closed-loop system of Fig. 9.2 (a) with $\mathbf{y}_1 = \mathbf{0}$ in the same way as above. As the transformed system has the structure drawn in Fig. 9.2 (c), the stability and even the output $\tilde{\mathbf{y}}(p)$ would depend also on the high-frequency uncertainties.

Bibliographical notes

The use of simplified models for the design of multivariable controllers has been the subject of several papers. The robustness of the Smith predictor controller was investigated by Chotai et al. (1984) whereas the model structure described in Section 9.2 has been proposed by Donati and Vallauri (1984). Owens and Chotai (1983, 1986) developed graphical means for assessing the effects of model simplifications on the closed-loop performance. Daniel and Kouvaritakis (1983) considered the use of normal approximate transfer functions $\bar{\mathbf{G}}$, which make the design process very easy.

10 ROBUST MULTIVARIABLE PI-CONTROLLERS

Although sophisticated controller structures and design algorithms have been elaborated, the conventional PID-controller is frequently used in process control. It proved to be a simple and realistic means to meet basic control aims such as command tracking and disturbance rejection. Moreover, in single-loop control the PID-controller is significantly robust against variations and uncertainties of the plant dynamics. Its parameters can be adjusted by on-plant tuning without developing an accurate model of the plant. This design technique is often used in the engineering practice to bring control loops into operation.

For these reasons, it is desirable to extend the PID-controller's field of application to multi-input multi-output plants. The first mentioned property, the attainability of asymptotic regulation, can be directly taken over to multivariable control. As the integral part of the controller can be interpreted as the internal model of the external step signals, asymptotic stability of the closed-loop system implies set-point tracking and disturbance rejection due to the Internal Model principle. However, the extension of the robustness properties is more involved, particularly, since *quantitative* results are necessary if the robustness is to be utilised in the controller design.

This chapter is devoted to the robustness of multivariable PI-controllers with regard to unknown-but-bounded plant uncertainties. In Section 10.1 we consider the existence of robust PI-controllers and derive a condition on the model accuracy. Interestingly, this condition refers merely to the knowledge of the static plant behaviour. Section 10.2 applies the results of Section 8.4 and 8.5 to PI-controllers and presents industrial application studies. Whereas all these results presuppose an analytic plant model the tuning guidelines presented in Section 10.3 can be used whenever some coarse information about the plant step responses is available. A further practical application illustrates this design technique.

10.1 Existence of robust multivariable PI-controllers

In this section we characterise the largest amount of uncertainties that may be tolerated by multivariable PI-controllers. We restrict ourselves to the stability requirement on the closed-loop original system and call the controller robust if it ensures the closed-loop stability for all uncertainties under consideration.

The plant is assumed to be linear and stable, so that it can be described by some model of the form

$$\begin{aligned} \dot{\mathbf{x}} &= \mathbf{A}\mathbf{x} + \mathbf{B}\mathbf{u}, \qquad \mathbf{x}(0) = \mathbf{x}_0 \\ \mathbf{y} &= \mathbf{C}\mathbf{x} + \mathbf{D}\mathbf{u} + \mathbf{d} \end{aligned} \tag{1}$$

or, equivalently, by

$$\mathbf{y}(t) = \mathbf{G}(t) * \mathbf{u}(t) + \mathbf{y}_0(\mathbf{x}_0, t) + \mathbf{d}(t) \tag{2}$$

where

$$\begin{aligned} \mathbf{G}(t) &= \mathbf{D}\delta(t) + \mathbf{C} \exp \mathbf{A}t\, \mathbf{B} \\ \mathbf{y}_0(\mathbf{x}_0, t) &= \mathbf{C} \exp \mathbf{A}t\, \mathbf{x}_0 . \end{aligned} \tag{3}$$

For the simplicity of our treatment we assume that $\dim \mathbf{y} = \dim \mathbf{u} = m$. The multivariable PI-controller is described by

$$\begin{aligned} \dot{\mathbf{x}}_r &= \mathbf{y} - \mathbf{v} \\ \mathbf{u} &= -\mathbf{K}_P(\mathbf{y} - \mathbf{v}) - \mathbf{K}_I\mathbf{x}_r . \end{aligned} \tag{4}$$

If the controller matrices $\mathbf{K}_P$ and $\mathbf{K}_I$ are chosen so that the closed-loop system (1), (4) is stable, asymptotic regulation occurs for all step commands

$$\mathbf{v}(t) = \bar{\mathbf{v}}\sigma(t) \tag{5}$$

and step disturbances

$$\mathbf{d}(t) = \bar{\mathbf{d}}\sigma(t) \tag{6}$$

or, more generally, for asymptotically constant external signals.

In the following, we want to investigate which information about the original plant (1) must be available for the controller design. We begin with the derivation of a necessary stability condition for the system (1), (4).

Necessary condition for closed-loop stability

Combining eqns (1) and (4) we get

$$\begin{aligned} \dot{\bar{\mathbf{x}}} &= \begin{pmatrix} \mathbf{A} - \mathbf{B}\tilde{\mathbf{K}}_P\mathbf{C} & -\mathbf{B}\tilde{\mathbf{K}}_I \\ \tilde{\mathbf{M}}\mathbf{C} & -\mathbf{D}\tilde{\mathbf{K}}_I \end{pmatrix} \bar{\mathbf{x}} + \begin{pmatrix} \mathbf{B}\tilde{\mathbf{K}}_P \\ -\tilde{\mathbf{M}} \end{pmatrix} \mathbf{v} + \begin{pmatrix} -\mathbf{B}\tilde{\mathbf{K}}_P \\ \tilde{\mathbf{M}} \end{pmatrix} \mathbf{d} \\ \mathbf{y} &= (\tilde{\mathbf{M}}\mathbf{C} \quad -\mathbf{D}\tilde{\mathbf{K}}_I)\, \mathbf{x} + \mathbf{D}\tilde{\mathbf{K}}_P\mathbf{v} + \tilde{\mathbf{M}}\mathbf{d} \end{aligned} \tag{7}$$

with

$$\tilde{\mathbf{M}} = (\mathbf{I} + \mathbf{D}\mathbf{K}_P)^{-1}$$
$$\tilde{\mathbf{K}}_P = (\mathbf{I} + \mathbf{K}_P\mathbf{D})^{-1}\,\mathbf{K}_P, \qquad \tilde{\mathbf{K}}_I = (\mathbf{I} + \mathbf{K}_P\mathbf{D})^{-1}\,\mathbf{K}_I \tag{8}$$

(cf eqn (8.2-16) for our special controller structure). The stability of the closed-loop original system claims that all eigenvalues of the matrix

$$\bar{\mathbf{A}} = \begin{pmatrix} \mathbf{A} - \mathbf{B}\tilde{\mathbf{K}}_P\mathbf{C} & -\mathbf{B}\tilde{\mathbf{K}}_I \\ \tilde{\mathbf{M}}\mathbf{C} & -\mathbf{D}\tilde{\mathbf{K}}_I \end{pmatrix} \tag{9}$$

possess negative real parts. Hence, we have

$$\det(-\bar{\mathbf{A}}) = \prod \lambda_i[-\bar{\mathbf{A}}] > 0, \tag{10}$$

where the identity $\det(\mathbf{A}) = \prod \lambda_i[\mathbf{A}]$ has been used. Due to the determinant identity

$$\det \begin{pmatrix} \mathbf{R} & \mathbf{S} \\ \mathbf{T} & \mathbf{U} \end{pmatrix} = \det \mathbf{R} \det(\mathbf{U} - \mathbf{T}\mathbf{R}^{-1}\mathbf{S}), \tag{11}$$

provided that $\mathbf{R}$ is invertible, eqns (9) and (10) lead to

$$\det(-\mathbf{A} + \mathbf{B}\tilde{\mathbf{K}}_P\mathbf{C}) \det\big(+\mathbf{D}\tilde{\mathbf{K}}_I + \tilde{\mathbf{M}}\mathbf{C}(-\mathbf{A} + \mathbf{B}\tilde{\mathbf{K}}_P\mathbf{C})^{-1}\,\mathbf{B}\tilde{\mathbf{K}}_I\big) > 0. \tag{12}$$

If we have only an I-controller ($\mathbf{K}_P = \mathbf{0}$) eqn (12) yields

$$\det(-\mathbf{A}) \det\big((\mathbf{D} - \mathbf{C}\mathbf{A}^{-1}\mathbf{B})\,\mathbf{K}_I\big) > 0$$

and, because $\mathbf{A}$ is assumed to be stable,

$$\det \mathbf{K}_s\mathbf{K}_I > 0, \tag{13}$$

where $\mathbf{K}_s$ represents the static transmission matrix (2.1-11) of the plant (1).

Lemma 10.1. A necessary condition for the stability of the closed-loop system (1), (4) with $\mathbf{K}_P = \mathbf{0}$ is given by eqn (13).

The condition (13) can be referred to as the *negative-feedback-condition* for multivariable I-control systems. For single-loop systems it claims $k_s k_I > 0$, ie the controller (4) should counteract the control error. The general version $\det \mathbf{K}_s \det \mathbf{K}_I > 0$ answers the question what 'negative feedback' means for multivariable controllers.

For example, if we have

$$\mathbf{K}_s = \begin{pmatrix} -1 & -0.8 \\ -2 & 3 \end{pmatrix}$$

it is actually a problem to decide which controller does ensure the negative feedback. The lemma tells us that the controller

$$\mathbf{K}_I = \begin{pmatrix} -1 & 0.2 \\ 3 & 2 \end{pmatrix}$$

does.

A similar result can be received for PI-controllers if the proportional part of the controller is assumed to lead to a stable closed-loop system (which is a very reasonable assumption!). Then $(\mathbf{A} - \mathbf{B}\tilde{\mathbf{K}}_P\mathbf{C})$ is stable and eqn (12) yields, after some calculations,

$$\det \tilde{\mathbf{K}}_s\tilde{\mathbf{K}}_I > 0 \tag{14}$$

where

$$\tilde{\mathbf{K}}_s = (\mathbf{I} + \mathbf{K}_s\mathbf{K}_P)^{-1}\,\mathbf{K}_s \tag{15}$$

represents the static reinforcement of the plant (1) subject to the proportional part of the controller (4).

Lemma 10.2 Assume that the plant is stable and that the proportional part of the controller is known to produce a stable P-controlled system. Then the condition (14) is necessary for the controller (4) to ensure the stability of the closed-loop system (1), (4),

Necessary model accuracy

As the necessary stability conditions refer only to the static behaviour of the plant (1), we may confine our considerations to the static plant model

$$\mathbf{y} = \mathbf{K}_s\mathbf{u}\,. \tag{16}$$

The matrix $\mathbf{K}_s$ is assumed to belong to a set of possible representatives $\mathbf{K}_s \in \mathcal{K}_s$. Then the controller has to be chosen so as to satisfy the conditions (13) or (14), respectively, for all elements of $\mathcal{K}_s$. We want to investigate this for pure I-controllers first. Then

$$\det \mathbf{K}_s\mathbf{K}_I > 0 \qquad \text{for all} \qquad \mathbf{K}_s \in \mathcal{K}_s \tag{17}$$

is required. More specifically, if $\mathbf{K}_s$ is decomposed according to

$$\mathbf{K}_s = \hat{\mathbf{K}}_s + \delta\mathbf{K}_s \tag{18}$$

and an upper bound $\bar{\mathbf{K}}$ of $|\delta\mathbf{K}_s|$ is known

$$\delta\mathbf{K}_s| \leqq \bar{\mathbf{K}} \tag{19}$$

then the requirement (17) can be transformed into

$$\det(\hat{\mathbf{K}}_s\mathbf{K}_I + \delta\mathbf{K}_s\mathbf{K}_I) > 0\,.$$

As $\mathbf{K}_s = \hat{\mathbf{K}}_s$ is a member of $\mathcal{K}_s$, the relation

$$\det(\hat{\mathbf{K}}_s\mathbf{K}_I) > 0 \tag{20}$$

represents a necessary condition on $\mathbf{K}_I$ for the robust stability of the closed-loop original system. Combining the last two inequalities we get

$$\det(\mathbf{I} + \delta\mathbf{K}_s\hat{\mathbf{K}}_s^{-1}) > 0 \qquad \text{for all} \qquad |\delta\mathbf{K}_s| \leqq \bar{\mathbf{K}}\,. \tag{21}$$

This equation is a condition on the accuracy of the knowledge of the static reinforcement of the plant. It claims

$$\sup_{|\delta \mathbf{K}_s| \leq \bar{\mathbf{K}}} \varrho[\delta \mathbf{K}_s \hat{\mathbf{K}}_s^{-1}] < 1$$

which can be analysed in the same way as eqn (8.3-2) for $\mathbf{G}_2$ from (8.3-13). Accordingly, a sufficient condition for (21) is given by

$$\lambda_p[\bar{\mathbf{K}} |\hat{\mathbf{K}}_s^{-1}|] < 1. \tag{22}$$

Moreover, from (8.3-22) and the restriction to real-valued matrices $\delta \mathbf{K}_s$ the supremum is known to be assumed for

$$\delta \tilde{\mathbf{K}}_s = \operatorname{diag} l_i \bar{\mathbf{K}} \operatorname{diag} r_i \tag{23}$$

with l_i, $r_i = +1$ or -1. Hence, eqn (21) is equivalent to

$$\sup_{|\delta \mathbf{K}_s| \leq \bar{\mathbf{K}}} \varrho[\delta \mathbf{K}_s \hat{\mathbf{K}}_s^{-1}] = \inf_{l_i, r_i \in \{+1, -1\}} \sigma_{\max}\left[\operatorname{diag} \frac{1}{r_1} \hat{\mathbf{K}}_s^{-1} \operatorname{diag} \frac{1}{l_i}\right] \times \sigma_{\max}[\operatorname{diag} l_i \bar{\mathbf{K}} \operatorname{diag} r_i] < 1. \tag{24}$$

For PI-controllers the condition (14) yields

$$\det (\mathbf{I} + \hat{\mathbf{K}}_s \mathbf{K}_P)^{-1} \hat{\mathbf{K}}_s \mathbf{K}_I > 0 \tag{25}$$

and, by an easy calculation,

$$\det \left(\mathbf{I} + \delta \mathbf{K}_s \mathbf{K}_P (\mathbf{I} + \hat{\mathbf{K}}_s \mathbf{K}_P)^{-1}\right) \det (\mathbf{I} + \delta \mathbf{K}_s \hat{\mathbf{K}}_s^{-1}) > 0. \tag{26}$$

That is, the model must satisfy the same necessary condition (21). Moreover, the choice of the controller matrix $\mathbf{K}_P$ has to ensure that

$$\det \left(\mathbf{I} + \delta \mathbf{K}_s \mathbf{K}_P (\mathbf{I} + \hat{\mathbf{K}}_s \mathbf{K}_P)^{-1}\right) > 0 \qquad \text{for all} \qquad |\delta \mathbf{K}_s| \leq \bar{\mathbf{K}} \tag{27}$$

holds.

Lemma 10.3 Assume that the original plant (1) is stable and that its static reinforcement can be described by eqns (18) and (19) with some known matrices $\hat{\mathbf{K}}_s$ and $\bar{\mathbf{K}}$. A necessary condition for the existence of multivariable PI-controllers that ensure robust stability of the closed-loop system is given by eqn (24). This condition is satisfied if eqn (22) holds.

In practical applications we always check the condition (22) first because this condition is simpler than eqn (24). In what follows we want to show that the conditions (22) and (24) are also sufficient for the existence of a robust PI-controller. We do this by presenting a method for the construction of a robust low-gain PI-feedback.

A stability condition for the low-gain feedback system

The following investigations are motivated by the idea that the robust stability of the closed-loop system (1), (4) can be ensured by a PI-controller with a low controller gain (low norm of the controller matrices). As the only unstable poles of the open loop are brought about by the integral part of the controller, such a feedback may stabilise these poles while not destabilising the uncertain plant poles. Using

$$\mathbf{K}_{\mathrm{I}} = \alpha \hat{\mathbf{K}}_{\mathrm{I}}, \qquad \mathbf{K}_{\mathrm{P}} = \beta \hat{\mathbf{K}}_{\mathrm{P}} \tag{28}$$

we have to choose the matrix $\hat{\mathbf{K}}_{\mathrm{I}}$ so that the root loci of the integrator poles move into the left-half complex plane at least for small positive values of the scalar α. If we use α and β small enough, the root loci of the plant poles can not cross the imaginary axis and the uncertain closed-loop system is stable (Fig. 10.1).

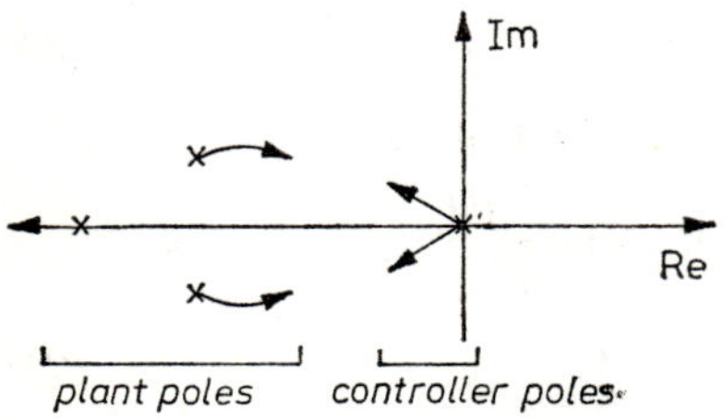

Fig. 10.1 Root loci for low-gain feedback (small scalar α)

To establish conditions under which the root loci behave in this way, we specify the matrix $\bar{\mathbf{A}}$ from eqn (9) for pure I-control with $\mathbf{K}_{\mathrm{I}}$ from eqn (28)

$$\bar{\mathbf{A}}(\alpha) = \begin{pmatrix} \mathbf{A} & -\alpha \mathbf{B}\hat{\mathbf{K}}_{\mathrm{I}} \\ \mathbf{C} & -\alpha \mathbf{D}\hat{\mathbf{K}}_{\mathrm{I}} \end{pmatrix}. \tag{29}$$

Our aim is to make the sensitivity vector of $\lambda_1[\mathbf{A}(\alpha)], \ldots, \lambda_m[\mathbf{A}(\alpha)]$ point into the left-half complex plane

$$\mathrm{Re}\left[\frac{\partial \lambda_i}{\partial \alpha}\right] < 0 \qquad (i = 1, 2, \ldots, m) \tag{30}$$

at $\alpha = 0$, where $\lambda_i[\mathbf{A}(0)] = 0$ (Fig. 10.1). In order to use eqn (3.2-12)

$$\frac{\partial \lambda}{\partial \alpha} = \frac{\tilde{\mathbf{v}}' \dfrac{\partial \mathbf{A}}{\partial \alpha} \tilde{\mathbf{u}}}{\tilde{\mathbf{v}}'\tilde{\mathbf{u}}} \tag{31}$$

we derive from eqn (29)

$$\frac{\partial \bar{\mathbf{A}}}{\partial \alpha} = \begin{pmatrix} \mathbf{0} & -\mathbf{B}\hat{\mathbf{K}}_{\mathrm{I}} \\ \mathbf{0} & -\mathbf{D}\hat{\mathbf{K}}_{\mathrm{I}} \end{pmatrix}.$$

The eigenvectors $\tilde{\mathbf{u}}$ and $\tilde{\mathbf{v}}$ of $\bar{\mathbf{A}}(0)$ that belong to $\lambda_i[\mathbf{A}(0)] = 0$ result from the equations $\bar{\mathbf{A}}(0)\,\tilde{\mathbf{u}} = \mathbf{0}$ and $\tilde{\mathbf{v}}'\bar{\mathbf{A}}(0) = \mathbf{0}$ which yield

$$\tilde{\mathbf{u}} = \begin{pmatrix} \mathbf{0} \\ \mathbf{u}_0 \end{pmatrix} \quad \text{and} \quad \tilde{\mathbf{v}}' = (-\tilde{\mathbf{v}}_0'\mathbf{C}\mathbf{A}^{-1}, \tilde{\mathbf{v}}_0').$$

The m-vectors $\tilde{\mathbf{u}}_0$ and $\tilde{\mathbf{v}}_0$ may be chosen arbitrarily. Then eqn (31) leads to

$$\frac{\partial \lambda_i}{\partial \alpha} = \frac{-\tilde{\mathbf{v}}_0{}'\mathbf{K}_s\hat{\mathbf{K}}_I\tilde{\mathbf{u}}_0}{\tilde{\mathbf{v}}_0{}'\tilde{\mathbf{u}}_0}. \tag{32}$$

As the conditions (30), (32) must be satisfied for all $\tilde{\mathbf{u}}_0$ and $\tilde{\mathbf{v}}_0$, we have to ensure that

$$\mathrm{Re}\,\big[\lambda_i[\mathbf{K}_s\hat{\mathbf{K}}_I]\big] > 0 \qquad (i = 1, \ldots, m). \tag{33}$$

Lemma 10.4 There exist positive scalars $\bar{\alpha}$ and $\bar{\beta}$ such that the closed-loop system (1), (4), (28) is stable for all $\alpha \in (o, \bar{\alpha})$ and $\mathbf{b} \in (o, \bar{\beta})$ if and only if $\hat{\mathbf{K}}_I$ satisfies the condition (33).

Condition (33) improves the negative-feedback condition (13). If (33) is satisfied then stability can be ensured by using a sufficiently low loop gain. For instance, if we use $\mathbf{K}_I$ in the numerical example above as $\hat{\mathbf{K}}_I$ then $\mathbf{K}_s\hat{\mathbf{K}}_I$ meets the requirement (33). Therefore, we can be sure to get a stable closed-loop system if we make α small.

Construction of a robust PI-controller

With $\mathbf{K}_s$ from eqns (18) and (19), the condition (33) claims

$$\mathrm{Re}\,\big[\lambda_i[\hat{\mathbf{K}}_s\hat{\mathbf{K}}_I + \delta\mathbf{K}_s\hat{\mathbf{K}}_I]\big] > 0 \qquad \text{for all} \qquad |\delta\mathbf{K}_s| \leqq \bar{\mathbf{K}}. \tag{34}$$

For $\delta\mathbf{K}_s = \mathbf{0}$ this equation suggests to use

$$\hat{\mathbf{K}}_I = \hat{\mathbf{K}}_s{}^{-1} \tag{35}$$

since then all eigenvalues equal $+1$ and satisfy the condition (34) in the 'best possible' way. Then eqn (34) changes to

$$\mathrm{Re}\,\big[\lambda_i[\mathbf{I} + \delta\mathbf{K}_s\hat{\mathbf{K}}_s{}^{-1}]\big] = 1 + \mathrm{Re}\,\big[\lambda_i[\delta\mathbf{K}_s\hat{\mathbf{K}}_s{}^{-1}]\big] > 0. \tag{36}$$

This condition is satisfied for all $|\delta\mathbf{K}_s| \leqq \bar{\mathbf{K}}$ if and only if eqn (24) holds. From this we conclude that the condition (24) is also sufficient for the existence of a robust PI-controller.

Theorem 10.1. (*Necessary and sufficient condition on model accuracy*). A multivariable PI-controller that ensures the stability of the closed-loop system (1), (4) for all stable plants whose static behaviour is described by eqns (16), (18) and (19) exists if and only if the condition (24) is satisfied.

In addition, a sufficient existence condition is given by eqn (22).

Theorem 10.2. (*Construction of robust low-gain PI-controllers*) Assume that the plant (1) is stable and that its static reinforcement $\mathbf{K}_s$ can be described by eqns (18) and (19) with some known matrices $\hat{\mathbf{K}}_s$ and $\bar{\mathbf{K}}$. If the condition (24) is satisfied there exist positive scalars $\bar{\alpha}$ and $\bar{\beta}$ such that the closed-loop system (1), (4), (28) with $\hat{\mathbf{K}}_I$ from eqn (35) is stable for all $\alpha \in (0, \bar{\alpha})$ and $\beta \in (0, \bar{\beta})$.

10.2 Robustness of PI-controlled systems

In this section we assume that the designer has access to an analytic plant model or reliably measured plant step responses which enable him to set up an approximate model and give some uncertainty bounds. On the basis of the results presented in Sections 8.2, 8.4 and 8.5 we evaluate the robust stability and the I/O-behaviour of the closed-loop original system in the time domain. The frequency domain counterpart is straightforward. For brevity of notation, we use the parallel model structure of Fig. 5.2(a) and assume that $\mathbf{d} = \mathbf{0}$ and $\mathbf{w} = \mathbf{0}$. Then the approximate model has the form

$$\begin{aligned} \dot{\hat{\mathbf{x}}} &= \hat{\mathbf{A}}\hat{\mathbf{x}} + \hat{\mathbf{B}}\mathbf{u} \\ \mathbf{y} &= \hat{\mathbf{C}}\mathbf{x} + \hat{\mathbf{D}}\mathbf{u} + \mathbf{s} \\ \mathbf{z} &= \mathbf{u}. \end{aligned} \tag{1}$$

An upper bound of the uncertainties is given by

$$|\mathbf{s}(t)| \leqq \mathbf{V}_2(t) * |\mathbf{z}(t)|. \tag{2}$$

The PI-controller is described by eqn (10.1-4)

$$\begin{aligned} \dot{\mathbf{x}}_r &= \mathbf{y} - \mathbf{v} \\ \mathbf{u} &= -\mathbf{K}_P(\mathbf{y} - \mathbf{v}) - \mathbf{K}_I\mathbf{x}_r. \end{aligned} \tag{3}$$

Then the closed-loop approximate model is represented by

$$\begin{aligned} \dot{\bar{\mathbf{x}}} &= \begin{pmatrix} \hat{\mathbf{A}} - \hat{\mathbf{B}}\tilde{\mathbf{K}}_P\hat{\mathbf{C}} & -\hat{\mathbf{B}}\tilde{\mathbf{K}}_I \\ \tilde{\mathbf{M}}\hat{\mathbf{C}} & -\hat{\mathbf{D}}\tilde{\mathbf{K}}_I \end{pmatrix} \bar{\mathbf{x}} + \begin{pmatrix} \hat{\mathbf{B}}\tilde{\mathbf{K}}_P \\ -\tilde{\mathbf{M}} \end{pmatrix} \mathbf{v} + \begin{pmatrix} -\hat{\mathbf{B}}\tilde{\mathbf{K}}_P \\ \tilde{\mathbf{M}} \end{pmatrix} \mathbf{s} \\ \mathbf{y} &= (\tilde{\mathbf{M}}\hat{\mathbf{C}} \quad -\hat{\mathbf{D}}\tilde{\mathbf{K}}_I)\, \bar{\mathbf{x}} + \hat{\mathbf{D}}\tilde{\mathbf{K}}_P\mathbf{v} + \tilde{\mathbf{M}}\mathbf{s} \\ \mathbf{z} &= (-\tilde{\mathbf{K}}_P\hat{\mathbf{C}} \quad -\tilde{\mathbf{K}}_I)\, \bar{\mathbf{x}} + \tilde{\mathbf{K}}_P\mathbf{v} - \tilde{\mathbf{K}}_P\mathbf{s} \end{aligned} \tag{4}$$

with

$$\begin{aligned} \tilde{\mathbf{M}} &= (\mathbf{I} + \hat{\mathbf{D}}\mathbf{K}_P)^{-1} \\ \tilde{\mathbf{K}}_P &= (\mathbf{I} + \mathbf{K}_P\hat{\mathbf{D}})^{-1}\, \mathbf{K}_P, \qquad \tilde{\mathbf{K}}_I = (\mathbf{I} + \mathbf{K}_P\hat{\mathbf{D}})^{-1}\, \mathbf{K}_I \\ \bar{\mathbf{x}} &= \begin{pmatrix} \mathbf{x} \\ \mathbf{x}_r \end{pmatrix}. \end{aligned} \tag{5}$$

Due to Theorem 8.11 we get the stability condition

$$\lambda_P \left[\left(|\tilde{\mathbf{K}}_P| + \int_0^\infty \left| (-\tilde{\mathbf{K}}_P\hat{\mathbf{C}} \quad -\tilde{\mathbf{K}}_I) \exp \bar{\mathbf{A}}t \begin{pmatrix} -\hat{\mathbf{B}}\tilde{\mathbf{K}}_P \\ \tilde{\mathbf{M}} \end{pmatrix} \right| \mathrm{d}t \right) \int_0^\infty \mathbf{V}_2(t)\, \mathrm{d}t \right] < 1 \tag{6}$$

where $\bar{\mathbf{A}}$ is the system matrix in eqn (4).

The approximate $\hat{\mathbf{y}}$ of the closed-loop system output is given by

$$\dot{\bar{\mathbf{x}}} = \bar{\mathbf{A}}\bar{\mathbf{x}} + \begin{pmatrix} \hat{\mathbf{B}}\tilde{\mathbf{K}}_P \\ -\tilde{\mathbf{M}} \end{pmatrix} \mathbf{v}$$
$$\hat{\mathbf{y}} = (\tilde{\mathbf{M}}\hat{\mathbf{C}} \quad -\hat{\mathbf{D}}\tilde{\mathbf{K}}_I)\,\bar{\mathbf{x}} + \hat{\mathbf{D}}\tilde{\mathbf{K}}_P\mathbf{v} \tag{7}$$

and the approximation error is bounded by

$$|\mathbf{y} - \hat{\mathbf{y}}| \leqq \mathbf{V}_{\mathbf{yv}}(t) * |\mathbf{v}(t)| \tag{8}$$

where

$$\mathbf{V}_{\mathbf{yv}} = \left|\tilde{\mathbf{M}}\delta(t) + (\tilde{\mathbf{M}}\hat{\mathbf{C}} \quad -\hat{\mathbf{D}}\tilde{\mathbf{K}}_I) \exp \bar{\mathbf{A}}\mathbf{t} \begin{pmatrix} -\hat{\mathbf{B}}\tilde{\mathbf{K}}_P \\ \tilde{\mathbf{M}} \end{pmatrix}\right| * \mathbf{V}_2 * \tilde{\mathbf{G}}$$
$$* \left|\tilde{\mathbf{K}}_P\delta(t) + (-\tilde{\mathbf{K}}_P\hat{\mathbf{C}} \quad -\tilde{\mathbf{K}}_I) \exp \bar{\mathbf{A}}\mathbf{t} \begin{pmatrix} \hat{\mathbf{B}}\tilde{\mathbf{K}}_P \\ -\tilde{\mathbf{M}} \end{pmatrix}\right| \tag{9}$$
$$\tilde{\mathbf{G}} = \delta(t)\,\mathbf{I} + \left|\tilde{\mathbf{K}}_P\delta(t) + (-\tilde{\mathbf{K}}_P\hat{\mathbf{C}} \quad -\tilde{\mathbf{K}}_I) \exp \bar{\mathbf{A}}t \begin{pmatrix} -\hat{\mathbf{B}}\tilde{\mathbf{K}}_P \\ \tilde{\mathbf{M}} \end{pmatrix}\right| * \mathbf{V}_2 * \tilde{\mathbf{G}}.$$

In particular, for step input $\mathbf{v} = \bar{\mathbf{v}}\delta(t)$ we have

$$|\mathbf{y} - \hat{\mathbf{y}}| \leqq \int_0^t \mathbf{V}_{\mathbf{yv}}(\tau)\,\mathrm{d}\tau |\bar{\mathbf{v}}|. \tag{10}$$

Because of $\mathbf{V}_{\mathbf{yv}}(t) \geqq \mathbf{0}$ for all t, the right-hand side of eqn (10) is monotonically increasing. However, as we know that the system is asymptotically stable and that

$$\mathbf{y}(\infty) = \hat{\mathbf{y}}(\infty) = \bar{\mathbf{v}} \tag{11}$$

holds, we can reduce the width of the tolerance band. Eqns (8) and (11) yield

$$|\mathbf{y}(t) - \hat{\mathbf{y}}(t)| \leqq \int_t^\infty \mathbf{V}_{\mathbf{yv}}(\tau)\,\mathrm{d}\tau\,|\bar{\mathbf{v}}| = \left(\int_0^\infty \mathbf{V}_{\mathbf{yv}}(t)\,\mathrm{d}t - \int_0^t \mathbf{V}_{\mathbf{yv}}(\tau)\,\mathrm{d}\tau\right)|\bar{\mathbf{v}}|. \tag{12}$$

As both the eqns (10) and (12) hold, we can use the smaller bound to construct the tolerance band

$$|y_i - \hat{y}_i| \leqq \min\left[\left(\int_0^t \mathbf{V}_{\mathbf{yv}}(\tau)\,\mathrm{d}\tau\,|\bar{\mathbf{v}}|\right)_i,\ \left(\left(\int_0^\infty \mathbf{V}_{\mathbf{yv}}(\tau)\,\mathrm{d}\tau - \int_0^t \mathbf{V}_{\mathbf{yv}}(\tau)\,\mathrm{d}\tau\right)|\bar{\mathbf{v}}|\right)_i\right] \tag{13}$$

where $(.)_i$ denotes the i-th component of the indicated vector. The right-hand side of eqn (13) is smaller than the bound given by eqn (10) for large t and vanishes for $t \to \infty$. Hence, the property of asymptotic regulation is reproduced by the tolerance bands.

Evaluation of PI-controlled boiler

Our first example should give an impression of the tolerance band described by eqns (7) and (13). It concerns the feedback control of a boiler with two control inputs and outputs. As this control problem will be explained in more detail in the next section we are satisfied here with knowing the parameters of the approximate model (1) (Table 10.1). The uncertainty is assumed to be less than 10% of the nominal model, ie $\mathbf{V}_2(t) = 0.1\,|\mathbf{G}(t)|$ where $\mathbf{G}(t)$ is the impulse response matrix of the approximate model (1).

Table 10.1 Parameters of the approximate model (1)

$$\mathbf{A} = \left(\begin{array}{cc|cc|cc|cc} -0.037 & 0 & & & & & & \\ 0.1 & -0.1 & & & & 0 & & \\ \hline & & -0.0385 & 0 & & & & \\ & & 0.05 & -0.05 & & & & \\ \hline & & & & -0.0385 & 0 & & \\ & & & & 0.1 & -0.1 & & \\ \hline & 0 & & & & & -0.05 & 0 \\ & & & & & & 0.025 & -0.025 \end{array}\right)$$

$$\mathbf{B} = \left(\begin{array}{cc} 0.037 & 0 \\ 0 & 0 \\ \hline 0 & 0.0385 \\ 0 & 0 \\ \hline 0.0385 & 0 \\ 0 & 0 \\ \hline 0 & 0.05 \\ 0 & 0 \end{array}\right)$$

$$\mathbf{C} = \left(\begin{array}{cc|cc|cc|cc} 0 & 2.2 & 0 & 0.033 & 0 & 0 & 0 & 0 \\ 0 & 0 & 0 & 0 & -0.26 & 0 & 0 & 0.09 \end{array}\right)$$

$$\mathbf{D} = \begin{pmatrix} 0 & 0 \\ 0 & 0 \end{pmatrix}$$

For the PI-controller with

$$\mathbf{K}_\mathrm{P} = \beta \begin{pmatrix} 1 & 0 \\ 0 & 1 \end{pmatrix}, \quad \mathbf{K}_\mathrm{I} = \alpha \begin{pmatrix} -0.0262 & -0.0346 \\ 0.0351 & -0.00582 \end{pmatrix}$$

where $\mathbf{K}_\mathrm{I}$ is chosen as in eqn (10.1-35) eqns (7) and (13) yield the tolerance band for the step response between $v_2(t)$ and $y_2(t)$ shown in Fig. 10.2 for different values of α and β. These bands illustrate that a design compromise has to be made. For low controller gains the band is small but reaches its final value slowly. If the gain is increased the response becomes quicker,

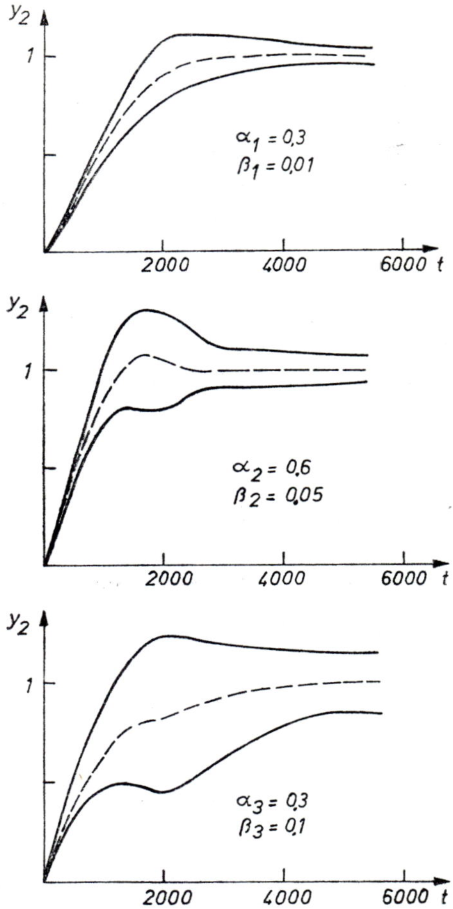

Fig. 10.2 Tolerance band for the step response $y_2(t)$ of the closed-loop original system for different values of α and β ($v_1 = O$, $v_2 = \sigma(t)$).

but the width of the band grows simultaneously. If the controller with parameters α_2 and β_2 is selected, the tolerance bands of the whole step response matrix of the closed-loop system has the form shown in Fig. 10.3.

Design of a PI-controller for a chemical process

As second example we consider the control of a process in the chemical industry that produces a solution of ammonia and urea. The aim is to get constant concentrations of both substances within the final solution although their concentrations in the intermediate products, from which the solution is made, vary. The input vector of the two-input two-output plant consists of the quantity of the second component flowing through an input valve and the temperature of the initial solution.

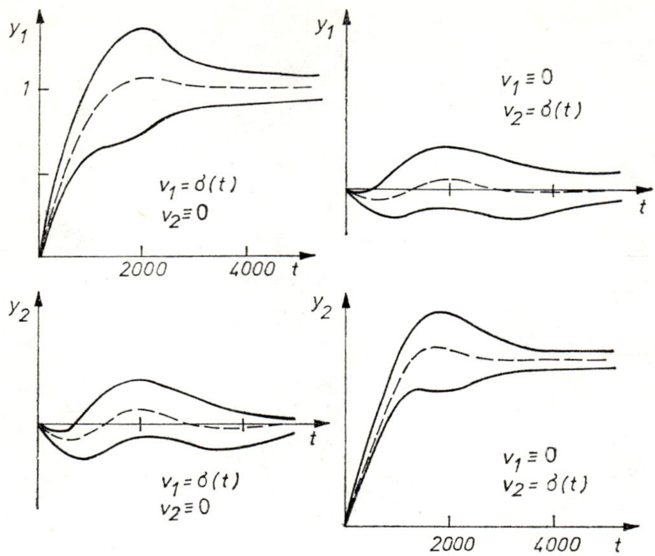

Fig. 10.3 Tolerance band for the step response matrix of the closed-loop original system with $\alpha = \alpha_2$ and $\beta = \beta_2$

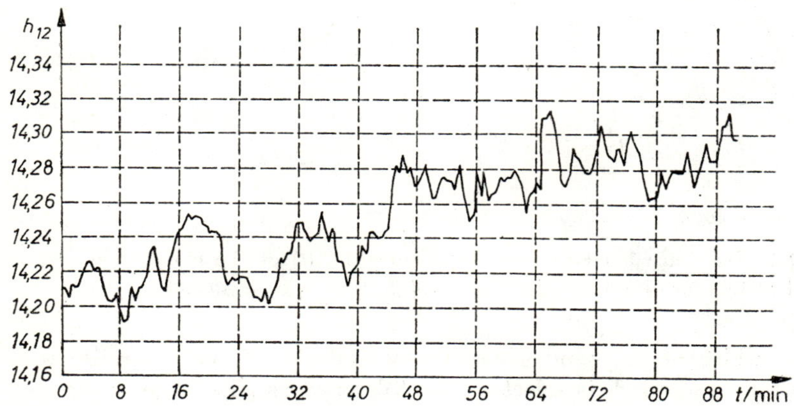

Fig. 10.4 Measured signal of the plant step response $h_{12}(t)$

In order to set up a plant model, experiments have been made to measure the step responses. Although these experiments have been disturbed by heavy unavoidable stochastic signals (Fig. 10.4), a detailed analysis, which we skip here, leads to low order approximate models and to the certainty that the differences between the model outputs and the measurement signals can be regarded as a pure stochastic plant disturbance $\mathbf{d}(t)$. What we have to investigate is the robustness of the controller against uncertainties that are caused by the time-varying behaviour of the plant. Experiments at different operating points and under different operational conditions yield a set $\mathcal{S}$ of substantially deviating models. The uncertainties regard both the static reinforcement $\mathbf{K}_s$ and the dynamical behaviour. To give an impression different delay-lag models of the element $h_{11}(t)$ of the step response matrix are listed in Table 10.2.

Table 10.2 Parameters of the models describing the step response $h_{11}(t)$

No.	Form of the model	Parameters		
	$\dfrac{k_{s11}}{(1+pT_1)(1+pT_2)}$	T_1	T_2	k_{s11}
(1)		10 min	27 min	1.68
(2)		19 min	19 min	1.87
	$\dfrac{k_{s11}}{(1+pT_1)}\,e^{-pT_t}$	T_t	T_1	k_{s11}
(3)		11 min	33 min	1.68
(4)		9 min	32 min	1.87
(5)		10 min	28 min	1.86
(6)		8 min	30 min	1.62
(7)		15 min	50 min	2.12
(8)		13 min	36 min	2.02
(9)		14 min	30 min	1.45
(10)		17 min	31 min	2.03
(11)		18 min	48 min	1.96
(12)		18 min	32 min	1.46
(13)		18 min	38 min	1.78

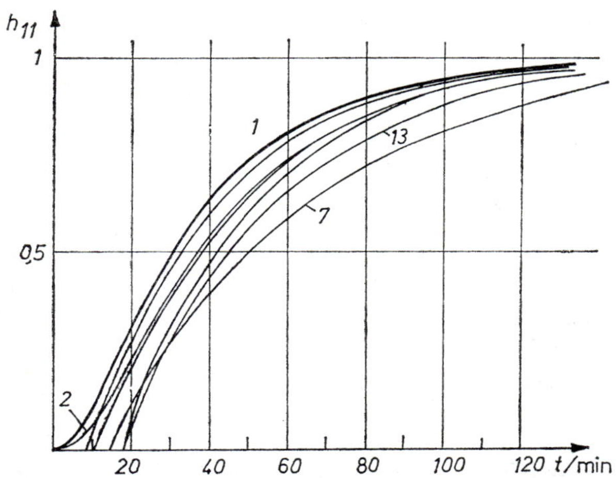

Fig. 10.5 Step responses $h_{11}(t)$ of the model for different parameters (cf Table 10.2) after having been scaled with regard to their final values

The static plant characteristic can be described by

$$\hat{\mathbf{K}}_s = \begin{pmatrix} 2.2 & 0.033 \\ -0.26 & 0.09 \end{pmatrix}, \qquad \overline{\mathbf{K}} = \begin{pmatrix} 0.7 & 0.01 \\ 0.1 & 0.03 \end{pmatrix}.$$

The existence condition (10.1-22) for a robust PI-controller is satisfied $\lambda_p[\overline{\mathbf{K}}\,|\mathbf{K}_s^{-1}|] = 0.1 < 1$. The range of variation of the dynamical behaviour is shown in Fig. 10.5.

After having carried out the experiments we have a set $\mathcal{S}$ of plant models at our disposal. One element of this set is used as approximate plant model (1) (Table 10.3).

Table 10.3 Parameters of the approximate plant model

$$\hat{\mathbf{A}} = \left(\begin{array}{cc|cc|cc|cc} -.00606 & 0 & & & & & & \\ .00166 & -.00166 & & & & 0 & & \\ \hline & & -.00641 & & & & & \\ & & .00084 & -.00084 & & & & \\ \hline & & & & -.00641 & & & \\ & & & & .00166 & -.00166 & & \\ \hline & 0 & & & & & -.00084 & \\ & & & & & & .00040 & -.00040 \end{array}\right)$$

$$\hat{\mathbf{B}} = \left(\begin{array}{cc} .00606 & 0 \\ 0 & 0 \\ \hline 0 & .00641 \\ 0 & 0 \\ \hline .00641 & 0 \\ 0 & 0 \\ \hline 0 & .00084 \\ 0 & 0 \end{array}\right)$$

$$\hat{\mathbf{C}} = \left(\begin{array}{cc|cc|cc|cc} 0 & 2.2 & 0 & .033 & 0 & 0 & 0 & 0 \\ 0 & 0 & 0 & 0 & 0 & -.26 & 0 & .09 \end{array}\right)$$

$$\hat{\mathbf{D}} = \begin{pmatrix} 0 & 0 \\ 0 & 0 \end{pmatrix}$$

The PI-controller is used with

$$\mathbf{u} = -\beta_1 \hat{\mathbf{K}}_s^{-1}\mathbf{y} + \beta_2 \hat{\mathbf{K}}_s^{-1}\mathbf{v} - \alpha \hat{\mathbf{K}}_s^{-1}\mathbf{x}_r \tag{14}$$

instead of eqn (3). For $\beta_1 = 0.51$, $\beta_2 = 1.18$, and $\alpha = 0.0195$ the closed-loop approximate model (7) has a satisfactory behaviour (solid lines in Fig. 10.6).

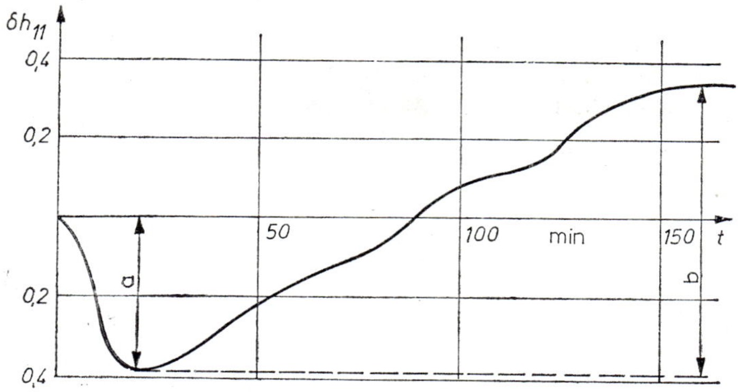

Fig. 10.6 Step response of the uncertainty model with impulse response $\delta g_{11}(t) = g_{11}(t) - \hat{g}_{11}(t)$ where $\hat{g}_{11}(t)$ and $g_{11}(t)$ correspond to the parameter sets (1) and (8) of Table 10.2

Absolute variation: $a + b = 1.12$

The stability robustness is analysed by eqn (6). As explained in Section 8.4, $\int_0^\infty \mathbf{V}_2(t)\, dt$ can be determined in a graphical way by determining the maximum absolute variation of the differences $\delta\mathbf{H}(t)$ between the step responses of the approximate model and the other models of the set $\mathcal{S}$. In Fig. 10.6 the total variation of $\delta h_{11}(t)$ amounts to 1.12. As we need the maximum values of the total variations we compare the approximate model with all other models from $\mathcal{S}$ and get

$$\int_0^\infty \mathbf{V}_2(t)\, dt = \begin{pmatrix} 1.4 & 0.02 \\ 0.2 & 0.04 \end{pmatrix}. \tag{15}$$

The closed-loop approximate model gives

$$\int_0^\infty |\bar{\mathbf{G}}_{zs}(t)|\, dt = \begin{pmatrix} 0.57 & 0.205 \\ 1.62 & 13.7 \end{pmatrix}.$$

Hence, the Perron root in eqn (6) is 0.48, which proves the stability robustness of the closed-loop system.

The I/O-behaviour is approximately given by eqn (7). As the Perron root in eqn (6) is about 0.5, the tolerance bands received by means of eqn (13) are very broad. That is, a global characterisation of all models from $\mathcal{S}$ by the approximate model and an upper error bound is too coarse for this example because such an uncertainty description includes many other plants too. Therefore, we assess the closed-loop system behaviour by determining the family of step responses for different (or all) models of $\mathcal{S}$. The

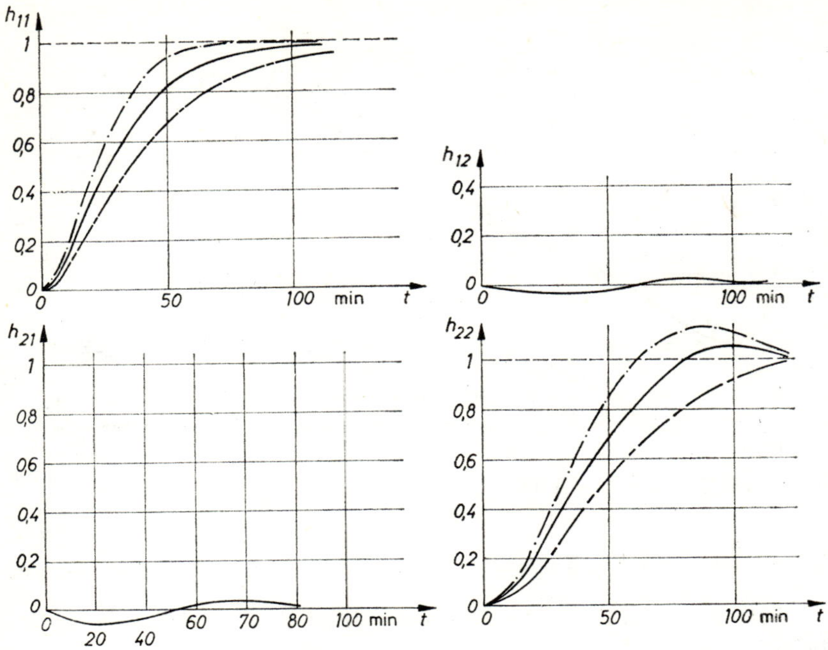

Fig. 10.7 Step response matrix of the closed-loop original sytem (simulation results)
——— closed-loop approximate model
— — —, —·—·— performance for other models (other operating conditions)

results in Fig. 10.7 show that under different operating conditions the original system may have a greater overshoot or may be more slowly than the approximate model. The cross-coupling remain neglectable. All the curves satisfy the design specifications and prove the controller to be robust. Fig. 10.8 shows experimental results with the closed-loop system. As the controller has been designed for the purpose of good command tracking, the stochastic disturbances with high frequencies cannot be completely attenuated.

10.3 Tuning of multivariable PI-controllers

A design technique which is fundamentally based on the robustness properties of the PI-controllers is on-line tuning. No exact quantitative model of the plant is assumed to be known. Instead, some measured trajectories of the plant are directly used for determining the controller matrices. Because of the unavoidable disturbances and measurement errors these trajectories do not by any means represent a complete description of the plant and the controller has to endure substantial model uncertainties.

Tuning means looking for suitable controller parameters by means of experiments with the plant and the tentative closed-loop system. The

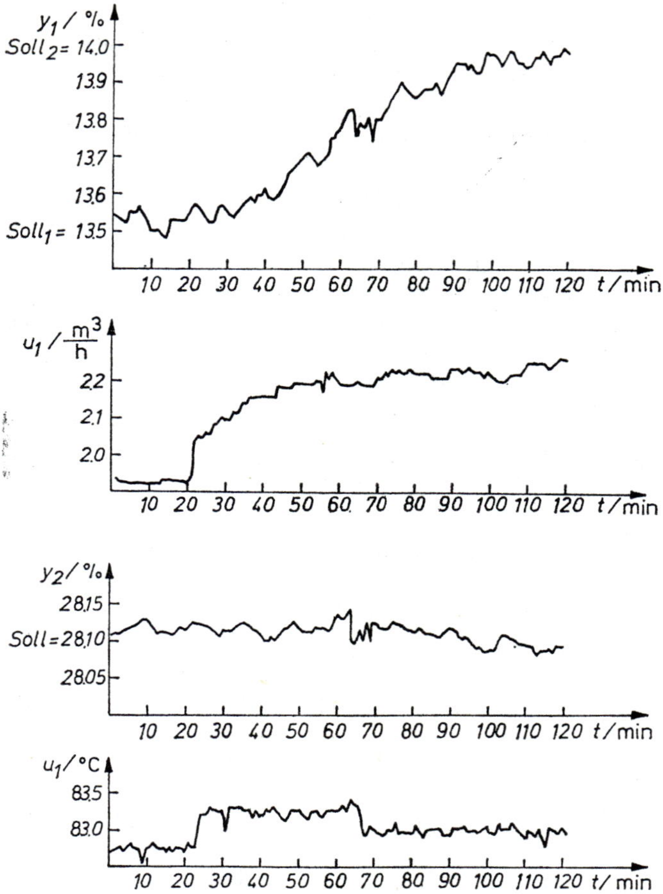

Fig. 10.8 Response to command step of v_1 at $t = 20$ min (experimental results)

Ziegler-Nichols rules for tuning single-loop systems may serve as an example to motivate this design technique. They require that a P-controller is attached to the plant and the controller gain k_P is increased until undamped oscillations occur in the closed-loop system. The oscillation time T is measured. The corresponding controller gain is denoted by $k_P{}^0$. Then suitable parameters of the PID-controller

$$u(p) = \left(k_P + \frac{1}{pT_n} + pT_v\right)(v(p) - y(p))$$

are given by

P-controller:	$k_P = 0.5\, k_P{}^0$,	$T_n = \infty$,	$T_v = 0$
PI-controller:	$k_P = 0.45\, k_P{}^0$,	$T_n = 0.85\, T$,	$T_v = 0$
PID-controller:	$k_P = 0.6\, k_P{}^0$,	$T_n = 0.5\, T$,	$T_v = 0.12\, T$

The corresponding closed-loop system has a damping factor of about 0.25, and its unit step response with regard to disturbance steps exhibits a peak value of 0.4. Therefore, these rules are suitable for slowly changing disturbances.

The Ziegler-Nichols rules are based on the determination of the plant characteristics at the stability limit. However, by virtue of the cross-couplings within or the complexity of the plant it is practically impossible or technologically not permitted to drive multivariable processes towards instability. Instead, the first step in multivariable tuning regulation must be a reasonable reduction of the freedom in the choice of the controller parameters in order to get a clear view of which parameter modifications are possible without endangering the closed-loop stability. Further problems are to test the existence of a robust regulator by means of the acquired information about the plant and to establish rules for experiments which help to improve the closed-loop system behaviour.

In general, multivariable controller tuning proceeds in the following steps:

(1) Make experiments with the plant in order to measure important plant characteristics.
(2) Check the existence of a robust controller.
(3) Fix as many controller parameters as possible and establish rules for the on-plant tuning in Step 4.
(4) Attach the controller to the plant. Using the guidelines for the modification of the controller parameters and making experiments with the closed-loop system, improve the system behaviour.

Theoretical investigations have to prove that in each step of the tuning process the closed-loop original system remains stable.

In the following we assume that the experiments in Step 1 concern the measurement of plant step responses. Input signals

$$\mathbf{u}(t) = \mathbf{u}^i \sigma(t), \qquad (i = 1, \ldots, m) \tag{1}$$

are applied where the constant amplitude vectors $\mathbf{u}^i$ are linearly independent and the trajectories

$$\mathbf{y}^i(t) = \int_0^t \mathbf{G}(\tau)\, \mathbf{u}^i\, \mathrm{d}\tau + \delta\mathbf{y}^i(t) \qquad (i = 1, \ldots, m) \tag{2}$$

are measured. The term

$$\mathbf{H}(t) = \int_0^t \mathbf{G}(\tau)\, \mathrm{d}\tau \tag{3}$$

represents the step response of the system (10.1-2). The measurement error $\delta\mathbf{y}^i(t)$ is caused by the plant disturbance $\mathbf{d}(t)$ and nonzero initial conditions

$\mathbf{x}_0$. Lumping all measured signals together we get ($\dim \mathbf{y}^i = \dim \mathbf{u}^i = m$)

$$\mathbf{H}(t) = \left(\mathbf{y}^1(t) \quad \mathbf{y}^2(t) \ldots \mathbf{y}^m(t)\right) (\mathbf{u}^1\, \mathbf{u}^2 \ldots \mathbf{u}^m)^{-1} \\ - \left(\delta\mathbf{y}^1(t) \quad \delta\mathbf{y}^2(t) \ldots \delta\mathbf{y}^m(t)\right) \quad (\mathbf{u}^1\, \mathbf{u}^2 \ldots \mathbf{u}^m)^{-1}. \tag{4}$$

The first term in eqn (4) is used as approximation $\hat{\mathbf{H}}(t)$ of the plant step response matrix $\mathbf{H}(t)$. The second term corresponds to the model error $\delta\mathbf{H}(t)$, which is assumed to be bounded by some matrix function $\bar{\mathbf{H}}(t)$

$$|\delta\mathbf{H}(t)| \leqq \bar{\mathbf{H}}(t). \tag{5}$$

In particular, the experiments yield the static characteristics of the plant

$$\begin{aligned} \hat{\mathbf{K}}_s &= \lim_{t\to\infty} \hat{\mathbf{H}}(t) \\ \bar{\mathbf{K}} &= \lim_{t\to\infty} \bar{\mathbf{H}}(t). \end{aligned} \tag{6}$$

Hence,

$$\mathbf{K}_s = \hat{\mathbf{K}}_s + \delta\mathbf{K}_s \quad \text{with} \quad |\delta\mathbf{K}_s| \leqq \bar{\mathbf{K}} \tag{7}$$

holds.

Tuning by means of the static plant model

At first we assume that we have only measured the final values of the step responses. Then we only know $\hat{\mathbf{K}}_s$ and $\bar{\mathbf{K}}$ in eqn (7). According to Theorem 10.1 the measurement error $\bar{\mathbf{K}}$ may not be arbitrarily large.

Corollary 10.1. A solution to the tuning problem exists only if the measurement error satisfies the condition (10.1-24).

If this condition is satisfied, Theorem 10.2 ensures that the stability of the closed-loop original system can be ensured in all steps of the following tuning algorithm.

Algorithm 1

(1) Make experiments with the plant to measure the static reinforcement $\hat{\mathbf{K}}_s$. Estimate an upper bound $\bar{\mathbf{K}}$ of the uncertainty $\delta\mathbf{K}_s$.

(2) Check the condition (10.1-24). If $\hat{\mathbf{K}}_s^{-1}$ does not exist or if (10.1-24) is violated then there exists no PI-controller for the plant or the experiment has been too coarse.

(3) Use the controller matrices (10.1-28) where $\hat{\mathbf{K}}_I$ is given by (10.1-35) and $\hat{\mathbf{K}}_P$ can be chosen arbitrarily. Use small values of α and β and connect the controller (10.1-4) to the plant.

(4) Make experiments with the closed-loop system. Increase α and β slowly with the controller attached to the plant until a satisfactory transient behaviour is received.

$\hat{\mathbf{K}}_P$ can, in principle, be chosen arbitrarily. Since $\hat{\mathbf{K}}_s$ is the only information of the plant performance

$$\hat{\mathbf{K}}_P = \hat{\mathbf{K}}_s^{-1} \tag{8}$$

is a reasonable controller matrix.
Alternatively, $\hat{\mathbf{K}}_P$ can be fixed in accordance with the tangents at the step responses for $t = 0$

$$\hat{\mathbf{K}}_P = \dot{\hat{\mathbf{H}}}(0)^{-1}$$

if $\dot{\hat{\mathbf{H}}}(0)$ can be reliably measured and is invertible.

The dynamic plant model received by the experiments

In our further investigations we want to utilise the information about the dynamical behaviour of the plant as it is received by the experiments described above. That is, we know $\hat{\mathbf{H}}(t)$ and $\overline{\mathbf{H}}(t)$ from eqns (4) and (5). Since the plant is stable, the step response reaches its static value $\mathbf{K}_s$ asymptotically. This property is not reproduced by eqn (5) because $\overline{\mathbf{K}} = \overline{\mathbf{H}}(\infty) > \mathbf{0}$ admits a never ending movement of the system. Therefore, we assume to know a time constant T_2 that characterises how the system approaches its steady state for large time, ie

$$|\mathbf{H}(t) - \mathbf{K}_s| \leqq |\mathbf{H}(T_1) - \mathbf{K}_s| \exp\left(-(t - T_1)/T_2\right) \quad \text{for} \quad t > T_1 . \tag{9}$$

T_1 and T_2 describe the least possible stability degree of the plant. They might be determined by physical considerations of the plant. Then, the set of representatives of the original plant contains all systems (1) whose step response matrix $\mathbf{H}(t)$ satisfies the eqns (5) and (9) for given $\hat{\mathbf{H}}(t)$.

It is impossible to determine the impulse response matrix $\mathbf{G}(t)$ from the bounds of the step response matrix without making further strong assumptions because the information about the plant used so far does not prevent the step response from oscillating quickly within the band of $\mathbf{H}(t)$. Therefore, we restrict our further considerations on multivariable I-controllers for which the given information suffices. The impossibility to determine $\mathbf{G}(t)$ distinguishes the tuning problem from the robustness analysis of Section 10.2.

In order to set up a model of the form (5.3-23), (5.3-27), we decompose the closed-loop system in the following way. From eqn (10.1-4) with $\mathbf{K}_P = \mathbf{0}$ we get

$$\mathbf{u}(t) = \int_0^t \mathbf{K}_I\left(\mathbf{v}(\tau) - \mathbf{y}(\tau)\right) \mathrm{d}\tau = \mathbf{I} * \tilde{\mathbf{u}}(t) \tag{10}$$

with

$$\tilde{\mathbf{u}}(t) = \mathbf{K}_I\left(\mathbf{v}(t) - \mathbf{y}(t)\right). \tag{11}$$

Eqn (10.1-2) yields (for $\mathbf{d}(t) = \mathbf{0}$ and $\mathbf{x}_0 = \mathbf{0}$)

$$\begin{aligned}\mathbf{y}(t) &= \mathbf{G}(t) * \mathbf{u}(t) = \mathbf{H}(t) * \tilde{\mathbf{u}}(t) \\ &= \hat{\mathbf{H}}(t) * \tilde{\mathbf{u}}(t) + \left(\delta\mathbf{H}(t) - \delta\mathbf{K}_s\right) * \tilde{\mathbf{u}}(t) + \mathbf{I} * \delta\mathbf{K}_s\tilde{\mathbf{u}}(t).\end{aligned}$$

Using

$$\begin{aligned}\mathbf{f}_d(t) &= \left(\delta\mathbf{H}(t) - \delta\mathbf{K}_s\right) * \tilde{\mathbf{u}}(t) \\ \mathbf{f}_s(t) &= \delta\mathbf{K}_s\tilde{\mathbf{u}}(t)\end{aligned} \tag{12}$$

we have

$$\begin{aligned}\mathbf{y}(t) &= \hat{\mathbf{H}}(t) * \tilde{\mathbf{u}}(t) + \mathbf{f}_d(t) + \mathbf{I} * \mathbf{f}_s(t) \\ \mathbf{z}(t) &= \tilde{\mathbf{u}}(t)\end{aligned} \tag{13}$$

as approximate model of the plant (Fig. 10.9). Eqns (12), (13) and

$$\mathbf{s}(t) = \begin{pmatrix}\mathbf{f}_d(t) \\ \mathbf{f}_s(t)\end{pmatrix} \tag{14}$$

constitute a model of the form prescribed by eqns (5.3-23) and (5.3-27), where we made use of the freedom in choosing the model structure in order to get an uncertainty model whose upper bound $\mathbf{V}_2(t)$ can be evaluated by the available information. Due to eqns (11) and (14), $\mathbf{V}_2(t)$ in

$$|\mathbf{s}(t)| \leqq \mathbf{V}_2(t) * |\mathbf{z}(t)| \tag{15}$$

has to satisfy the relation

$$\mathbf{V}_2(t) \geqq \begin{pmatrix}|\delta\mathbf{H}(t) - \delta\mathbf{K}_s| \\ |\delta\mathbf{K}_s|\ \delta(t)\end{pmatrix}.$$

For the first row $\mathbf{V}_{21}(t)$ of $\mathbf{V}_2(t)$ we have, using eqn (9),

$$|\delta\mathbf{H}(t) - \delta\mathbf{K}_s| \leqq \bar{\mathbf{H}}(t) + \bar{\mathbf{K}} = \mathbf{V}_{21}(t) \quad \text{for} \quad 0 \leqq t < T_1 \tag{16}$$

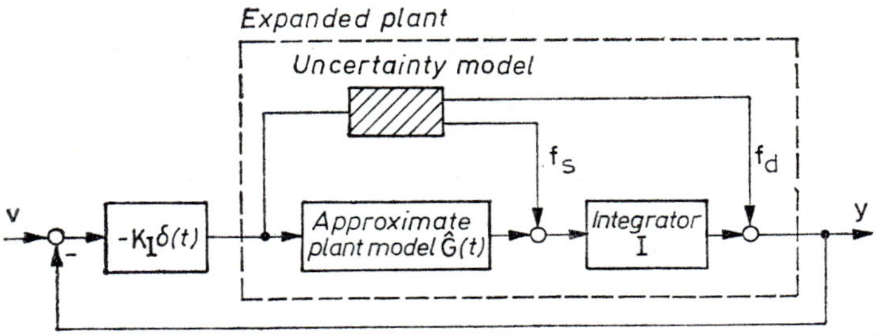

Fig. 10.9 Closed-loop original system (11), (12), (13) (the indicated matrices describe the impulse response of the corresponding blocks)

and

$$\begin{aligned}|\delta\mathbf{H}(t) - \delta\mathbf{K}_s| &\leqq |\mathbf{H}(t) - \mathbf{K}_s + \hat{\mathbf{H}}(t) - \hat{\mathbf{K}}_s| \\ &\leqq |\mathbf{H}(t) - \mathbf{K}_s| + |\hat{\mathbf{H}}(t) - \hat{\mathbf{K}}_s| \\ &\leqq |\mathbf{H}(T_1) - \mathbf{K}_s| \exp\left(-(t - T_1)/T_2\right) + |\hat{\mathbf{H}}(t) - \hat{\mathbf{K}}_s| \\ &= \mathbf{V}_{21}(t) \qquad \text{for} \qquad t \geqq T_1 . \end{aligned} \tag{17}$$

Note that there is some candidate of the original plant for which all relations hold with the equality sign. Because of eqn (6) we get

$$\mathbf{V}_2 = \begin{pmatrix} \mathbf{V}_{21}(t) \\ \overline{\mathbf{K}}\delta(t) \end{pmatrix}. \tag{18}$$

Analysis of the closed-loop system

The resulting closed-loop system (11), (13), (14), (15), (18) can be analysed in the way described in Sections 8.3 and 8.4.

(1) The closed-loop approximate model (11), (13) is given by

$$\begin{aligned}\mathbf{y}(t) &= \tilde{\mathbf{G}}_1 * \hat{\mathbf{H}}\mathbf{K}_\mathrm{I} * \mathbf{v} + \tilde{\mathbf{G}}_1 * \mathbf{f}_\mathrm{d} + \tilde{\mathbf{G}}_1 * \mathbf{I} * \mathbf{f}_\mathrm{s} \\ \tilde{\mathbf{u}}(t) &= \tilde{\mathbf{G}}_2\mathbf{K}_\mathrm{I} * \mathbf{v} - \tilde{\mathbf{G}}_2\mathbf{K}_\mathrm{I} * \mathbf{f}_\mathrm{d} - \tilde{\mathbf{G}}_2\mathbf{K}_\mathrm{I} * \mathbf{I} * \mathbf{f}_\mathrm{s}\end{aligned} \tag{19}$$

with

$$\begin{aligned}\tilde{\mathbf{G}}_1(t) &= \delta(t)\,\mathrm{I} - \hat{\mathbf{H}}(t)\,\mathbf{K}_\mathrm{I} * \tilde{\mathbf{G}}_1(t) \\ \tilde{\mathbf{G}}_2(t) &= \delta(t)\,\mathrm{I} - \mathbf{K}_\mathrm{I}\hat{\mathbf{H}}(t) * \tilde{\mathbf{G}}_2(t) .\end{aligned} \tag{20}$$

By using $\mathbf{K}_\mathrm{I}$ from eqns (10.1-28), (10.1-35) the system (19) can be made stable by using small α.

(2) The closed-loop original system is stable if

$$\lambda_\mathrm{p}\left[\int_0^\infty |\tilde{\mathbf{G}}_2\mathbf{K}_\mathrm{I}|\,\mathrm{d}t \cdot \int_0^\infty \mathbf{V}_{21}\,\mathrm{d}t + \int_0^\infty |\tilde{\mathbf{G}}_2\mathbf{K}_\mathrm{I} * \mathbf{I}|\,\mathrm{d}t \cdot \overline{\mathbf{K}}\right] < 1 \tag{21}$$

is satisfied. It can be proved that eqn (21) holds for sufficiently small α provided that the inequality (10.1-22) is satisfied. Instead of the rather lengthy proof we make the following explanation. Look at Fig. 10.9, ignore the uncertainty model and excite the closed-loop approximate model at $\mathbf{f}_\mathrm{s}$ with step functions. As the system is stable, the system becomes quiet if $\mathbf{u} = -\hat{\mathbf{K}}_\mathrm{s}^{-1}\mathbf{f}_\mathrm{s}$. If the system behaviour is slow enough, ie $\alpha \leqq \bar{\alpha}$ for some $\bar{\alpha}$, then

$$\int_0^\infty |\tilde{\mathbf{G}}_2\mathbf{K}_\mathrm{I} * \mathbf{I}|\,\mathrm{d}t = \hat{\mathbf{K}}_\mathrm{s}^{-1} \quad \text{for} \quad \alpha \leqq \bar{\alpha}. \tag{22}$$

A similar consideration concerning the input $\mathbf{f}_\mathrm{d}$ yields

$$\int_0^\infty |\tilde{\mathbf{G}}_2\mathbf{K}_\mathrm{I}|\,\mathrm{d}t \xrightarrow[\alpha\to 0]{} 0. \tag{23}$$

Hence, eqn (21) can be satisfied by choosing a sufficiently small α.

(3) The trajectory of the closed-loop original system can be approximated by

$$\hat{\mathbf{y}}(t) = \tilde{\mathbf{G}}_1 * \hat{\mathbf{H}}\mathbf{K}_\mathrm{I} * \mathbf{v} \tag{24}$$

where the approximation error is bounded by

$$|\mathbf{y}(t) - \hat{\mathbf{y}}(t)| \leq |\tilde{\mathbf{G}}_1 \; \tilde{\mathbf{G}}_1| * \mathbf{I} * \mathbf{V}_2 * \tilde{\mathbf{G}} * |\tilde{\mathbf{G}}_2\mathbf{K}_\mathrm{I}| * |\mathbf{v}| \tag{25}$$

with

$$\tilde{\mathbf{G}}(t) = \delta(t)\,\mathbf{I} + |\tilde{\mathbf{G}}_2\mathbf{K}_\mathrm{I} \; \tilde{\mathbf{G}}_2\mathbf{K}_\mathrm{I} * \mathbf{I}| * \mathbf{V}_2 * \tilde{\mathbf{G}}. \tag{26}$$

As step commands (10.1-5) are considered, the tolerance band can be reduced as described in eqn (10.2-13). Note that the width of the tolerance band increases if the scalar α is made larger.

Tuning by means of the dynamic plant model

The following tuning algorithm exploits all the information that is received by measuring the whole step responses:

Algorithm 2

(1) Make experiments with the plant in order to determine $\hat{\mathbf{H}}(t)$ and $\overline{\mathbf{H}}(t)$ of an envelope of the step responses. Evaluate T_1 and T_2 so that eqn (9) holds.

(2) Check the condition (10.1-22). If $\hat{\mathbf{K}}_\mathrm{s}^{-1}$ does not exist or if (10.1-22) is violated then this tuning algorithm cannot be applied or the experiment has been too coarse.

(3) Use an I-controller with controller matrix $\mathbf{K}_\mathrm{I}$ from eqns (10.1-28), (10.1-35). Determine α such that the stability condition (21) is satisfied and the I/O-behaviour is satisfactory.

(4) Apply the controller to the plant. The transient behaviour may be improved by changing the tuning parameter α or using an additional P-controller whose gain is initially very small.

In comparison with Algorithm 1, this algorithm enables the design engineer to completely determine the controller matrix $\mathbf{K}_\mathrm{I}$. Step 3 answers the question how to choose the tuning parameter so as to ensure the stability of the closed-loop system even in the first closed-loop experiment. Moreover, the I/O-behaviour of the closed-loop system can be evaluated. This provides the possibility to search an initial value of the tuning parameter for which the closed-loop original system is not only theoretically stable but has a substantial stability degree. Finally, it is worth mentioning that the Algorithm 2 does not necessitate other experiments with the plant but that it takes advantage of more information received by such experiments.

In order to evaluate the *computational requirements* of the proposed algorithm note that the functions $\hat{\mathbf{H}}(t)$ and $\bar{\mathbf{H}}(t)$ may be available in form of analytic terms, graphs or data files. In any case, the application of the tuning algorithm necessitates very easy calculations or graphical manipulations (Fig. 10.10). The conditions (10.1-22) and (21) can be checked by determining the Perron root according to the simple algorithm given in Appendix 1. All convolutions can be carried out with data columns using numerical integration methods. Eqn (20) can be resolved by a transformation into its discrete-time counterpart, which may be solved recursively for the given data series.

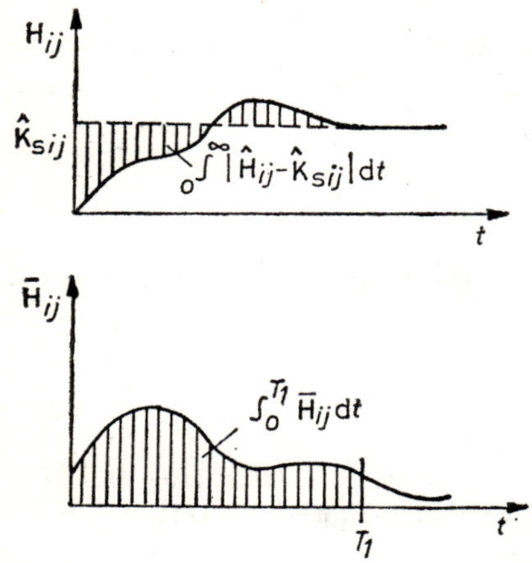

Fig. 10.10 Graphical determination of the main terms occuring in $\int_0^\infty \mathbf{V}_{21}(t)\,dt$

Example: Control of a steam boiler

To demonstrate the tuning algorithm let us study the design of a multivariable I-controller for a once-through forced-flow steam generator. Although from the engineering point of view such a plant is well known, precise quantitative models of the form (10.1-1) are not available. We have to regulate the steam pressure p and temperature T subject to unknown load changes. Actuators are valves for fresh water supply $\dot{Q}_w$ and air supply $\dot{Q}_a$.

In accordance with the engineering experience the following assumption can be made:

- Near the operating point, the I/O-behaviour of the boiler can be described by some linear model.
- Due to an already installed controller the plant is asymptotically stable.
- The control problem can be solved by a PI-controller.

Then Algorithm 2 proceeds as follows:

(1) The step responses of the plant are measured. To reduce the uncertainties we have performed the experiments three times. After selecting large stochastic measurement errors the tolerance bands shown in Fig. 10.11 have been received. The middle of the bands is used as $\hat{\mathbf{H}}(t)$ and half the width as $\overline{\mathbf{H}}(t)$. T_1 and T_2 are given as $T_1 = 1800$ s, $T_2 = 100$ s.

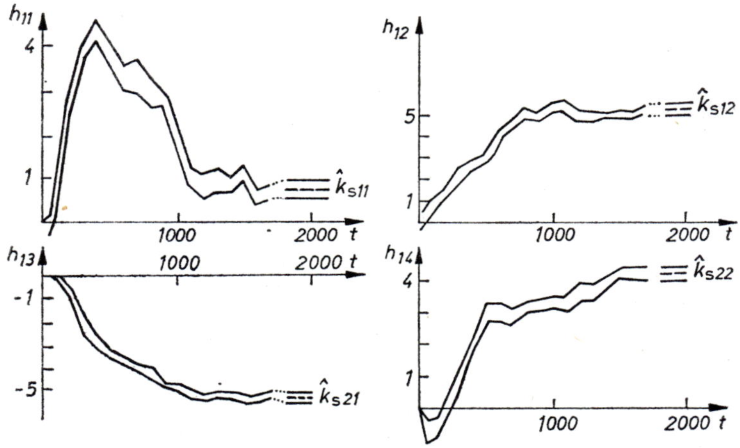

Fig. 10.11 Tolerance bands of the step response matrix of the boiler as received by experiments

(2) The condition (10.1-22) is satisfied for

$$\hat{\mathbf{K}}_s = \begin{pmatrix} 0.9 & 5.3 \\ -5.2 & 4.1 \end{pmatrix}, \qquad \overline{\mathbf{K}} = \begin{pmatrix} 0.2 & 0.3 \\ 0.2 & 0.3 \end{pmatrix}.$$

(3) The controller matrix is given by

$$\mathbf{K}_I = \alpha \begin{pmatrix} 0.129 & -0.167 \\ 0.164 & 0.028 \end{pmatrix}.$$

For $\alpha = 0.11$ the stability condition (21) is satisfied and the envelopes of the closed-loop system suggest that all trajectories they include have a satisfactory behaviour (Fig. 10.12). Therefore the I-controller can be implemented.

(4) Improvements of the closed-loop original system are possible by increasing α or by using an additional P-controller with initially small tuning factor β.

Our results concerning tuning regulation show that no exact identification of the process is necessary if the controller has to satisfy moderate design objectives. The approximately measured step responses provide

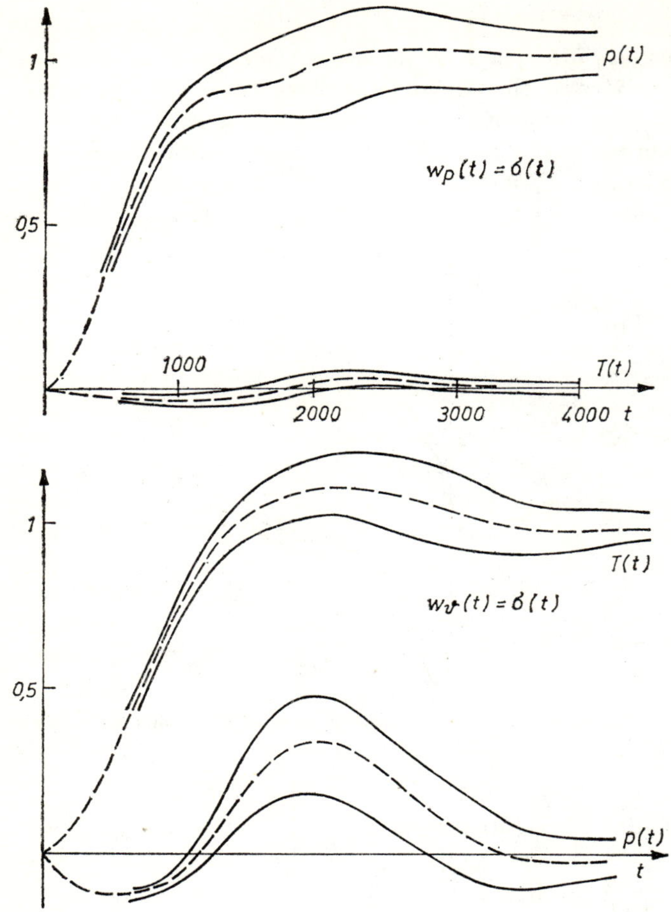

Fig. 10.12 Tolerance bands of the command step responses of the closed-loop original system

enough information about the plant to construct an I- or PI-controller that ensures the robust stability of the closed-loop original system and, moreover, a suitable I/O-behaviour.

As the tuning rules rely heavily on the robustness of feedback controllers the methods for quantitatively assessing the robustness provide a means for elaborating such rules. However, the whole tuning process includes the implementation of the controller, the exploitation of the information about the closed-loop original system received for the controller in reality, and the on-plant improvement of the controller parameters. Therefore, the off-line results received by the help of the model have only to ensure a reasonable 'first choice' of the controller together with a way for on-plant modifications. This contrasts with the design method based on the robustness analysis of Section 10.2. There, the aim is to get a completely specified controller that, if attached to the plant, meets the control objectives at once.

Bibliographical notes

Necessary stability conditions for PI-control systems and their implications on the accuracy of the plant model have been investigated independently by Lunze (1982b, 1985), Morari (1983), and Grosdidier et al. (1984).

The tuning problem for single-loop systems was solved several decades ago, eg by Ziegler and Nichols (1942). Corresponding tuning rules can be found in textbooks or manuals, eg in (Reinisch 1979). The extension to multivariable control has been initiated by Davison (1976b), who assumed an exact measurement of the static reinforcement $\mathbf{K}_s$ of the plant. Results along this line have been published by Penttinen and Koivo (1980), Davison et al. (1980), Porter (1981, 1982), Owens and Chotai (1981) and others, some of whom explained the possibility of including proportional controller actions. The generalisation to uncertain information about $\mathbf{K}_s$ and the Algorithm 2 have been described by Lunze (1982d, 1985). Easily computable off-line estimates of $\bar{\alpha}$ so that the closed-loop stability is ensured at least for $\bar{\alpha} \in (0, \alpha)$ are given by Lunze (1985) and Owens and Chotai (1986). An extension to infinite-dimensional systems has been described by Logemann and Owens (1987).

The examples used in this chapter are explained in more detail in (Lunze, 1984a), (Lunze and v. Kurnatowski, 1987) and (Lunze, 1985), respectively. The experiments for the boiler control problem of Section 10.3 have been described in (Wilfert et al. 1978).

11 ROBUST DECENTRALISED CONTROL

11.1 Robustness issues of decentralised control

Multivariable controllers are used if strong cross-couplings within the plant have to be compensated for. However, there are practical reasons that make restrictions on the controller structure necessary or reasonable. In the extreme case the controller is split into several local feedbacks and represents a *decentralised controller*.

Decentralised controllers are encountered in different situations. On the one hand, in many frequency domain design techniques decoupling compensators are used to decompose the control problem of an m-input m-output plant into m independent single-loop problems. Then, the expanded plant, which includes the decoupling compensator, is controlled by a decentralised controller. The result, however, represents a multivariable feedback (Fig. 11.1 (a)). On the other hand, large-scale systems necessitate controller structure constraints because, for example, the on-line information transmission between the plant and the controller has to be reduced or the reliability of the control scheme must be improved (Fig. 11.1 (b)). In contrast to multivariable controllers with voluntarily introduced constraints, the control stations have often more than one control variable.

Both situations pose the same theoretical problems. The inputs $\tilde{\mathbf{u}}$ or $\mathbf{u}$ and output $\mathbf{y}$ have to be decomposed into subvectors $\mathbf{u}_i$ and $\mathbf{y}_i$ so that the plant is described by

$$\begin{pmatrix} \mathbf{y}_1 \\ \cdot \\ \cdot \\ \mathbf{y}_N \end{pmatrix} = \begin{pmatrix} \mathbf{G}_{11} & \mathbf{G}_{12} & \cdots & \mathbf{G}_{1N} \\ \mathbf{G}_{21} & \mathbf{G}_{22} & \cdots & \mathbf{G}_{2N} \\ \cdot & \cdot & \cdots & \cdot \\ \mathbf{G}_{N1} & \mathbf{G}_{N2} & \cdots & \mathbf{G}_{NN} \end{pmatrix} \begin{pmatrix} \mathbf{u}_1 \\ \cdot \\ \cdot \\ \mathbf{u}_N \end{pmatrix}, \quad (\dim \mathbf{u}_i = \dim \mathbf{y}_i = m_i). \tag{1}$$

The controller is restricted to consist of N independent parts

$$\mathbf{u}_i(p) = -\mathbf{K}_i(p)\,\mathbf{y}_i(p) + \mathbf{K}_i(p)\,\mathbf{v}_i(p)\,, \qquad (i = 1, \ldots, N). \tag{2}$$

Besides the robustness against plant uncertainties, decentralised controllers bring about additional robustness problems. As the local controllers are not simultaneously connected to the plant and there is even no given sequence of their implementation, the closed-loop system must exhibit a stable performance with reasonable dynamical properties even if several

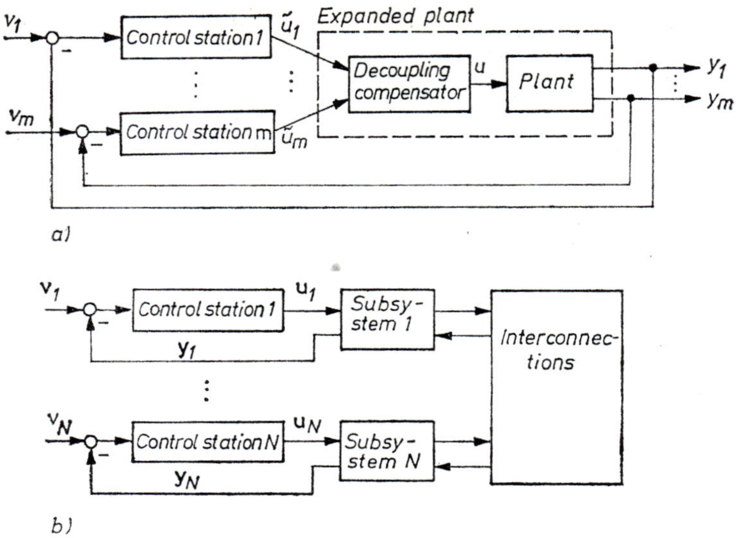

Fig. 11.1 Feedback system with controller structure constraints

(a) Centralised multivariable controller consisting of a decoupling compensator and m independent control loops

(b) Decentralised control of an interconnected system

control stations are missing. This kind of stability robustness has been called 'integrity' in multivariable control and 'connective stability' in decentralised control where the latter is often used with regard to breakdowns of subsystem interconnections rather than of the controller input or output.

In Section 11.2 we are concerned with the question in which way the couplings $\mathbf{G}_{ij}$ must be restricted so that integrity can be ensured. Section 11.3 is devoted to a design technique in which all local controllers are constructed independently for the isolated subsystem $\mathbf{G}_{ii}$ and made robust against the neglected cross-couplings $\mathbf{G}_{ij}$ of the plant.

11.2 Existence of robust decentralised PI-controllers

We consider a linear stable plant (11.1-1) with a decentralised PI-controller

$$\begin{aligned} \dot{\mathbf{x}}_{\mathrm{r}i} &= \mathbf{y}_i - \mathbf{v}_i \\ \mathbf{u}_i &= -e_i \mathbf{K}_{\mathrm{P}i}(\mathbf{y}_i - \mathbf{v}_i) + e_i \mathbf{K}_{\mathrm{I}i}\mathbf{x}_{\mathrm{r}i} \end{aligned} \tag{1}$$

where the scalar $e_i \in \{0, 1\}$ indicates whether the control station i is ($e_i = 1$) or is not ($e_i = 0$) in operation. Obviously,

$$\mathbf{e} = (e_1, \ldots, e_N)' \in \mathscr{E} = \left\{\mathbf{e} \text{ with } e_i \in \{0, 1\}\right\} \tag{2}$$

holds.

We want to investigate under what conditions controllers (1) exist such that the closed-loop system (11.1-1), (1) is I/O-stable for all $\mathbf{e} \in \mathcal{E}$.

To this end we use the existence condition of Lemma 10.1 for PI-controllers from the point of view of a certain subsystem, say subsystem 1. The system controlled by control station 1 includes all other control stations that are in operation. Assuming $\mathbf{v}_i = \mathbf{0}$ we have to determine the static model

$$\mathbf{y}_1 = \overline{\mathbf{K}}_{s11}(\mathbf{e})\,\mathbf{u}_1 . \tag{3}$$

Decomposing the static reinforcement $\mathbf{K}_s$ of the plant (11.1-1)

$$\begin{pmatrix} \mathbf{y}_1 \\ \cdot \\ \cdot \\ \cdot \\ \mathbf{y}_N \end{pmatrix} = \begin{pmatrix} \mathbf{K}_{s11} & \mathbf{K}_{s12} & \cdots & \mathbf{K}_{s1N} \\ \cdot & \cdot & & \cdot \\ \cdot & \cdot & & \cdot \\ \cdot & \cdot & & \cdot \\ \mathbf{K}_{sN1} & \mathbf{K}_{sN2} & \cdots & \mathbf{K}_{sNN} \end{pmatrix} \begin{pmatrix} \mathbf{u}_1 \\ \cdot \\ \cdot \\ \cdot \\ \mathbf{u}_N \end{pmatrix} \tag{4}$$

we have

$$\begin{pmatrix} \mathbf{y}_2 \\ \cdot \\ \cdot \\ \cdot \\ \mathbf{y}_N \end{pmatrix} = \begin{pmatrix} \mathbf{K}_{s22} & \mathbf{K}_{s23} & \cdots & \mathbf{K}_{s2N} \\ \cdot & \cdot & & \cdot \\ \cdot & \cdot & & \cdot \\ \cdot & \cdot & & \cdot \\ \mathbf{K}_{sN2} & \mathbf{K}_{sN3} & \cdots & \mathbf{K}_{sNN} \end{pmatrix} \begin{pmatrix} \mathbf{u}_2 \\ \cdot \\ \cdot \\ \cdot \\ \mathbf{u}_N \end{pmatrix} + \begin{pmatrix} \mathbf{K}_{s21} \\ \cdot \\ \cdot \\ \cdot \\ \mathbf{K}_{sN1} \end{pmatrix} \mathbf{u}_1 . \tag{5}$$

At first, we consider the stable closed-loop system with $e_2 = e_3 = \cdots = e_N = 1$. The equilibrium state is assumed for $\dot{\mathbf{x}}_{ri} = \mathbf{0} = \mathbf{y}_i$. From eqn (5) we get

$$\begin{pmatrix} \mathbf{u}_2 \\ \cdot \\ \cdot \\ \cdot \\ \mathbf{u}_N \end{pmatrix} = - \begin{pmatrix} \mathbf{K}_{s22} & \mathbf{K}_{s23} & \cdots & \mathbf{K}_{s2N} \\ \cdot & \cdot & \cdots & \cdot \\ \mathbf{K}_{sN2} & \mathbf{K}_{sN3} & \cdots & \mathbf{K}_{sNN} \end{pmatrix}^{-1} \begin{pmatrix} \mathbf{K}_{s21} \\ \cdot \\ \cdot \\ \cdot \\ \mathbf{K}_{sN1} \end{pmatrix} \mathbf{u}_1 . \tag{6}$$

Then from eqns (3), (4) we get

$$\overline{\mathbf{K}}_{s11}(\mathbf{e}) = \mathbf{K}_{s11} - \delta\mathbf{K}_{s11}(\mathbf{e}) \tag{7}$$

with

$$\delta\mathbf{K}_{s11}(0, 1, 1, \ldots, 1) = -(\mathbf{K}_{s12} \ldots \mathbf{K}_{s1N}) \begin{pmatrix} \mathbf{K}_{s22} & \cdots & \mathbf{K}_{s2N} \\ \cdot & \cdots & \cdot \\ \mathbf{K}_{sN2} & \cdots & \mathbf{K}_{sNN} \end{pmatrix}^{-1} \begin{pmatrix} \mathbf{K}_{s21} \\ \cdot \\ \cdot \\ \mathbf{K}_{sN1} \end{pmatrix} . \tag{8}$$

If we have $e_i = 0$ for some indices i, $\mathbf{u}_i = \mathbf{0}$ and $\mathbf{y}_i \neq \mathbf{0}$ hold. We have to delete the i-th rows and columns in eqn (5) before proceeding in the same way as above. Then, in eqn (8) the elements with at least one index 'i' are missing. This result can be brought into the more convenient form

$$\begin{aligned}&\delta\mathbf{K}_{s11}(e_1, e_2, \ldots, e_N)\\ &= -(\mathbf{K}_{s12}\ldots\mathbf{K}_{s1N})\begin{pmatrix}\mathbf{K}_{s22} & e_2\mathbf{K}_{s23} & \ldots & e_2\mathbf{K}_{s2N}\\ e_3\mathbf{K}_{s32} & \mathbf{K}_{s33} & \ldots & e_3\mathbf{K}_{s3N}\\ \ldots & \ldots & \ldots & \ldots\\ e_N\mathbf{K}_{sN2} & e_N\mathbf{K}_{sN3} & \ldots & \mathbf{K}_{sNN}\end{pmatrix}^{-1}\begin{pmatrix}e_2\mathbf{K}_{s21}\\ e_s\mathbf{K}_{s31}\\ \\ e_N\mathbf{K}_{sN1}\end{pmatrix}. \qquad (9)\end{aligned}$$

$\overline{\mathbf{K}}_{s11}(\mathbf{e})$ is given by eqns (7) and (9). Note that it depends on e_i but not on the numerical values of the controller parameters of the control stations $2, \ldots, N$.

We have to choose the controller 1 so as to satisfy the negative feedback condition (10.1-13)

$$\det\left(\overline{\mathbf{K}}_{s11}(\mathbf{e})\,\mathbf{K}_{I1}\right) > 0 \quad \text{for all} \quad \mathbf{e} \in \mathscr{E}. \qquad (10)$$

By assuming that the signs of $\mathbf{u}_1$ and $\mathbf{y}_1$ are defined so as to ensure $\det(\mathbf{K}_{s11}) > 0$, eqn (10) for $\mathbf{e} = \mathbf{0}$ claims $\det(\mathbf{K}_{I1}) > 0$. Hence, eqn (10) is equivalent to $\det\left(\overline{\mathbf{K}}_{s11}(\mathbf{e})\right) = \det\left(\mathbf{K}_{s11} + \delta\mathbf{K}_{s11}(\mathbf{e})\right) > 0$, which can be reformulated by means of the determinant identity (10.1-11) as

$$\det\begin{pmatrix}\mathbf{K}_{s11} & \mathbf{K}_{s12} & \mathbf{K}_{s13} & \ldots & \mathbf{K}_{s1N}\\ e_2\mathbf{K}_{s21} & \mathbf{K}_{s22} & e_2\mathbf{K}_{s23} & \ldots & e_2\mathbf{K}_{s2N}\\ \ldots & \ldots & \ldots & \ldots & \ldots\\ e_N\mathbf{K}_{sN1} & e_N\mathbf{K}_{sN2} & e_N\mathbf{K}_{sN3} & \ldots & \mathbf{K}_{sNN}\end{pmatrix} > 0 \quad \text{for all} \quad \mathbf{e} \in \mathscr{E}. \qquad (11)$$

As corresponding relations have to hold for the other control stations, eqn (11) has to be satisfied with $e_1\mathbf{K}_{s1i}$ $(i = 2, \ldots, N)$ in the first row too.

Theorem 11.1 Consider a stable linear system with static characteristic (4). Assume that $\det(\mathbf{K}_{sii}) > 0$. Then a necessary condition for the existence of decentralised PI-controllers (1) that ensure closed-loop stability for all $\mathbf{e} \in \mathscr{E}$ is given by the requirement that the determinants of $\mathbf{K}_s$ and all matrices that result from $\mathbf{K}_s$ by deleting corresponding block rows and block columns have to be positive.

For systems with $\dim \mathbf{y}_i = \dim \mathbf{u}_i = 1$ Theorem 11.1 demands that all principal minors of $\mathbf{K}_s$ are positive. This condition is sufficient as well because the controller (1) with $k_{Pi} = 0$ and

$$k_{Ii} = -\alpha_i k_{sii}^{-1} \qquad (12)$$

satisfies the condition (10.1-33) in connection with $\bar{k}_{sii}(\mathbf{e})$ for all $\mathbf{e} \in \mathscr{E}$ whenever Theorem 11.1 or, equivalently, $\det\left(\bar{k}_{sii}(\mathbf{e})\,k_{Ii}\right) = \bar{k}_{sii}(\mathbf{e})\,k_{Ii} > 0$ holds (cf eqn (11)).

Theorem 11.2 Consider a stable linear system with $\dim \mathbf{y}_i = \dim \mathbf{u}_i = 1$. Assume that $k_{sii} > 0$ $(i = 1, \ldots, N)$. Then a necessary and sufficient condition for the existence of decentralised PI-controllers (1) that ensure closed-loop stability for all $\mathbf{e} \in \mathcal{E}$ is given by the requirement that all principal minors of $\mathbf{K}_s$ are positive.

The conditions of Theorems 11.1 and 11.2 describe the necessary autonomy of the subsystems. They are satisfied at least for systems with low static cross-couplings $\mathbf{K}_{sij}$ $(j \neq i)$. More details about matrices with positive principal minors can be found in (Berman and Plemmons, 1978, Chapter 10).

Sufficient conditions for systems with dim $\mathbf{y}_i$ = dim $\mathbf{u}_i > 1$ can be received by means of Theorem 10.1 if $\sigma\overline{\mathbf{K}}_{sii}(\mathbf{e})$ is bounded from above by some matrix $\overline{\mathbf{K}}$. We leave this as an exercise and go on with an example.

Existence of decentralised voltage controllers

In large interconnected electric power systems the bus voltages $V_{ei} = y_i$ of the feeding nodes are to be maintained at optimum values. This control task should be carried out by a decentralised PI-controller which acts on the command inputs $V_{ci} = u_i$ of the terminal voltage controller of the synchronous generators and has to maintain the closed-loop stability even if some feeding nodes are not controlled at times (Fig. 11.2).

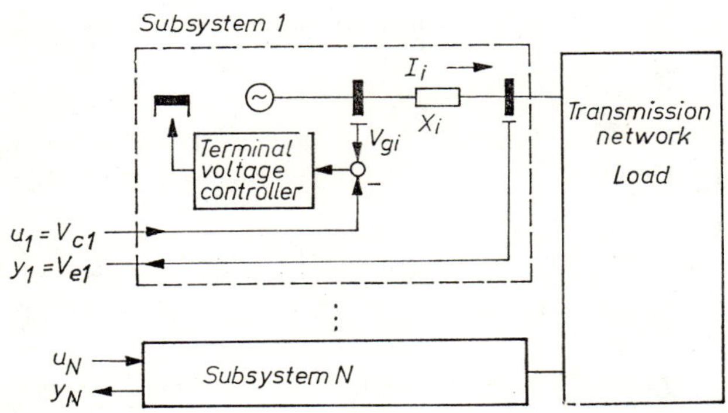

Fig. 11.2 Electric power system as the plant of the decentralised voltage control problem

At first, we consider the plant under the ideal assumption that we can split the system completely into a pure active and a pure reactive power system. Then, all admittances are reactive. To set up the static plant model we use the mesh theorem $V_{gi} = I_i X_i + V_{ei}$ where I_i is the reactive current and X_i the admittance of the transformer between the synchronous machine and the feeding node. In control engineering terms we have

$$u_i = I_i X_i + y_i. \tag{13}$$

The transmission network including the load can be described by

$$\begin{pmatrix} I_1 \\ \cdot \\ \cdot \\ I_N \end{pmatrix} = \mathbf{Y} \begin{pmatrix} V_{e1} \\ \cdot \\ \cdot \\ V_{eN} \end{pmatrix} \tag{14}$$

with $\mathbf{Y}$ denoting the admittance matrix. Eqns (13) and (14) yield $\mathbf{u} = (\mathbf{I} + \operatorname{diag} X_i \mathbf{Y})\,\mathbf{y}$. Hence, we have

$$\mathbf{K}_s^{-1} = \mathbf{I} + \operatorname{diag} X_i\, \mathbf{Y}. \tag{15}$$

The admittance matrix $\mathbf{Y} = (y_{ij})$ has the properties

$$y_{ij} \leqq 0 \quad \text{for} \quad i \neq j, \qquad y_{ii} > \sum_{j \neq i} |y_{ij}| \geqq 0. \tag{16}$$

Hence, $\mathbf{Y}$ is an M-matrix and so does the matrix $\mathbf{K}_s^{-1}$ (cf Appendix 1). By Theorem A1.7, the plant satisfies the existence condition of Theorem 11.2.

Corollary 11.1 For the 'ideal' reactive power system, in which interconnections between changes in the real and reactive power flows are ignored, decentralised PI-controllers (1) that ensure the closed-loop stability for all $\mathbf{e} \in \mathscr{E}$ do exist.

The existence is proved for an arbitrarily large number N of subsystems and is not affected by switching transmission lines on or off, or by changing the generator parameters.

If we release our 'ideal' assumptions, the matrix $\mathbf{K}_s^{-1}$ need not be an M-matrix. However, since the interactions of real and reactive power flow changes are actually very small, $\mathbf{K}_s^{-1}$ can be approximated by some M-matrix $\mathbf{A} = \hat{\mathbf{K}}_s^{-1}$

$$\mathbf{K}_s = \hat{\mathbf{K}}_s + \delta \mathbf{K}_s. \tag{17}$$

Due to M-matrix properties the following lemma can be proved.

Lemma 11.1 Assume that the matrix $\mathbf{K}_s$ can be represented by eqn (17) where $\hat{\mathbf{K}}_s^{-1} = \mathbf{A}$ is an M-matrix. Then all principle minors of $\mathbf{K}_s$ are positive if

$$\lambda_p[|\delta \mathbf{K}_s|\, \tilde{\mathbf{A}}] < 1 \tag{18}$$

holds where $\tilde{\mathbf{A}} = (\tilde{a}_{ij})$ can be determined from $\mathbf{A} = (a_{ij})$ and $\mathbf{K}_s = (k_{sij})$ by $\tilde{a}_{ii} = a_{ii}$ and $\tilde{a}_{ij} = k_{sij}/(k_{sii}k_{sjj} - k_{sij}k_{sji})$ for $i \neq j$.

For the matrix given in Table 11.1, which has been received by load-flow calculations of a large power system with 20 feeding nodes, $\mathbf{K}_s^{-1} = (\bar{k}_{ij})$ has 26 positive off-diagonal elements, eg $\bar{k}_{1,19} = 0.02099$, $\bar{k}_{1,20} = 0.0629$, $\bar{k}_{2,12} = 0.0215$. Replacing these elements by zero, an M-matrix $\mathbf{A}$ can be found. Then, eqn (18) yields $\lambda_p[|\delta \mathbf{K}_s|\, \tilde{\mathbf{A}}] = 0.545 < 1$ proving the existence of robust decentralised controllers for the real power system.

Table 11.1 Matrix $\mathbf{K}_s$ of the power system

0.5236	0.1237	0,0945	0.0368	0.0645	0.0416	0.0062	0.0055	0.0125	0.0099
0.1210	0.5012	0.0810	0.0470	0.040	0.1052	0.0060	0.0051	0.0093	0.0072
0.1656	0.1447	0.4373	0.0242	0.080	0.0376	0.0060	0.0052	0.0125	0.0099
0.0708	0.0968	0.0287	0.5420	0.026	0.111	0.0108	0.0009	0.0125	0.0097
0.1260	0.0798	0.0898	0.0250	0.3753	0.0292	0.0114	0.0101	0.0268	0.0020
0.0832	0.2277	0.0455	0.1155	0.0317	0.3118	0.0112	0.0093	0.0120	0.0096
0.0273	0.0282	0.0158	0.0245	0.027	0.0242	0.3030	0.0530	0.0120	0.0106
0.0235	0.0236	0.0138	0.0199	0.0236	0.0200	0.0527	0.230	0.0103	0.0090
0.0898	0.0691	0.0532	0.0443	0.1025	0.0414	0.0190	0.0168	0.3713	0.0964
0.1043	0.0789	0.0630	0.0515	0.1154	0.0487	0.0257	0.0220	0.1430	0.3389
0.0325	0.0481	0.0155	0.0620	0.0181	0.0560	0.076	0.0605	0.0080	0.0067
0.0857	0.0746	0.0423	0.0671	0.0670	0.0570	0.0123	0.0108	0.0363	0.0354
0.0401	0.0527	0.0161	0.0748	0.0152	0.0604	0.0083	0.0068	0.0073	0.0057
0.0794	0.0561	0.0532	0.0259	0.1065	0.0280	0.0320	0.0292	0.0370	0.0292
0.0794	0.0561	0.0532	0.0259	0.1065	0.0280	0.0320	0,0292	0.0370	0.0292
0.1955	0.1160	0.0636	0.1146	0.0550	0.1019	0.0143	0.0122	0.0213	0.0165
0.0584	0.1028	0.0296	0.0556	0.0314	0.0960	0.0157	0.0130	0.0143	0.0124
0.1188	0.1773	0.0501	0.2476	0.043	0.2156	0.0273	0.0222	0.0190	0.0150
0.1054	0.0818	0.0570	0.0645	0.1005	0.0580	0.0150	0.0132	0.0387	0.0320
0.0760	0.0636	0.0397	0.0550	0.0645	0.0470	0.0120	0.0105	0.041	0.0550

It is worth mentioning that checking the condition of Theorem 11.2 represents a mathematical problem of its own. We have solved it for a large system by exploiting a robustness property according to which not only the ideal approximate but also the real system meets the existence condition.

The controller design

As mentioned in the derivation of Theorem 11.2 decentralised PI-controllers (1), (12) can be found by using the low-gain feedback technique similar to that explained in Section 10.1 whenever the condition of Theorem 11.2 is satisfied. This or more comfortable design procedures, which have been elaborated in the theory of decentralised control, exploit the complete knowledge about the whole plant. In the next section, however, we want to consider a method for designing the control stations independently for the isolated subsystems.

11.3 Design of robust decentralised controllers

If we ignore the couplings among the subsystems, the problem of designing a decentralised controller (11.1-2) falls into N independent design tasks for the N control stations. We want to investigate under what conditions the resulting local feedbacks are compatible in the sense that they satisfy

0.0112	0.0447	0.0235	0.0106	0.0106	0.0596	0.0338	0.0140	0.0143	0.0311
0.0147	0.0210	0.0274	0.0073	0.0073	0.0289	0.0536	0.0174	0.0104	0.0245
0.0085	0.0214	0.0148	0.0122	0.0122	0.0249	0.0272	0.0088	0.0129	0.0270
0.0388	0.040	0.0791	0.0071	0.0071	0.0587	0.0606	0.0496	0.0172	0.0434
0.0110	0.0380	0.0156	0.0272	0.0272	0.0241	0.0326	0.0086	0.0257	0.0496
0.0363	0.0344	0.0664	0.0079	0.0079	0.0486	0.1077	0.0448	0.0150	0.0384
0.1077	0.0167	0.0197	0.0193	0.0193	0.0147	0.0380	0.0122	0.0090	0.0220
0.0857	0.0143	0.0160	0.0177	0.0177	0.0123	0.0317	0.0100	0.0080	0.0190
0.0180	0.0783	0.0277	0.0367	0.0367	0.0350	0.0560	0.0138	0.0377	0.1203
0.0230	0.1147	0.0323	0.0430	0.0430	0.0393	0.0730	0.0162	0.0460	0.2413
0.4450	0.0167	0.0557	0.0097	0.0097	0.0303	0.0350	0.0369	0.0080	0.0197
0.0180	0.3217	0.0413	0.0190	0.0190	0.0467	0.0583	0.0196	0.0760	0.2177
0.0327	0.0227	0.8190	0.0040	0.0040	0.0430	0.0337	0.0465	0.0103	0.0253
0.0227	0.0423	0.0167	0.2563	0.1260	0.0220	0.0473	0.0092	0.0253	0.0623
0.0227	0.0423	0.0167	0.1260	0.2563	0.0220	0.0473	0.0092	0.0253	0.0623
0.0417	0.0593	0.0990	0.0127	0.0127	0.5223	0.1173	0.0703	0.0283	0.0663
0.0207	0.0313	0.0333	0.0117	0.0117	0.0510	0.5223	0.0203	0.0147	0.0367
0.1157	0.0563	0.247	0.0120	0.0120	0.1510	0.1070	0.7802	0.0253	0.0630
0.0360	0.1697	0.0403	0.0257	0.0257	0.0490	0.0597	0.0194	0.1730	0.1497
0.0157	0.1617	0.0343	0.0207	0.0207	0.0387	0.0503	0.0164	0.0500	0.4353

the design objectives of the overall system under the influence of the cross-couplings within the system.

To ignore the cross-couplings during the design means to use

$$\hat{\mathbf{y}} = \operatorname{diag} \mathbf{G}_{ii}\mathbf{u} \tag{1}$$

as approximate model. A corresponding decomposition of eqn (11.1-1) is (see Fig. 11.3):

$$\begin{aligned} \mathbf{y} &= \operatorname{diag} \mathbf{G}_{ii}\mathbf{u} + \mathbf{s} \\ \mathbf{z} &= \mathbf{u} \end{aligned} \tag{2}$$

and

$$\mathbf{s} = \begin{pmatrix} \mathbf{0} & \mathbf{G}_{12} & \dots & \mathbf{G}_{1N} \\ \mathbf{G}_{21} & \mathbf{0} & \dots & \mathbf{G}_{2N} \\ \dots & \dots & \dots & \dots \\ \mathbf{G}_{N1} & \mathbf{G}_{N2} & \dots & \mathbf{0} \end{pmatrix} \mathbf{z}. \tag{3}$$

The closed-loop approximate model (2), (11.1-2) is

$$\begin{aligned} \mathbf{y} &= \operatorname{diag} (\mathbf{I} + \mathbf{G}_{ii}\mathbf{K}_i)^{-1} \mathbf{G}_{ii}\mathbf{K}_i\mathbf{v} + \operatorname{diag} (\mathbf{I} + \mathbf{G}_{ii}\mathbf{K}_i)^{-1} \mathbf{s} \\ \mathbf{z} &= \operatorname{diag} (\mathbf{I} + \mathbf{K}_i\mathbf{G}_{ii})^{-1} \mathbf{K}_i\mathbf{v} - \operatorname{diag} (\mathbf{I} + \mathbf{K}_i\mathbf{G}_{ii})^{-1} \mathbf{K}_i\mathbf{s}. \end{aligned} \tag{4}$$

If we use an upper bound of the 'uncertainty model' (3)

$$|\mathbf{s}| \leqq \mathbf{V}_2 \, |\mathbf{z}| \tag{5}$$

with

$$\mathbf{V}_2 = \begin{pmatrix} 0 & |\mathbf{G}_{12}| & \dots & |\mathbf{G}_{1N}| \\ \dots & \dots & \dots & \dots \\ |\mathbf{G}_{N1}| & |\mathbf{G}_{N2}| & \dots & 0 \end{pmatrix}, \tag{6}$$

the closed-loop system performance under the influence of the cross-couplings can be easily evaluated by means of the results presented in Sections 8.3 and 8.5.

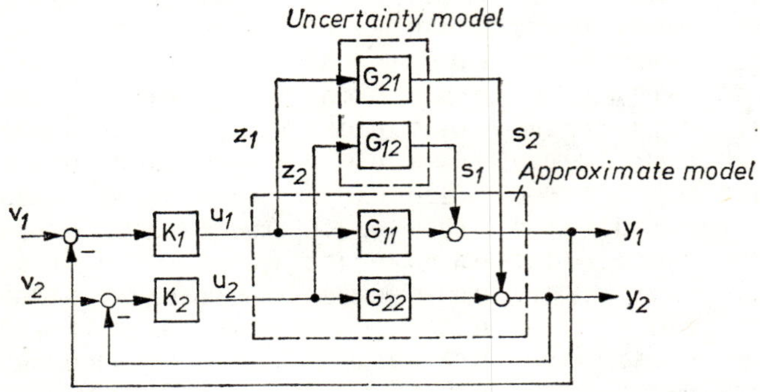

Fig. 11.3 Decomposed decentralised control system with $N = 2$ subsystems

The system is stable if

$$d(p) = \lambda_p[\mathbf{V}_2 \operatorname{diag} |(\mathbf{I} + \mathbf{K}_i\mathbf{G}_{ii})^{-1} \mathbf{K}_i|] < 1, \quad p \in \mathcal{D} \tag{7}$$

holds. This inequality bounds the strength of the cross-couplings so that they may not destabilise the overall system. The stability condition (7) remains valid if the couplings $|\mathbf{G}_{ij}|$ are smaller than $\mathbf{V}_{2ij}$ and even if $\mathbf{s}_i = \mathbf{0}$ holds for some indices i. Moreover, to break the system at some $\mathbf{s}_i$-signal has the same influence on the closed-loop stability as the disconnection of the control station i from the plant (Fig. 11.3). Hence, eqn (7) proves the integrity property of the controller.

Theorem 11.3 Consider a stable linear plant (11.1-1) under decentralised control (11.1-2). Assume that the control stations ensure the stability of the closed-loop subsystems (1), (11.1-2), or, equivalently, (4). If eqn (7) is satisfied then the closed-loop overall system (11.1-1), (11.1-2) is stable and remains stable if cross-couplings within the plant are decreased or broken down or if some control stations are disconnected from the plant.

Note that we have only to check a single condition (7) although we prove the stability under 2^N operating conditions $\mathbf{e} \in \mathcal{E}$.

For $\mathbf{s} = \mathbf{0}$ the closed-loop approximate model (4) splits into

$$\hat{\mathbf{y}}_i = (\mathbf{I} + \mathbf{G}_{ii}\mathbf{K}_i)^{-1} \mathbf{G}_{ii}\mathbf{K}_i\mathbf{v}_i \quad (i = 1, \dots, N) \tag{8}$$

describing the closed-loop isolated subsystems and, approximately, the closed-loop overall system. The model error is bounded by

$$|\mathbf{y} - \hat{\mathbf{y}}| \leqq \text{diag}\,|(\mathbf{I} + \mathbf{G}_{ii}\mathbf{K}_i)^{-1}|\,\mathbf{V}_2\big(\mathbf{I} + \text{diag}\,|(\mathbf{I} + \mathbf{K}_i\mathbf{G}_{ii})^{-1}\,\mathbf{K}_i|\,\mathbf{V}_2\big)^{-1} \times \text{diag}\,|(\mathbf{I} + \mathbf{K}_i\mathbf{G}_{ii})^{-1}\,\mathbf{K}_i|\,|\mathbf{v}|\,, \tag{9}$$

This error bound remains valid for decreased or vanishing cross-couplings or disconnected control stations (for which $\hat{\mathbf{y}}_i = 0$).

Theorem 11.4 The closed-loop overall system (11.1-1), (11.1-2) can be approximated by eqn (8). Under the assumptions and the condition of Theorem 11.3 the approximation error is bounded from above by eqn (9).

An immediate extension of the results of Theorems 11.3 and 11.4 to uncertain plants with $\mathbf{G}(p) = \hat{\mathbf{G}}(p) + \delta\mathbf{G}(p)$ can be received if eqn (3) is replaced by

$$\mathbf{s} = \begin{pmatrix} \delta\mathbf{G}_{11} & \hat{\mathbf{G}}_{12} + \delta\mathbf{G}_{12} & \dots & \hat{\mathbf{G}}_{1N} + \delta\mathbf{G}_{1N} \\ \cdots & \cdots & \cdots & \cdots \\ \hat{\mathbf{G}}_{N1} + \delta\mathbf{G}_{N1} & \hat{\mathbf{G}}_{N2} + \delta\mathbf{G}_{N2} & \dots & \delta\mathbf{G}_{NN} \end{pmatrix} \mathbf{z}$$

and the bound $\mathbf{V}_2$ in eqn (5) is chosen accordingly.

Time domain design

A similar design procedure can be elaborated in the time domain. Whereas the plant model and the controller are usually described in the state-space, the neglected cross-couplings have to be bounded in the form $|\mathbf{s}(t)| \leqq \mathbf{V}_2(t) * |\mathbf{z}(t)|$ with $\mathbf{V}_2(t)$ representing an upper bound of the module of the impulse response matrix of the system (3). Then, results analogous to those of Theorems 11.3 and 11.4 can be derived.

Specifications for single-input single-output subsystems

For systems with $\dim \mathbf{y}_i = \dim \mathbf{u}_i = 1$, $(i = 1, \dots, N)$ the stability condition (7) can be reformulated as an equivalent condition on the open-loop system. By Theorem A1.5 the condition (7) is equivalent to the assertion that $\mathbf{I} - \mathbf{V}_2\,\text{diag}\,|(1 + k_i g_{ii})^{-1}\,k_i|$ is an M-matrix. Multiplying this matrix with $\text{diag}\,|(1 + k_i g_{ii})|$ from the right the resulting matrix

$$\begin{pmatrix} |1 + k_1 g_{11}| & -|g_{12}k_2| & \dots & -|g_{1N}k_N| \\ -|g_{21}k_1| & |1 + k_2 g_{22}| & \dots & -|g_{2N}k_N| \\ \cdots & \cdots & \cdots & \cdots \\ -|g_{N1}k_1| & -|g_{N2}k_2| & \dots & |1 + k_N g_{NN}| \end{pmatrix} \tag{10}$$

is an M-matrix for $p \in \mathcal{D}$ if and only if eqn (7) holds (Theorem A1.4). Obviously, this matrix (10) can be received from the return-difference matrix $\mathbf{F}(p) = \mathbf{I} + \text{diag}\,k_i\mathbf{G}$ by introducing the symbol $|.|$ and the corresponding signs. It represents an M-matrix if and only if the return difference matrix is generalised diagonal dominant according to the definition below.

Definition 11.1 A complex (m, m)-matrix $\mathbf{A}$ is said to be ***generalised diagonal dominant*** if the matrix

$$\begin{pmatrix} |a_{11}| & -|a_{12}| & \dots & -|a_{1m}| \\ \dots & \dots & \dots & \dots \\ -|a_{m1}| & -|a_{m2}| & \dots & |a_{mm}| \end{pmatrix}$$

is an M-matrix or, equivalently,

$$d(\mathbf{A}) = \lambda_{\mathrm{p}} \left[\begin{pmatrix} 0 & |a_{12}| & \dots & |a_{1m}| \\ |a_{21}| & 0 & \dots & |a_{2m}| \\ \dots & \dots & \dots & \dots \\ |a_{m1}| & |a_{m2}| & \dots & 0 \end{pmatrix} \operatorname{diag} \frac{1}{|a_{ii}|} \right] < 1 \qquad (11)$$

holds.
$d\big(\mathbf{G}(p)\big)$ is called coupling factor of the plant $\mathbf{G}(p)$.

Therefore, a decentralised controller that makes the return-difference matrix diagonally dominant possesses the integrity property. The Perron root on the left-hand side of eqn (7) is referred to as the coupling factor. It must be less than unity for all p on the Nyquist contour.

The Direct Nyquist Array method

This interpretation of the stability condition (7) opens the way towards an extension of the Direct Nyquist Array method for designing multivariable feedback controllers. The extensions refer to the use of the generalised version of the diagonal dominance as described in Definition 11.1 and the explicit characterisation of the open and closed loops in terms of the cross-couplings within the plant.

The Direct Nyquist Array method proceeds in two steps:

(1) A decoupling compensator is designed so as to decrease the cross-couplings within the expanded plant (Fig. 11.1 (a)).
(2) Decentralised single-loop controllers are designed so as to ensure closed-loop stability and a suitable dynamical behaviour.

We skip step (1) with reference to the literature cited below and focus our attention on the second step with $\mathbf{G}(p)$ including the decoupling compensator. The expanded plant is assumed to be stable. The controllers $k_i(p)$ have to be chosen so that the closed-loop subsystems (8) are stable, the stability condition (7) is satisfied and the I/O-behaviour meets the given specifications. The first requirement is automatically satisfied if the single-loop feedbacks $k_i(p)$ are independently designed for the isolated subsystems $g_{ii}(p)$. The second and third requirements can be taken into account as follows.

First, the closed-loop overall system is stable if eqn (7) is satisfied. Then, the generalised version of the Gershgorin theorem enables us to assess the stability in terms of $k_i(p)$.

Lemma 11.2 For the eigenvalues of a generalised diagonal dominant matrix **A**

$$|\lambda_i[\mathbf{A}] - a_{ii}| \leq d\,|a_{ii}| \tag{13}$$

holds, where d is defined in eqn (11).

Applying this result to the return-difference matrix $\mathbf{F}(p)$ the characteristic functions $f_i(p)$, which are defined by $\det\big(f_i(p)\,\mathbf{I} - \mathbf{F}(p)\big) = 0$, are known to lie within the bands around $1 + g_{ii}k_i(p)$ swept out by circles with radius $d\,|1 + g_{ii}k_i(p)|$ with d from eqn (7). Consequently, the closed-loop original system is stable if the bands around $g_{ii}k_i(p)$ with radius $d\,|g_{ii}k_i|$ do not encircle the point $(-1 + \mathrm{j}0)$. These bands are called *generalised Gershgorin bands*. They are drawn around the Nyquist plot of the open-loop subsystems $g_{ii}k_i$. The distance of these bands on the real axis from the point -1 determine the gain margin of the i-th control loop.

Second, the I/O-behaviour of the overall system can be checked by means of eqns (8) and (9). If a certain controller $k_i(p)$ should be improved, the knowledge about the other controllers $k_j(p)$ with $(j \neq i)$ can be used to characterise the relationship $\tilde{g}_{ii}(p)$ between u_i and y_i with all controllers k_j attached to the plant (Fig. 11.4(a)). Although $\tilde{g}_{ii}(p)$ could be determined exactly, an approximate description can be easily derived from the results of Section 8.3 and 8.5. We do this for the relation $\tilde{g}_{11}(p)$ between u_1 and y_1 assuming that eqn (7) is satisfied with $k_1 = 0$.

Eqn (2) yields

$$y_1 = g_{11}u_1 + (1\ 0\ \ldots\ 0)\ \mathbf{s} \tag{14}$$

and

$$\hat{y}_1 = g_{11}u_1\,. \tag{15}$$

Instead of eqn (4) we have

$$\mathbf{z} = \begin{pmatrix} 1 \\ 0 \\ \vdots \\ 0 \end{pmatrix} u_1 + \begin{pmatrix} 0 & & & \\ & \dfrac{k_2}{1 + k_2 g_{22}} & & \\ & & \ddots & \\ & & & \dfrac{k_N}{1 + k_N g_{NN}} \end{pmatrix} \mathbf{s}$$

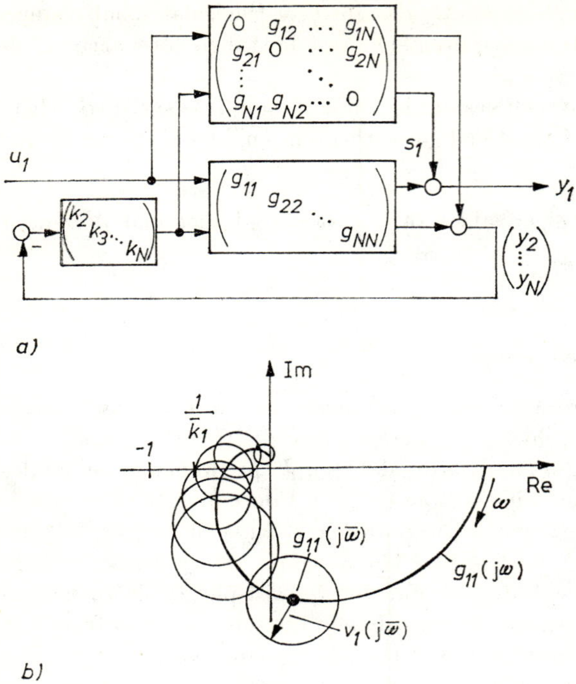

Fig. 11.4 Resulting plant of controller 1
(a) Components of the resulting plant
(b) Characterisation of $\tilde{g}_{11}(j\omega)$ by a tolerance band around $g_{11}(j\omega)$

since $k_1 = 0$ and $v_i = 0$ hold. With eqns (5) and (6) we receive

$$|y_1 - \hat{y}_1| \leqq v_1(p)\, |u_1| \tag{16}$$

with

$$v_1(p) = (1\ 0 \ldots 0)\, \mathbf{V}_2 \left(\mathbf{I} - \begin{pmatrix} 0 & & \\ & \left|\frac{k_2}{1+k_2 g_{22}}\right| & \\ & & \ddots \\ & & & \left|\frac{k_N}{1+k_N g_{NN}}\right| \end{pmatrix} \mathbf{V}_2 \right)^{-1} \begin{pmatrix} 1 \\ 0 \\ \vdots \\ 0 \end{pmatrix}. \tag{17}$$

That is, the frequency response $\tilde{g}_{11}(j\omega)$ lies within a band around the plant transfer function $g_{11}(j\omega)$, where the circles have radius $v_1(j\omega)$. This band is called *generalised Ostrowski band* (Fig. 11.4(b)). It lies inside the generalised Gershgorin band. The controller $k_1(p)$ can be designed for this band as it is usually designed for the Nyquist plot of the plant. The open-loop transfer function is received by 'multiplication' of the band by $k_i(p)$. As long as the resulting band does not encircle the point $-1 + j0$, the return-difference matrix remains to be generalised diagonal dominant.

These results suggest the following steps

(1) The control laws $k_i(p)$ are specified for the isolated subsystems (1) by single-loop controller design methods the stability and a well suited dynamical I/O-behaviour of the closed-loop isolated subsystem (8) being the design requirements.
(2) The diagonal dominance of the return-difference matrix is checked by eqn (7). A more detailed analysis of the stability is possible if the generalised Gershgorin bands are drawn.
(3) The controllers are adjusted successively while using the tolerance band (15)—(17) to characterise the plant of the corresponding controller i.

Design of the decentralised voltage controller

We continue with the example of Section 11.1 and demonstrate the application of the design procedure above by considering an overall system with $N = 5$ subsystems. The plant is described by a 45-th order state-space model. To get a global impression, Fig. 11.5 shows the step response. To ensure asymptotic regulation for stepwise voltage setpoint or load changes we use decentralised PI-controllers

$$k_i(p) = \alpha_i \left(\tilde{k}_i + \frac{\tilde{k}_i}{pT_i}\right). \tag{18}$$

(1) The separate single-loop design yields $T_i = 1$ $(i = 1, \ldots, 5)$ and $\tilde{k}_1 = 1$, $\tilde{k}_2 = 0.7$, $\tilde{k}_3 = 0.8$, $\tilde{k}_4 = 0.7$ and $\tilde{k}_5 = 1$.
(2) The stability condition (7) is satisfied. In fact, the scalars α_i could be simultaneously increased until $\bar{\alpha}_1 = 1.8$, $\bar{\alpha}_2 = 1.4$, $\bar{\alpha}_3 = 1.3$, $\bar{\alpha}_4 = 1.3$ and $\bar{\alpha}_5 = 2.2$ without violating the diagonal dominance condition.
(3) The tolerance bands for the resulting plants which include $N - 1$ controllers determine a broader interval of α_i. Fig. 11.6 shows the tolerance band for $\tilde{g}_{22}\left(\tilde{k}_2 + \dfrac{\tilde{k}_2}{j\omega T_2}\right)$. The middle of the band is given by the Nyquist plot of $g_{22}(p)\left(\tilde{k}_2 + \dfrac{\tilde{k}_2}{j\omega T_2}\right)$ while the radius of the smaller circles is determined according to eqn (16) (for subsystem 2). The cut of the real axis shows that α_2 can be increased till $\bar{\alpha}_2 = 2.8$ without endangering the diagonal dominance property of the return-difference matrix. Note that this increase of α_2 is possible because all other controllers remain fixed.

The step response of the closed-loop overall system is shown in Fig. 11.7. They satisfy the requirements on the dynamical I/O-behaviour prescribed for the voltage control problem under consideration.

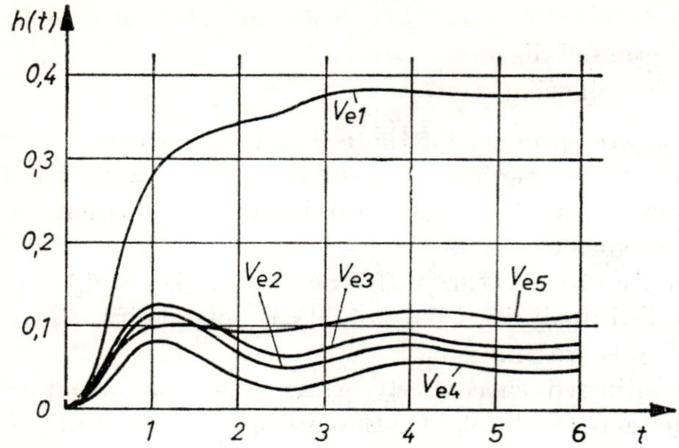

Fig. 11.5 Step response of the plant $\left(V_{c1} = \sigma(t)\right)$

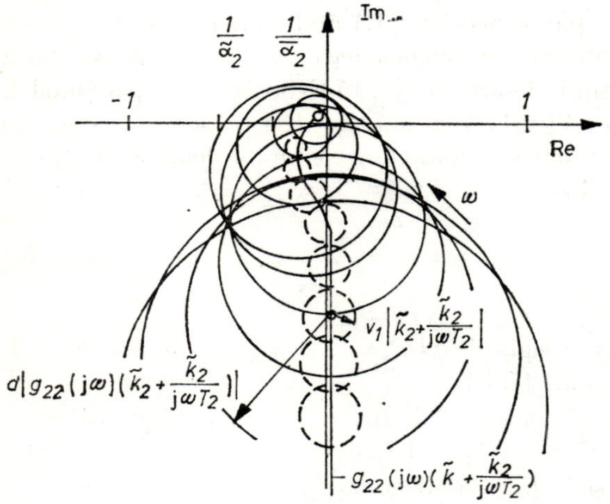

Fig. 11.6 Generalised Gershgorin and Ostrowski bands for g_{11}

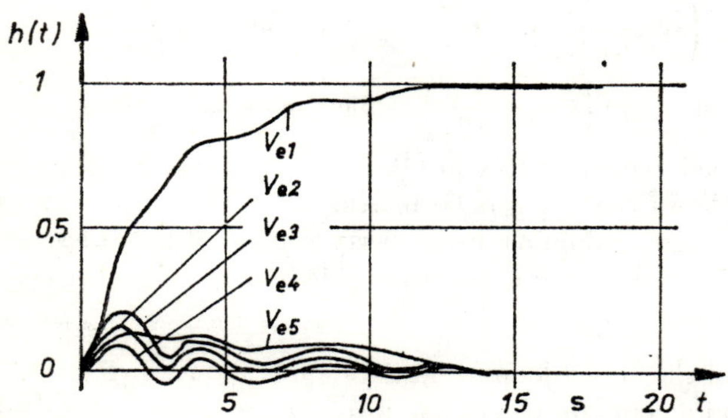

Fig. 11.7 Step response of the closed-loop system $\left(v_1 = \sigma(t)\right)$

Bibliographical notes

Decentralised control gave rise to an extensive research on this special topic of control system theory. Thorough surveys of the results are given by Siljak (1978), Sandell et al. (1978), Lunze (1908 b), Singh (1981) and Litz (1983). In this chapter, we referred only to decentralised control problems whose solution rely strongly on robustness issues of feedback systems. Therefore, only the aggregation-decomposition method of Siljak (1978) and stability considerations for structurally perturbed systems have a close connection to our treatment whereas the majority of the design procedures for decentralised control uses the complete model of the overall plant and does not refer to robustness properties as a primary design aim.

The existence of decentralised PI-controllers has been investigated by Davison (1978) in connection with the sequential design where the control stations are attached to the plant in a given order. The generalisation described in Section 11.2 is due to Lunze (1983 a, b) and Grosidier et al. (1984). Sufficient existence conditions on the basis of high-gain feedback have been derived by Locatelli et al. (1983). The application example is taken from (Gamaleja et al., 1984).

The results derived in Section 11.3 represent a straightforward extension of the Nyquist Array Method for designing multivariable controllers, as proposed by Rosenbrock (1974), to decentralised controllers, where each control station is a multivariable feedback itself. Furthermore, it shows the way to the elaboration of similar design methods in the time domain on the same conceptual basis. Results along this line have been described by several authors, eg by Nwokah (1980, 1984), Arkun et al. (1984), and Kidd (1984, 1986). In essence, these papers use the same notion of generalised diagonal dominance but differ in the tightness of the tolerance bands used and the way in which the bounds have to be determined. Other authors, eg Limebeer (1982), Hung and Limebeer (1984), or Bennett and Baras (1980) generalised the dominance condition by the use of a matrix norm. They end up with auxiliary single-loop subsystems which can be investigated as before. However, the results are generally more conservative than those given here.

Nwokah (1983) combined the decomposition of the design problem for diagonal dominant systems with the quantitative feedback design method described in Section 8.1 as a means for solving the N independent single-loop problems.

12 TRENDS IN ROBUST FEEDBACK CONTROL

The main feature of the theory of robust multivariable feedback control is the explicit consideration of model uncertainties. Robustness indicators which have been found by qualitative robustness investigations help to get a clear view of the conditions under which the closed loop is particularly robust and can be expected to tolerate given classes of uncertainties. Methods for considering robustness issues in a quantitative way enable us to make reliable assertions about the system performance and to design controllers that satisfy the design objectives despite of all uncertainties of a completely specified class.

Robustness analysis and design addresses a broad range of problems. They tackle quite different questions that arise, in essence, due to existing or deliberately introduced parameter variations or model uncertainties. As we have seen, the design of controllers that are robust against plant dynamics variations, the ensurance of integrity or connective stability of the closed-loop system, and the tuning of multivariable controllers on the basis of experimental data are examples that fall well into the scope of robustness investigations. For all of these topics, robust control theory has elaborated new theoretical results as well as practice-relevant analysis and design methods.

Besides these problems, which are directly concerned with feedback controller design, robust control theory has impact on other fields of control theory and systems analysis in which the effects of severe uncertainties of one system part on the performance of the whole system play an important role. This will be briefly outlined for two of such topics: structural problems of control system synthesis, and modelling and control of large scale systems. Until now, these fields have made only little advantage of robustness considerations but must be expected to be the subject of extensive robustness investigations in near future.

Structural problems of feedback control include questions concerning the information about the plant which must be available on each stage of design and implementation. As corresponding investigations cannot be based on particular numerical values of the system parameters, robust control theory can contribute to solve such problems.

At first, for the design phase of the controller, requirements on the necessary a-priori information about the plant can be derived by means of the existence conditions for robust controllers. For example, a static plant model and some information about the stability property of the plant proved to be sufficient for the design of stable PI-controlled systems. 'Parasitic' dynamical elements can be neglected in the plant model as long as they satisfy some structural conditions. These results contribute to bring the requirements on the model accuracy in accordance with the strength of the design objectives.

Secondly, for the implementation of the controller, robustness considerations yield guidelines on choosing appropriate measurement signals and actuators. The existence test of decentralised controllers ensuring closed-loop integrity may remind the reader of a result that contributes to the solution of such a structural synthesis problem.

The modelling and control of large scale systems are dominated by dimensionality and uncertainty problems. Therefore, robustness issues have to be introduced and utilised. In the modelling stage, each subsystem can only be described by an approximate model. Robustness considerations are necessary to assess the influence of the error model, which reflects the interaction with the other subsystems. In the stability analysis of interconnected systems, strong model simplifications are deliberately introduced in order to decrease the complexity of the stability test. Many methods use, in one or another way, upper bounds of the I/O-performance of the subsystems that are represented by 'comparison systems' as explained in Section 5.3. In the design of decentralised controllers, robustness considerations may help to reduce the information about the whole composite plant that must be available at the subsystem level in order to ensure the design objectives to be satisfied.

The main lines of current and future research in robust control aim at improving the robustness analysis and design methods and at developing novel answers to control systems synthesis problems that result from uncertainties. The critical evaluations, which have accompanied the presentation of all the methods in the book, and the bibliographical notes have already given some ideas of the basic problems that remain to be solved. They can be classified as follows.

New forms of the description of uncertain dynamical systems have to be found which enable the designer to express all the available information about the plant, in dependence upon its form and amount, by appropriately utilising the freedom of this model form. They have to be both tractable mathematically and interpretable in engineering terms. While the former should prevent conservatism of the analysis results, the latter has to diminish the conservatism of the modelling step. Towards this end, the model has to include as much structural information about the source of uncertainties as possible. A combination of the two model forms described in Sections 5.2 and 5.3 may be considered as a first step in this direction.

The existence of robust controllers remains a vital problem. On the one hand, the known existence conditions are merely stability-oriented and have to be extended so as to regard principal issues of the closed-loop I/O-performance. On the other hand, the necessity of and systematic ways for the introduction of dynamical elements into the controller in order to ensure the required robustness remain to be elaborated. The results presented in Section 7.1 can only be considered as starting point of such investigations.

The reduction of the conservatism of the analysis methods is a further open problem. Unlike most of the design methods, which do not incorporate the robustness requirement as a primary design aim but merely check the robustness for a given controller, new principles have to include the robustness issues directly into the choice of the controller structure and parameters. We should, therefore, keep trace of new concepts such as the H^{∞}-approach or algebraic methods, which are being rapidly developed now.

Extensions to nonlinear or time varying systems are necessary for many practical applications, but also for a deeper understanding of robustness as a 'structural' property of feedback systems. Some methods are naturally extendable because their basic ideas refer to a much larger class than that of linear systems, but important questions such as 'Which classes of nonlinear systems can be described by sufficiently tight upper bounds or sectors?' have not been answered clearly enough.

Packages for computer-aided design have to be equipped with algorithms for robustness assessment. This necessitates, among others, the development of numerically reliable and efficient procedures for optimisation problems that are incorporated in the determination of tight bounds of the closed-loop system uncertainties or new design steps as the determination of all rather than a single controller for a given control task.

APPENDICES

Appendix 1 Nonnegative and M-matrices

In this appendix we review some important properties of nonnegative and M-matrices, which are relevant to robustness analysis. For a more detailed theory see eg (Berman and Plemmons, 1973; Ortega and Rheinboldt, 1970; Gantmacher, 1958; Zurmühl, 1964; Bellman, 1960).

Nonnegative matrices

A matrix $\mathbf{A} = (a_{ij})$ is called *nonnegative* ($\mathbf{A} \geqq 0$) or positive ($\mathbf{A} > 0$) if all elements of $\mathbf{A}$ are real and nonnegative ($a_{ij} \geqq 0$) or positive ($a_{ij} > 0$) respectively. An (n, n)-matrix $\mathbf{A}$ is called reducible if there exists a permutation matrix $\mathbf{P}$ that transforms $\mathbf{A}$ into the form

$$\mathbf{PAP}' = \begin{pmatrix} \tilde{\mathbf{A}}_{11} & \tilde{\mathbf{A}}_{12} \\ 0 & \tilde{\mathbf{A}}_{22} \end{pmatrix}$$

where $\tilde{\mathbf{A}}_{11}$ and $\tilde{\mathbf{A}}_{22}$ are square. Otherwise, $\mathbf{A}$ is called *irreducible*.

Theorem A1.1 (*Frobenius-Perron Theorem*). Every irreducible nonnegative (n, n)-matrix $\mathbf{A}$ has a positive eigenvalue $\lambda_p[\mathbf{A}]$ that is a simple root of the characteristic equation and is not exceeded by the modules of all other eigenvalues of $\mathbf{A}$

$$\lambda_p[\mathbf{A}] \geqq |\lambda_i[\mathbf{A}]| . \tag{1}$$

The eigenvector corresponding to $\lambda_p[\mathbf{A}]$ is strictly positive and unique up to a multiplicative constant.

$\lambda_p[\mathbf{A}]$ is called the *Perron-root* of $\mathbf{A}$. Theorem A1.1 can be extended to reducible matrices. Then there exist eigenvalues $\lambda_i[\mathbf{A}]$ with $|\lambda_i[\mathbf{A}]| = \lambda_p[\mathbf{A}]$. All these eigenvalues are simple roots of the characteristic equation.

If $\mathbf{A}$ is irreducible then $\lambda_p[\mathbf{A}]$ can be determined by means of the following inclusion. For $\mathbf{x} = (x_1, \ldots, x_n)' > \mathbf{0}$ and $\mathbf{y} = (y_1, \ldots, y_n)' = \mathbf{Ax}$

$$\min_i \frac{y_i}{x_i} \leqq \lambda_p[\mathbf{A}] \leqq \max_i \frac{y_i}{x_i} \tag{2}$$

15 Lunze

holds. Used in an iterative way eqn (2) represents the basis for a simple algorithm for determining $\lambda_p[\mathbf{A}]$.

Theorem A1.2 For two (n, n)-matrices $\mathbf{A}$ and $\mathbf{B}$ with $\mathbf{A} \geqq |\mathbf{B}| \geqq \mathbf{0}$ the relation

$$\varrho[\mathbf{B}] \leqq \lambda_p[\mathbf{A}] \tag{3}$$

holds where $\varrho[\mathbf{B}]$ denotes the spectral radius of $\mathbf{B}$.

Corollaries: For an arbitrary (n, n)-matrix $\mathbf{C}$, which may not be nonnegative or positive, we get

$$\varrho[\mathbf{C}] \leqq \lambda_p[|\mathbf{C}|]. \tag{4}$$

If $\mathbf{A} \geqq \mathbf{B} \geqq \mathbf{0}$, Theorem A1.2 yields

$$\lambda_p[\mathbf{A}] \geqq \lambda_p[\mathbf{B}]. \tag{5}$$

M-matrices

Definition A1.1 An (n, n)-matrix $\mathbf{P} = (p_{ij})$ is said to be an *M-matrix* if $p_{ij} \leqq 0$ for all $i \neq j$ holds and all eigenvalues of $\mathbf{P}$ have positive real parts. □

Theorem A1.3 An (n, n)-matrix $\mathbf{P}$ with $p_{ij} \leqq 0$ for all $i \neq j$ is an M-matrix if and only if one of the following equivalent conditions hold:

(a) All eigenvalues of $\mathbf{P}$ have positive real parts.
(b) $\mathbf{P}$ is nonsingular and $\mathbf{P}^{-1}$ is nonnegative.
(c) All the leading principal minors of $\mathbf{P}$ are strictly positive

$$\det \begin{pmatrix} p_{11} & \cdots & p_{1k} \\ p_{k1} & \cdots & p_{kk} \end{pmatrix} > 0, \qquad (k = 1, \ldots, n). \tag{6}$$

(d) All the principal minors of $\mathbf{P}$ are strictly positive.
(e) There exists some vector $\mathbf{x}$ such that all elements of the vector $\mathbf{Px}$ are strictly positive.
(f) There exists a vector $\mathbf{y}'$ such that all elements of the vector $\mathbf{y}'\mathbf{P}$ are strictly positive.

Note that any of the conditions (a)—(f) is necessary and sufficient for the matrix $\mathbf{P}$ with $p_{ij} \leqq 0$ for all $i \neq j$ to be an M-matrix.

Theorem A1.4 Assume that $\mathbf{P}$ is an M-matrix.

(a) If $\mathbf{P}_1 \geqq \mathbf{P}$ and $\mathbf{P}_1$ satisfies the sign condition of Definition A1.1 then $\mathbf{P}_1$ is an M-matrix as well.
(b) $\operatorname{diag} d_i \mathbf{P}$ and $\mathbf{P} \operatorname{diag} d_i$ with $d_i > 0$ are M-matrices as well.

Theorem A1.5 If $\mathbf{A}$ is a nonnegative (n, n)-matrix then $\mathbf{P} = \mu\mathbf{I} - \mathbf{A}$ is an M-matrix if and only if

$$\mu > \lambda_p[\mathbf{A}]. \qquad \square$$

Theorem A1.6 Assume that $\mathbf{A}$ and $\mathbf{B}$ are square and nonnegative and that $\lambda_p[\mathbf{A}] < 1$. Then the following assertions are equivalent:

(a) $\mathbf{I} - \mathbf{A} - \mathbf{B}$ is an M-matrix.
(b) $\mathbf{I} - \mathbf{B}(\mathbf{I} - \mathbf{A})^{-1}$ is an M-matrix.
(c) $\mathbf{I} - (\mathbf{I} - \mathbf{A})^{-1}\mathbf{B}$ is an M-matrix.
(d) $\lambda_p[\mathbf{B}(\mathbf{I} - \mathbf{A})^{-1}] < 1$.
(e) $\lambda_p[(\mathbf{I} - \mathbf{A})^{-1}\mathbf{B}] < 1$.
(f) $\lambda_p[\mathbf{A} + \mathbf{B}] < 1$.

Proof. We prove that (a) $\Leftrightarrow$ (b). Then the other equivalences are straightforward. (a) $\Rightarrow$ (b): Since $\lambda_p[\mathbf{A}] < 1$, $(\mathbf{I} - \mathbf{A})$ is an M-matrix. Because of (A1.4-(b)) $\mathbf{B}(\mathbf{I} - \mathbf{A})^{-1} \geqq \mathbf{0}$, ie $\mathbf{P}_2 = \mathbf{I} - \mathbf{B}(\mathbf{I} - \mathbf{A})^{-1}$ satisfies the sign condition of Theorem A1.3. Because of (a) and (A1.3-(f)) there exists some $\mathbf{y}'$ with $\mathbf{y}'(\mathbf{I} - \mathbf{A} - \mathbf{B}) = \tilde{\mathbf{y}}' > \mathbf{0}$. Hence $\mathbf{0} < \tilde{\mathbf{y}}'(\mathbf{I} - \mathbf{A})^{-1} = \mathbf{y}'\big(\mathbf{I} - \mathbf{B}(\mathbf{I} - \mathbf{A})^{-1}\big)$ That is, $\mathbf{P}_2$ satisfies (A1.3-(f)) and, thus, represents an M-matrix.
(b) $\Leftarrow$ (a): As $\mathbf{A}$ and $\mathbf{B}$ are nonnegative, $\mathbf{P}_3 = \mathbf{I} - \mathbf{A} - \mathbf{B}$ satisfies the sign condition of Theorem A1.3. We have $\mathbf{0} \leqq (\mathbf{I} - \mathbf{A} - \mathbf{B})^{-1} = (\mathbf{I} - \mathbf{A})^{-1}(\mathbf{I} - \mathbf{B}(\mathbf{I} - \mathbf{A})^{-1})^{-1}$ where $(\mathbf{I} - \mathbf{A})^{-1}$ exists because $\lambda_p[\mathbf{A}] < 1$. Then Theorem A1.3(b) proves $\mathbf{P}_3$ to be an M-matrix. □

Theorem A1.7 Assume that $\mathbf{P}$ is an M-matrix. Then all the principal minors of $\mathbf{K} = \mathbf{P}^{-1}$ are positive.

Proof. We prove Theorem A1.7 for some leading principal minor. The proof for the other principal minors is analogous if corresponding column and row permutations have transformed the principal minor into a leading principal minor. Note that such permutations do not effect the M-matrix property of $\mathbf{P}$. Decompose $\mathbf{P}$ and $\mathbf{K}$ in the same manner

$$\mathbf{K} = \begin{pmatrix} \mathbf{K}_{11} & \mathbf{K}_{12} \\ \mathbf{K}_{21} & \mathbf{K}_{22} \end{pmatrix}, \qquad \mathbf{P} = \begin{pmatrix} \mathbf{P}_{11} & \mathbf{P}_{12} \\ \mathbf{P}_{21} & \mathbf{P}_{22} \end{pmatrix}$$

where $\mathbf{K}_{11}$ is the leading principal minor under consideration. Since $\mathbf{K} = \mathbf{P}^{-1}$ we have $\mathbf{K}_{11} = (\mathbf{P}_{11} - \mathbf{P}_{12}\mathbf{P}_{22}^{-1}\mathbf{P}_{21})^{-1}$, where $\mathbf{P}_{22}^{-1}$ exists since $\mathbf{P}$ is an M-matrix. This and Theorem A1.3 yield

$$\det \mathbf{K}_{11} = \frac{\det \mathbf{P}_{22}}{\det \mathbf{P}} > 0. \qquad \square$$

Appendix 2 Singular value decomposition

We consider (n, m)-matrices $\mathbf{A}$ with complex-valued elements and let $k = \min(n, m)$.

Definition. The *singular values* $\sigma_i[\mathbf{A}]$ are defined by

$$\sigma_i[\mathbf{A}] = \sqrt{\lambda_i[\mathbf{A}^*\mathbf{A}]}, \qquad i = 1, 2, \ldots, k \tag{1}$$

where $\lambda_i[.]$ denotes the eigenvalues of the indicated matrix.

Properties. The singular values exist uniquely for all complex-valued matrices. They are real and nonnegative. It is assumed that the singular values are ordered according to

$$\sigma_1[\mathbf{A}] \geqq \sigma_2[\mathbf{A}] \geqq \cdots \geqq \sigma_k[\mathbf{A}] \geqq 0. \tag{2}$$

The first k' singular values are strictly positive where $k' = \operatorname{rank} \mathbf{A}$. The maximum and minimum singular values are denoted by

$$\sigma_{\max}[\mathbf{A}] = \sigma_1[\mathbf{A}], \qquad \sigma_{\min}[\mathbf{A}] = \sigma_k[\mathbf{A}]. \tag{3}$$

$\sigma_{\max}[\mathbf{A}]$ is identical to the spectral norm of the matrix $\mathbf{A}$

$$\sigma_{\max}[\mathbf{A}] = \max_{\mathbf{x} \neq \mathbf{0}} \frac{\|\mathbf{Ax}\|_2}{\|\mathbf{x}\|_2} \tag{4}$$

where $\|.\|_2$ is the Euclidean vector norm. For the minimum singular value the relation

$$\sigma_{\min}[\mathbf{A}] = \min_{\mathbf{x} \neq \mathbf{0}} \frac{\|\mathbf{Ax}\|_2}{\|\mathbf{x}\|_2} = \begin{cases} \dfrac{1}{\|\mathbf{A}^{-1}\|_2} & \text{if } n = m, \quad \det \mathbf{A} \neq 0, \\ 0 & \text{otherwise} \end{cases} \tag{5}$$

holds.

Singular value decomposition. For all complex-valued matrices $\mathbf{A}$ there exists a decomposition

$$\mathbf{A} = \mathbf{UDV}^* = \sum_{i=1}^{k} \sigma_i \mathbf{u}_i \mathbf{v}_i^*, \tag{6}$$

where the (n, n)-matrix $\mathbf{U} = (\mathbf{u}_1, \ldots, \mathbf{u}_n)$ and the (m, m)-matrix $\mathbf{V} = (\mathbf{v}_1, \ldots, \mathbf{v}_m)$ are composed of the eigenvectors of $\mathbf{AA}^*$ and $\mathbf{A}^*\mathbf{A}$, respectively. The (n, m)-matrix $\mathbf{D}$ includes the singular values:

$$\text{for } n \geqq m \ \mathbf{D} = \begin{pmatrix} \sigma_1 & & \\ & \ddots & \\ & & \sigma_m \\ 0 & \ldots & 0 \end{pmatrix} \qquad \text{for } n < m \ \mathbf{D} = \begin{pmatrix} \sigma_1 & & & 0 \\ & \ddots & & \vdots \\ & & \sigma_n & 0 \end{pmatrix}. \tag{7}$$

$\mathbf{U}$ and $\mathbf{V}$ are unitary, ie $\mathbf{UU}^* = \mathbf{I}$ and $\mathbf{VV}^* = \mathbf{I}$. $\mathbf{UV}^*$ is unique. $\mathbf{u}_i$ and $\mathbf{v}_i$ are called output principal direction and input principal direction, respectively. $\mathbf{u}_1$ and $\mathbf{v}_1$ are referred to as the *major principal directions* of $\mathbf{A}$. For distinct singular values the vectors $\mathbf{u}_i$ and $\mathbf{v}_i$ are unique up to a scalar factor $e^{j\phi}$.

From eqn (6)

$$\mathbf{A}^*\mathbf{A} = \mathbf{VD}^*\mathbf{U}^*\mathbf{UDV}^* = \mathbf{V} \operatorname{diag} \sigma_i^2 \mathbf{V}^* \tag{8}$$

and

$$\mathbf{A}\mathbf{v}_i = \sigma_i[\mathbf{A}]\, \mathbf{u}_i \tag{9}$$

follow, where eqn (9) is received from $\mathbf{AV} = \mathbf{UD}$ by considering the i-th column of both sides of the equation.

Results. For quadratic (n, n)-matrices $\mathbf{A}$ the maximum singular value provides an upper bound of the spectral radius since it represents a matrix norm (cf eqn (4))

$$\varrho[\mathbf{A}] \leqq \sigma_{\max}[\mathbf{A}]. \tag{10}$$

Lemma A2.1 $\varrho[\mathbf{A}] = \sigma_{\max}[\mathbf{A}]$ **if and only if the major principal directions $\mathbf{u}_1$ and $\mathbf{v}_1$ of $\mathbf{A}$ are aligned, ie for some ϕ**

$$\mathbf{v}_1 = e^{j\phi}\mathbf{u}_1. \tag{11}$$

Proof. If eqn (11) holds eqn (9) leads to

$$\mathbf{A}\mathbf{v}_1 = \sigma_1[\mathbf{A}]\, e^{j\phi}\mathbf{v}_1.$$

Hence, $\lambda = \sigma_1[\mathbf{A}]\, e^{j\phi}$ is an eigenvalue of $\mathbf{A}$. As $|\lambda| = \sigma_1[\mathbf{A}] = \sigma_{\max}[\mathbf{A}]$, eqn (10) holds with equality sign. On the other hand, if eqn (10) holds as equality, there exist some vector $\mathbf{w}$ such that

$$\mathbf{A}\mathbf{w} = e^{j\psi}\sigma_{\max}[\mathbf{A}]\,\mathbf{w}$$

and

$$\mathbf{w}^*\mathbf{A}^* = e^{-j\psi}\sigma_{\max}[\mathbf{A}]\,\mathbf{w}^*.$$

Multiplication of both equations yields

$$\sigma_{\max}[\mathbf{A}]^2 = \frac{\mathbf{w}^*\mathbf{A}^*\mathbf{A}\mathbf{w}}{\mathbf{w}^*\mathbf{w}} = \frac{\mathbf{w}^*\mathbf{V}\operatorname{diag}\sigma_i^2\mathbf{V}^*\mathbf{w}}{\mathbf{w}^*\mathbf{V}^*\mathbf{V}\mathbf{w}} = \frac{\mathbf{z}^*\operatorname{diag}\sigma_i^2\mathbf{z}}{\mathbf{z}^*\mathbf{z}}$$

where $\mathbf{z} = \mathbf{V}^*\mathbf{w}$. Using the specific vector $\mathbf{z} = (1 \;\; 0 \;\; \ldots \;\; 0)'$ and, for the shortage of the proof, assuming $\sigma_{\max} > \sigma_2$ we get $\mathbf{w} = \mathbf{V}\mathbf{z} = \mathbf{v}_1$. Hence,

$$\mathbf{A}\mathbf{v}_1 = e^{j\psi}\sigma_{\max}[\mathbf{A}]\,\mathbf{v}_1$$

holds. On the other hand, eqn (9) yields

$$\mathbf{A}\mathbf{v}_1 = \sigma_{\max}[\mathbf{A}]\,u_1.$$

Consequently, we have $\mathbf{v}_1 = e^{-j\psi}\mathbf{u}_1$. □

For a more detailed survey see eg (Klema and Laub, 1980; Hung and MacFarlane 1982).

Appendix 3 The Nichols chart

The Nichols chart is a graphical means for determining the closed-loop frequency response $t(j\omega)$ for a given open loop frequency response $l(j\omega)$ according to

$$t(j\omega) = \frac{l(j\omega)}{1 + l(j\omega)}. \tag{1}$$

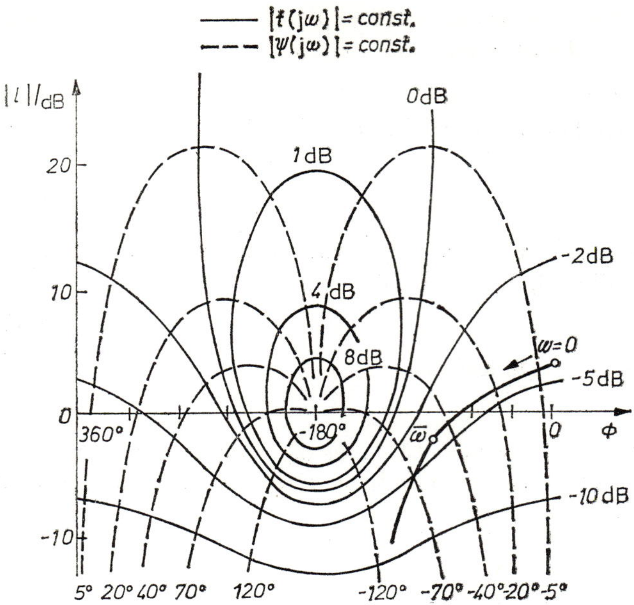

Fig. A3.1 The Nichols chart

Example:

$$l(j\omega) = \frac{1}{(1 + j\omega)(2 + j\omega)}$$

(see solid line)

for $\bar{\omega} = \sqrt{2}$: $l(j\bar{\omega}) = 0.71 = -3$ dB

$\phi(j\bar{\omega}) = -90°$

The Nichols chart gives for the closed loop

$t(j\bar{\omega}) = -4.8$ dB $= 0.57$

$\psi(j\bar{\omega}) = -55°$

For each frequency $\bar{\omega}$ the open-loop response $l(j\bar{\omega}) = |l(j\bar{\omega})|\, e^{j\phi(j\bar{\omega})}$ is drawn in the $|l|$-ϕ-coordinate system (abscissa in degree (°); ordinate in decibel (dB) where $|l|/\text{dB} = 20 \lg |l|$). Considering $\omega = 0 \ldots \infty$ a locus of $l(j\omega)$ is received (Fig. A3.1).

For each point $l(j\bar{\omega})$ in the Nichols chart the corresponding closed-loop frequency response

$$t(j\bar{\omega}) = |t(j\bar{\omega})|\, e^{j\psi(j\bar{\omega})}$$

can be determined. Equation (1) yields

$$t(j\bar{\omega}) = \frac{l(j\bar{\omega})}{\sqrt{1 + 2\,|l(j\bar{\omega})| \cos \phi(j\bar{\omega}) + |l(j\bar{\omega})|^2}} \tag{2}$$

$$\psi(j\bar{\omega}) = \arctan \frac{|l(j\bar{\omega})| \sin \phi(j\bar{\omega})}{|l(j\bar{\omega})| + \cos \phi(j\bar{\omega})}. \tag{3}$$

For fixed values of $|t(\mathrm{j}\bar{\omega})|$ and $\psi(\mathrm{j}\bar{\omega})$, say $|t(\mathrm{j}\bar{\omega})| = t_1$ and $\psi(\mathrm{j}\bar{\omega}) = \psi_1$, these equations give a relationship between $|l(\mathrm{j}\bar{\omega})|$ and $\phi(\mathrm{j}\bar{\omega})$. The geometric loci of these points $(|l|, \phi)$ are drawn as contour lines into the Nichols chart, t_1 and ψ_1 being the curve parameters.

To determine the closed-loop frequency response for a given frequency $\bar{\omega}$, $|l(\mathrm{j}\bar{\omega})|$ and $\phi(\mathrm{j}\bar{\omega})$ are calculated from the given plant and controller frequency responses and drawn into the Nichols chart. The parameters of the curves of the chart passing through the point $\big(|l(\mathrm{j}\bar{\omega})|, \phi(\mathrm{j}\bar{\omega})\big)$ give the values $|t(\mathrm{j}\bar{\omega})|$, $\psi(\mathrm{j}\bar{\omega})$ of the corresponding closed-loop frequency response $t(\mathrm{j}\omega)$ (Fig. A3.1).

Note that the curves focus on the point $(0, -180°)$ which corresponds to the point $(-1 + \mathrm{j}0)$ in the complex plane and that

$$|t(\mathrm{j}\omega)| \approx 0\ \mathrm{dB} \quad \text{if} \quad |l(\mathrm{j}\omega)| > 20\ \mathrm{dB}$$

$$|t(\mathrm{j}\omega)| \approx |l(\mathrm{j}\omega)| \quad \text{if} \quad |l(\mathrm{j}\omega)| < -20\ \mathrm{dB}.$$

REFERENCES

Ackermann J. (1980) 'Parameter space design of robust control systems', *IEEE Trans.* **AC-25** 1058—1072.

Ackermann J. (1983) *Abtastregelungen*, Band 2: *Entwurf robuster Systeme*, Springer Verlag, Berlin/Heidelberg/New York.

Ackermann J. (1984) 'Robustness against sensor failures', *Automatica* **20** 211—215.

Ackermann J. and Kaesbauer D. (1981) 'D-decomposition in the space of feedback gains for arbitrary pole regions', *IFAC-Congress*, Kyoto, paper 17.3.

Ackermann J. and Türk S. (1982) 'A common controller for a family of plant models', *IEEE Conf. Decision and Control*, Orlando, pp 240—244.

Alosh A. J. (1983) 'Stabilization of a class of plants with possible loss of outputs or actuator failures', *IEEE Trans* **AC-28** 231—233.

Anderson B. D. O. and Moore, J. B. (1971) *Linear Optimal Control*, Prentice-Hall, Englewood Cliffs.

Arkun Y.; Manousiouthakis B. and Putz P. (1984) 'Robust Nyquist array methodology: a new theoretical framework for analysis and design of robust multivariable feedback systems', *Intern. J. Control* **40** 603—629.

Ashworth M. J. (1982) *Feedback Design of Systems with Significant Uncertainties*, John Wiley and Sons, Chichester.

Aström K. J. (1980) 'Robustness of a design method based on assignment of poles and zeros', *IEEE Trans.* **AC-25** 588—591.

Bauer F. L. (1963) 'Optimally scaled matrices', *Numerische Mathematik* **5** 73—87.

Barmish B. R. (1982) 'On enlarging the class of stabilisable linear systems with uncertainties', 16-*th Asilomar Conf. on Circuits, Systems and Computers*, Pacific Grove, pp 364—368.

Barmish B. R.; Corless M. and Leitmann G. (1983) 'A new class of stabilizing controllers for uncertain dynamical systems', *SIAM J. Control and Optimization* **21** 246—255.

Barmish B. R. and Hollot C. V. (1984) 'Counterexample to a recent result on the stability of interval matrices by S. Bialas', *Intern. J. Control* **39** 1103—1104.

Barrett M. F. (1980) 'Conservatism with robustness tests for linear feedback control systems', *IEEE Conf. Decision and Control, Albuquerque*, 855—890.

Bellman R. (1960) *Introduction to Matrix Analysis*, McGraw-Hill, New York.

Bennett W. H. and Baras J. S. (1980) 'Block diagonal dominance and design of decentralized compensators', *IFAC-Symposium on Large Scale Systems, Toulouse*, pp 93—101.

Berman A. and Plemmons R. J. (1973) *Nonnegative Matrices in the Mathematical Sciences*, Academic Press, New York.

Bialas S. (1983) 'A necessary and sufficient condition for the stability of interval matrices', *Intern. J. Control* **37** 717–722.

Bode H. W. (1945) *Network Analysis and Feedback Amplifier Design*, Van Nostrand, New York.

Bux D. (1975) Anwendung und Entwurf konstanter linearer Zustandsregler bei linearen Systemen mit langsam veränderlichen Parametern (*Fortschrittsberichte der VDI-Zeitschriften Reihe* 8 Nr. 21), VDI-Verlag, Düsseldorf.

Carlucci D. and Vallauri M. (1981) 'Feedback control of linear multivariable systems with uncertain description in the frequency domain', *Intern. J. Control* **33** 903 to 912.

Charitonov W. L. (1978) 'On asymptotic stability of the equilibrium of a family of linear differential equations' (in Russian), *Differentialnoye Uravnenie* **14** 2086 to 2088.

Chen M. J. and Desoer C. A. (1982) 'Necessary and sufficient condition for robust stability of linear distributed feedback systems', *Intern. J. Control* **35** 255–267

Chotai, A.; Owens D. H.; Raya A. and Wang H. M. (1984) 'Design of Smith control scheme for time-delay systems based on plant step data', *Intern. J. Control* **40** 297–316.

Chow J. H. and Kokotovic P. V. (1976) 'A decomposition of near-optimum regulators for systems with slow and fast modes', *IEEE Trans.* **AC-21** 701–705.

Cruz J. B. (Ed.) (1973) *System Sensitivity Analysis*, Dowden, Hutchison and Ross, Stroudsburg.

Cruz J. B. and Perkins (1964) 'A new approach to the sensitivity problem in multivariable feedback system design', *IEEE Trans.* **AC-9** 216–233.

Daniel R. W. and Kouvaritakis B. (1983) 'The choice and use of normal approximations to transfer function matrices of multivariable control systems', *Intern. J. Control* **37** 1121–1133.

Daniel R. W. and Kouvaritakis B. (1985a) 'A new robust stability criterion for linear and nonlinear multivariable feedback systems', *Intern. J. Control* **41** 1349–1379.

Daniel R. W.; Kouvaritakis B. and Latchman H. (1985b) 'Principal direction alignment, a geometric framework for the complete solution to the μ-problem', *Report No OUEL* 1595/85, Univ. of Oxford.

Daoyi X. (1985) 'Simple criteria for stability of interval matrices', *Intern. J. Control* **41** 289–295.

Davison E. J. (1976a) 'The robust control of a servomechanism problem for linear time-invariant multivariable systems', *IEEE Trans.* **AC-21** 25–34 and **AC-22** 283.

Davison E. J. (1976b) 'Multivariable tuning regulators: the feedforward and robust control of a general servomechnism problem', *IEEE Trans.* **AC-21** 35–47.

Davison E. J. (1978) 'Decentralized robust control of unknown systems using tuning regulators', *IEEE Trans.* **AC-23** 276–289.

Davison E. J.; Taylor P. and Wright J. (1980) 'On the application of tuning regulation to obtain robust feedforward-feedback controllers of an industrial heat exchanger', *IEEE Trans.* **AC-25** 361–375.

Davison E. J. and Wang S. H. (1977) 'Properties and calculation of transmission zeros of linear multivariable systems' in Fallside F. (Ed.) *Control Systems Design by Pole-Zero Assignment*, Academic Press, London.

Desoer C. A.; Liu R.; Murray J. and Saeks R. (1980) 'Feedback system design: the fractional representation approach to analysis and synthesis', *IEEE Trans.* **AC-25** 399–412.

Desoer C. A. and Vidyasagar M. (1975) *Feedback Systems*: I/O-*Properties*, Academic Press, New York.

Donati F. and Carlucci D. (1975) 'Control of norm uncertain systems', *IEEE Trans.* **AC-20** 792–795.

Donati F. and Vallauri M. (1984) 'Guaranteed control of 'almost-linear' plants', *IEEE Trans.* **AC-29** 34–41.

Doyle J. C. (1978) 'Guaranteed margins for LQG-regulators', *IEEE Trans.* **AC-23** 756–767.

Doyle J. C. (1979) 'Achievable performance in multivariable feedback systems', *IEEE Conf. Decision Control, Fort Lauderdale*, pp 250–251.

Doyle J. (1982) 'Analysis of feedback systems with structured uncertainties', *Proc. IEE* **D-129**.

Doyle, J. C. (1986) 'Quantitative feedback theory and robust control', *American Control Conference*, Seattle, pp. 1691–1698.

Doyle J. C. and Stein G. (1979) 'Robustness with observers' *IEEE Trans.* **AC-24** 607–611.

Doyle J. C. and Stein G. (1981) 'Multivariable feedback design: concept for a classical/modern synthesis', *IEEE Trans.* **AC-26** 4–16.

El-Attar R. A. and Vidyasagar M. (1978) 'Optimal order reduction using the induced operator norm', *J. Franklin Institute* **306** 457–477.

El-Attar R. A. and Vidyasagar M. (1979) 'Subsystem simplification in large-scale systems analysis', *IEEE Trans.* **AC-24** 321–323.

Emre E. (1983) 'Simultaneous stabilization with fixed closed-loop characteristic polynomial', *IEEE Trans.* **AC-28** 103–104.

Evans R. J. and Xianya X. (1985) 'Robust regulator design', *Intern. J. Control* **41** 461–476.

Francis, B. A. (1987) *A Course in H^∞-Control Theory.* (Lecture Notes in Control and Information Sciences, vol. 88). Springer-Verlag, Berlin–Heidelberg–New York–London–Paris–Tokyo.

Francis B.; Helton J. W. and Zames G. (1984) 'H^∞-optimal feedback controllers for linear multivariable systems', *IEEE Trans.* **AC-29** 888–900.

Francis B. and Vidyasagar M. (1983) 'Algebraic and topological aspects of the regulator problem for lumped linear systems', *Automatica* **19** 87–90.

Frank P. M. (1976) *Empfindlichkeitsanalyse dynamischer Systeme.* Oldenbourg-Verlag, München.

Frank P. M.; Heger F. and Radischat J. (1982) 'Design of robust control systems by pole placement', *IASTED-Symposium, Davos.*

Franklin S. N. and Ackermann J. (1981) 'Robust flight control: a design example' *J. Guidance and Control* **4** 597–605.

Freeman E. A. (1973) 'Stability of linear constant multivariable systems: contraction mapping approach', *Proc. IEE* **120** 379–384.

Freudenberg L. S. and Looze D. P. (1986) 'An analysis of H^∞-optimization design method' *IEEE Trans.* **AC-31** 194–201.

Foo Y. K. and Postlethwaite I. (1984) 'An H^∞-minimax approach to the design of robust control systems', *Systems and Control Letters* **5** 81–88.

Gamaleja T.; Hoy C.; Lunze J.; Wilfert H. H. and Naumann K. (1984) 'Decentralized voltage control in electric power systems', *IFAC-Congress, Budapest.*

Gantmacher F. R. (1958) *Matrizenrechnung*, VEB Deutscher Verlag der Wissenschaften, Berlin.

Ghosh B. K. (1983) *Simultaneous Stabilization and Pole Placement of a Multimode Linear Dynamical System*, PhD Thesis, Harvard Univ., Cambridge.

Ghosh B. K. and Byrnes C. I. (1983) 'Simultaneous stabilization and simultaneous pole-placement by nonswitching dynamical compensation', *IEEE Trans.* **AC-28** 735–741.

Grimble M. J. (1986) 'Optimal H^∞ robustness and the relationship to LQG design problem', *Intern. J. Control* **43** 351–372.

Grimble M. J. and Owens T. J. (1986) 'On improving the robustness of LQ regulators', *IEEE Trans.* **AC-31** 54–55.

Goodwin G. and Sin K. (1984) *Adaptive Filtering, Prediction and Control*, Prentice-Hall Inc., Englewood Cliffs.

Gräser A.; Dickmann M. and Neddermeyer W. (1982) 'Robuster Regelkreisentwurf: Das Verfahren von Horowitz und seine Automatisierung', *Regelungstechnik* **30** 387–393 and 428–432.

Gräser A. and Tolle H. (1983) 'Robust controller design in the frequency domain for a multiaxial servohydraulic vibration test bench', *IEEE Conf. on Appl. of Adaptive and Multivariable Control, Hull*, pp 26–31.

Grosdidier P.; Morari M. and Holt B. R. (1984) 'Integral controllability, failure tolerance, robustness and the relative gain array', *American Control Conf., San Diego*, pp 1290–1295.

Gutman S. (1979) 'Uncertain dynamical systems — a Lyapunov minimax approach', *IEEE Trans.* **AC-24** 437–443 and **AC-25** 613.

Hassan M. F. and Singh M. G. (1979) 'Controllers for linear interconnected dynamical systems with prespecified degree of stability', *Intern. J. Syst. Sci.* **10** 339–350.

Heger F. (1984) Entwurf robuster Regler für Strecken mit großen Parametervariationen (*Fortschrittsberichte der VDI-Zeitschriften Reihe* 8 Nr. 71), VDI-Verlag, Düsseldorf.

Heinen J. A. (1984) 'Sufficient condition for stability of interval matrices', *Intern. J. Control* **39** 1323–1328.

Hinrichsen D. and Prichard A. J. (1985) 'Stability radii of linear systems', *Control Theory Centre Report No.* **131**, Univ. of Warwick.

Horowitz I. M. (1963) *Synthesis of Feedback Systems*, Academic Press, New York.

Horowitz I. M. (1975) 'A synthesis theory for linear time-varying feedback systems with plant uncertainties', *IEEE Trans.* **AC-20** 454–463.

Horowitz I. M. (1976) 'Synthesis of feedback systems with nonlinear time-varying plants to satisfy quantitative performance specifications', *Proc. IEEE* **64** 123–130.

Horowitz I. M. (1979) 'Quantitative synthesis of uncertain multiple input-output feedback systems', *Intern. J. Control* **30** 81–106.

Horowitz I. (1982) 'Quantitative feedback theory', *IEE Proc.* **D-129** 215–226.

Horowitz I. M. and Sidi M. (1972) 'Synthesis of feedback systems with large plant ignorance for prescribed time-domain tolerances', *Intern. J. Control* **16** 287–309.

Horowitz I. and Sidi M. (1978) 'Synthesis of linear time-varying, non-minimum phase systems with plant uncertainties', *Intern. J. Control* **27** 351–359.

Hung Y. S. and Limebeer D. J. N. (1984) 'Robust stability of additively perturbed interconnected systems', *IEEE Trans.* **AC-29** 1069–1075.

Hung Y. S. and MacFarlane A. G. J. (1982) *Multivariable Feedback: A Quasi-classical Approach*, Springer Verlag, Berlin/Heidelberg/New York.

Joshi S. M. (1984) 'On the design of robust LQ-regulators', *American Control Conf., San Diego*, pp 981–984.

Kalman R. E. (1960) 'On a general theory of control systems', *IFAC Congress, Moscow*, **vol 1**, pp 481–492.

Kalman R. E. (1964) 'When is a linear control system optimal?', *Trans ASME, J. Basic Engn.* **86** 51–60.

Kantor J. C. and Andres R. P. (1983) 'Characterization of 'allowable perturbations' for robust stability', *IEEE Trans.* **AC-28** 107–109.

Karl W. C.; Greschak J. P. and Veghese G. C. (1984) 'Comments on 'A necessary and sufficient condition for the stability of interval matrices' ', *Intern. J. Control* **39** 849–951.

Khalil H. K. (1981) 'On robustness of output feedback control methods to modelling errors', *IEEE Trans.* **AC-26** 524–526.

Khalil H. K. and Kokotovic P. V. (1979) 'Control of linear systems with multiparameter singular perturbations', *Automatica* **15** 197–207.

Kidd P. T. (1984) 'Extension of the direct Nyquist array design technique to uncertain multivariable systems subject to external disturbances', *Intern. J. Control* **40** 875–901.

Kidd P. T. (1986) 'Design of a controller for a multivariable system with varying parameters using the extended direct Nyquist array', *Intern. J. Control* **43** 901 to 920.

Kimura H. (1982) 'Perfect and subperfect regulation in linear multivariable control systems', *Automatica* **18** 125–145.

Klema V. C. and Laub A. J. (1980) 'The singular value decomposition: its computation and some applications', *IEEE Trans.* **AC-25** 164–176.

Kokotovic P. V.; O'Malley R. E. and Sannuti P. (1976) 'Singular perturbations and order reduction in control theory — an overview', *Automatica* **12** 123–132.

Korn U. and Wilfert H. H. (1982) *Mehrgrößenregelungen*, VEB Verlag Technik, Berlin.

Kosut R. L. (1982) 'Analysis of performance robustness for uncertain multivariable systems', *IEEE Conf. Decision and Control, Orlando*, pp 1289–1294.

Kouvaritakis B. and Latchman H. (1985) 'Necessary and sufficient stability criterion for systems with structured uncertainties: the major principal direction alignment principle', *Intern. J. Control* **42** 575–598.

Kouvaritakis B. and Shaked K. (1976) 'Asymptotic behaviour of root-loci of multivariable systems', *Intern. J. Control* **23** 297–340.

Krisham K. R. and Cruickshanks A. (1977) 'Frequency domain design of feedback systems for specified insensitivity of time-domain response to parameter variations', *Intern. J. Control* **25** 609–620.

Kwakernaak J. (1969) 'Optimal low sensitivity linear feedback systems', *Automatica* **5** 279–285.

Kwakernaak H. (1982) 'A condition for robust stabilizability', *System and Control Letters* **2** 1–5.

Kwakernaak H. (1984) 'Minimax frequency domain optimization of multivariable linear feedback systems', *IFAC-Congress*, Budapest.

Kwakernaak H. (1985) 'Minimax frequency domain performance and robustness optimization of linear feedback systems', *IEEE Trans.* **AC-29** 994–1004.

Kwakernaak H. and Sivan R. (1972) *Linear Optimal Control Systems*, Wiley, New York.

Lauckner G.; Lunze J. and Wilfert H. H. (1985) 'Entwurf robuster dezentraler Knotenspannungsregelungen in Elektroenergiesystemen', *Intern. Wiss. Kolloquium, Ilmenau*, pp 217–220.

Lee W. H.; Gully S. W.; Eterno J. S. and Sandell N. R. (1982) 'Structural information in robustness analysis', *American Control Conf., Arlington*, pp 1040–1045.

Lehtomaki N. A.; Castanon D.; Levy B.; Stein G.; Sandell N. R. and Athans M. (1981 a) 'Robustness tests utilizing the structure of modelling errors', *IEEE Conf. Decision and Control, San Diego*, pp 1173–1190.

Lehtomaki N. A.; Sandell N. R. and Athans M. (1981 b) 'Robustness results in linear-quadratic Gaussian based multivariable control design', *IEEE Trans.* **AC-26** 75–92.

Leitmann G. (1983) 'Deterministic control of uncertain systems', *4-th Intern. Conf. on Mathematical Modelling in Science and Technology, Zurich*, pp 1–9.

Leitmann G.; Ryan E. P. and Steinberg A. (1986) 'Feedback control of uncertain systems: robustness with respect to neglected actuator and sensor dynamics', *Intern. J. Control* **43** 1243–1256.

Limebeer D. J. N. (1982) 'The application of generalized diagonal dominance to linear system stability theory', *Intern. J. Control* **36** 185–212.

Litz L. (1983) *Dezentrale Regelung*, Oldenbourg-Verlag, München/Wien.

Locatelli A.; Romeo F.; Scattolini R. and Schavoni N. (1983) 'A parameter optimization approach to the design of reliable robust decentralized regulators', *IFAC-Symposium Large Scale Systems Theory and Applications, Warsaw*, pp 291–296.

Logemann, H. and Owens, D. H. (1987) 'Robust high-gain feedback control of infinite-dimensional minimum-phase systems', *IMA Journal of Mathematical Control and Information* **4** 195–220.

Logemann, H. and Owens, D. H. (1987) 'Low-gain feedback control of unknown infinite-dimensional systems', Univ. of Strathclyde, *Research Report No. 7.*

Looze D. P. (1983) 'A dual optimization procedure for linear quadratic robust control problems', *Automatica* **19** 299–302.

Lunze J. (1980 a) 'The estimation of the trajectory of linear interconnected systems by means of an approximate model and an error estimate', *Symposium Chemplant, Heviz*, pp 263–281.

Lunze J. (1980 b) 'Übersicht über die Verfahren zum Entwurf dezentraler Regler für lineare zeitinvariante Systeme', *Messen, Steuern, Regeln* **23** 315–322.

Lunze J. (1981 a) 'Ein Verfahren zum Entwurf von Mehrgrößenreglern für unvollständig bekannte Regelstrecken', *Messen, Steuern, Regeln* **24** 428–432.

Lunze J. (1981 b) 'An approximation approach to the I/O-behaviour in the analysis and design of nonlinear large-scale systems', *IFAC-Congress, Kyoto*, **vol 9**, pp 115–120.

Lunze J. (1982 a) 'Evaluation of the I/O-behaviour of large scale dynamical systems by means of a generalized majorization principle', *4-th Formator Symposium, Liblice.*

Lunze J. (1982 b) 'Notwendige Modellkenntnisse zum Entwurf robuster Mehrgrößenregler mit I-Charakter', *Messen Steuern, Regeln* **25** 608–612.

Lunze J. (1982 c) 'The design of robust feedback controllers for partly unknown systems by optimal control procedures', *Intern. J. Control* **36** 611–630.

Lunze J. (1982 d) *Entwurf robuster Mehrgrößenregler bei beschränkter Modellunsicherheit*, Akademie der Wissenschaften, ZKI Berlin.

Lunze J. (1983a) 'Decentralized design of robust decentralized controllers for incompletely unknown systems', *IFAC-Symposium on Large Scale Systems Theory and Applications, Warsaw.*

Lunze J. (1983b) 'Untersuchungen zur Autonomie der Teilregler einer dezentralen Regelung mit I-Charakter', *Messen, Steuern, Regeln* **26** 451—455.

Lunze J. (1983c) 'A majorization approach to the quantitative analysis of incompletely known large scale systems', *Zeitschrift für elektrische Informations- und Energietechnik* **13** 99—117.

Lunze J. (1984a) 'The design of robust feedback controllers in the time domain', *Intern. J. Control* **39** 1243—1260.

Lunze J. (1984b) 'Robustness tests for feedback control systems using multidimensional uncertainty bounds', *System and Control Letters* **4** 85—89.

Lunze J. (1984c) 'A unified approach for designing linear feedback controllers', *IFAC-Congress, Budapest.*

Lunze J. (1985) 'Determination of robust multivariable I-controllers by means of experiments and simulation', *Syst. Anal. Model. Simul.* **2** 227—249.

Lunze J. (1986) 'Robustness analysis of control systems by means of a structured uncertainty description', *Foundations of Control Engineering* **10** 201—213.

Lunze J. and v. Kurnatowski, B. (1987) 'Experimentelle Erprobung einer Einstellregel für PI-Mehrgrößenregler bei der Herstellung von Ammoniumnitrat-Harnstoff-Lösung', *Messern, Steuern, Regeln* **30** 2—6.

Molander P. and Willems J. C. (1980) 'Synthesis of state feedback control laws with specified gain and phase margin', *IEEE Trans.* **AC-25** 928—931.

Morari M. (1983) 'Robust stability of systems with integral control', 22-*nd IEEE Conf. Decision and Control, San Antonio.*

Noldus E. (1982) 'Design of robust state feedback laws', *Intern. J. Control* **35** 935—944.

Nwokah O. J. (1980) 'A recurrent issue on the extended Nyquist array', *Intern. J. Control* **31** 606—614.

Nwokah O. D. I. (1983) 'Synthesis of feedback systems for specified time domain insensitivity to interaction induced plant ignorance', *Intern. J. Control* **37** 421 to 428.

Nwokah O. D. I. (1984) 'Synthesis of controllers for uncertain multivariable plants for prescribed time-domain tolerances', *Intern. J. Control* **40** 1189—1209.

Ortega J. M. and Rheinboldt W. C. (1970) *Iterative Solution of Nonlinear Equations in Several Variables*, Academic Press, New York.

Owens D. H. (1979) 'Discrete first order models for multivariable process control', *Proc IEE* **D-126** 525—530.

Owens D. H. and Chotai A. (1981) 'Controllers for unknown multivariable systems using monotone modelling errors', *Proc IEE* **D-129** 57—69.

Owens D. H. and Chotai A. (1983) 'Robust controller design for linear dynamic systems using approximate models', *Proc IEE* **D-130** 45—54.

Owens D. H. and Chotai A. (1984) 'On eigenvalues, eigenvectors and singular values in robust stability analysis', *Intern. J. Control* **40** 285—296.

Owens D. H. and Chotai A. (1986) 'Approximate models in multivariable process control', *Proc IEE* **D-133** 1—12.

Patel R. V. and Munro N. (1982) *Multivariable System Theory and Design*, Pergamon Press, Oxford.

Penttinen J. and Koivo H. N. (1980) 'Multivariable tuning regulators for unknown systems', *Automatica* **16** 393—398.

Porter B. (1974) 'Singular perturbation methods in the design of stabilising feedback controllers for multivariable systems', *Intern. J. Control* **20** 689–692.

Porter B. (1981) 'Design of error-actuated controllers for unknown multivariable plants', *Electr. Letters* **17** 106–107.

Porter B. (1982) 'Design of tunable set point tracking controllers for linear multivariable plants', *Intern. J. Control* **35** 1107–1115.

Postlethwaite I.; Edmunds J. M. and MacFarlane A. G. J. (1981) 'Principal gains and principal phases in the analysis of linear multivariable feedback system', *IEEE Trans.* **AC-16** 32–46.

Postlethwaite I. and Foo Y. K. (1984) 'Representation of uncertainty and robustness tests for multivariable feedback systems', in Tzafestas S. G. (Ed.) *Multivariable Control,* Reidel Publ. Co., pp 151–160

Postlethwaite I. and Foo Y. K. (1985) 'Robustness with simultaneous pole and zero movement across the $j\omega$-axis', *Automatica* **21** 433–443.

Reinisch K. (1979) *Analyse und Synthese kontinuierlicher Steuerungssysteme,* VEB Verlag Technik, Berlin.

Reinschke, K. (1988) *Multivariable Control. A Graph-theoretic Approach.* (Lecture Notes in Control and Information Sciences, vol. 108). Springer-Verlag, Berlin. Heidelberg–New York–London–Paris–Tokyo and (Mathematical Research vol. 41) Akademie-Verlag Berlin.

Rosenbrock H. H. (1974) *Computer-Aided Control Systems Design,* Academic Press, London.

Safonov M. G. (1978) 'Tight bounds on the response of multivariable systems with component uncertainty', *16-th Allerton Conf. on Communication, Control and Computing, Monticello,* pp 451–460.

Safonov M. G. (1979) 'Frequency domain design of multivariable control systems for insensitivity to large plant modelling errors', *IEEE Conf. Decision and Control, Fort Lauderdale,* pp 247–249.

Safonov M. G. (1980) *Robustness of Multivariable Feedback Systems,* MIT Press, Cambridge.

Safonov M. G. (1982) 'Stability margins of diagonally perturbed multivariabel feedback systems', *Proc IEE* **D-129** 251–256.

Safonov M. G. and Athans M. (1977) 'Gain and phase margin for multi-loop LQG regulators', *IEEE Trans.* **AC-22** 173–179.

Sandell N. R.; Varaiya P.; Athans M. and Safonov M. G. (1978) 'Survey of decentralized control methods for large scale systems', *IEEE Trans.* **AC-23** 108–128.

Seneta E. (1981) *Non-negative Matrices and Markov Chains,* Springer-Verlag, Berlin.

Sezer M. E. and Siljak D. D. (1980) 'Robustness of suboptimal control: gain and phase margin', *IEEE Conf. Decision and Control, Albuquerque.*

Sezer M. E. and Siljak D. D. (1981) 'Sensitivity of large-scale control systems', *J. Franklin Institute* **312** 179–197.

Shaked U.; Horowitz I. and Golde S. (1976) 'Synthesis of multivariable basically non-interacting systems with significant plant uncertainties', *Automatica* **12** 61–71.

Shaked U. and Soroka E. (1985) 'On the stability robustness of the continuous-time LQG optimal control', *IEEE Trans.* **AC-30** 1039–1043.

Shi Z.-E. and Gao W.-B. (1986) 'A necessary and sufficient condition for the positive definiteness of interval symmetric matrices', *Intern. J. Control* **43** 325–328.

Singh M. G. (1981) *Decentralised Control,* North-Holland, Amsterdam.

Siljak D. D. (1978) *Large Scale Dynamic Systems. Stability and Structure*, North-Holland, New York.

Smith M. C. and Sondergeld K. P. (1987) 'On the order of stable compensators', *Automatica*, to appear.

Soh Y. C. and Evans R. J. (1985) 'Robust multivariable regulator design — special cases', *IEEE Conf. Decision and Control.*

Solodownikow W. W. (1971) *Stetige lineare Systeme*, Verlag Technik, Berlin.

Tchon K. (1983) 'On generic properties of linear systems: an overview', *Kybernetika* **19** 467—482.

Thompson P. M.; Stein G. and Athans M. (1983) 'Conic sectors for sampled-data feedback systems', *System and Control Letters* **3** 77—82.

Thorp J. S. (1980) 'Guaranteed stability of uncertain linear systems', *Joint Automatic Control Conference, San Francisco*, paper TA3-A.

Tolle H. (1983) *Mehrgrößenregelkreissynthese.* Band 1: *Grundlagen und Frequenzbereichsverfahren*, Band 2: *Entwurf im Zustandsraum*, R. Oldenbourg-Verlag, München/Wien.

Unbehauen H. (1985) *Regelungstechnik.* Band 3: *Identification, Adaption, Optimierung*, Friedr. Vieweg & Sohn, Braunschweig.

Vidyasagar M. (1984) 'The graph metric for unstable plants and robustness estimates for feedback stability', *IEEE Trans.* **AC-29** 403—417.

Vidyasagar M. and Viswanadham N. (1982) 'Algebraic design technique for reliable stabilization', *IEEE Trans.* **AC-27** 1085—1095.

Walke J. C.; Horowitz I. M. and Houpis C. H. (1984) 'Quantitative synthesis of highly uncertain multiple-input multiple-output flight control system for the forward swept using X-29 aircraft', *IEEE National Aerospace and Electronics Conf.*, pp 576—583.

Wilfert H. H.; Marschner H. H. and Hertel U. (1978) 'Experimentelle Identifikation eines braunkohlengefeuerten 210-MW-Kraftwerksblocks als Mehrgrößenregelstrecke', *Messen, Steuern, Regeln* **21** 567—572.

Willems J. L. and Willems J. C. (1983) 'Robust stabilization of uncertain systems', *SIAM J. Control and Optimization* **11** 352—374.

Wonham W. M. (1967) 'On pole assignment in multi-input controllable linear systems', *IEEE Trans.* **AC-12** 660—665.

Yaniv O. and Horowitz I. M. (1986) 'A quantitative design method for MIMO linear feedback systems having uncertain plants', *Intern. J. Control* **43** 401—421.

Yedavalli R. K. (1986) 'Stability analysis of interval matrices: another sufficient condition', *Intern. J. Control* **43** 767—772.

Yeh H. H.; Vanda S. and Ridgely D. B. (1985) 'Stability robustness measures utilizing structural information', *Intern. J. Control* **41** 365—387.

Yeung K. S. (1983) 'Linear system stability under parameter uncertainties', *Intern. J. Control* **38** 459—464.

Youla D. C.; Bongiorno J. J. and Jabr H. A. (1976) 'Modern Wiener-Hopf design of optimal controllers: part I, the single-input single-output case', *IEEE Trans.* **AC-21** 3—13.

Youla D. C.; Bongiorno J. J. and Lu C. N. (1974) 'Single-loop feedback-stabilization of linear multivariable dynamical plants', *Automatica* **10** 159—173.

Zafiriou, E. and M. Morari (1986) 'Design of the IMC filter by using the structured singular value approach', *American Control Conference*, Seattle, pp. 1—6.

Zames G. (1966) 'On the input-output stability of time-varying nonlinear feedback systems (part II)', *IEEE Trans.* **AC-11** 465—476.

Zames G. (1981) 'Feedback and optimal sensitivity: model reference transformation, multiplicative seminorms, and approximate inverses', *IEEE Trans.* **AC-26**, 301–320.

Zames G. and Francis B. (1983) 'Feedback, minimal sensitivity, and optimal robustness', *IEEE Trans.* **AC-28** 585–601.

Zhou C. S. and Deng J. L. (1986) 'The stability of the grey linear system', *Intern. J. Control* **43** 313–320.

Ziegler J. G. and Nichols N. B. (1942) 'Optimum settings for automatic controllers', *Trans. ASME, J. Dyn. Syst. Meas. and Control* **64** 759–768.

Zurmühl R. (1964) *Matrizen und ihre technischen Anwendungen*, Springer-Verlag, Berlin.

SYMBOLS

Mathematical symbols

In general, matrices and operators are denoted by roman boldface capital letters, vectors by boldface small letters, and scalars by italics.

$\mathbf{A}'$, $\mathbf{A}^*$	transpose of the real or complex matrix $\mathbf{A}$, conjugate transpose of the complex matrix $\mathbf{A}$
$\mathbf{I}$	unity matrix, unity operator
$\lambda_i[\mathbf{A}]$	eigenvalue of the matrix $\mathbf{A}$
$\lambda_p[\mathbf{A}]$	Perron root (maximum eigenvalue) of the nonnegative matrix $\mathbf{A}$ (cf Appendix 1)
$\varrho[\mathbf{A}]$	spectral radius of $\mathbf{A}$
$\sigma_i[\mathbf{A}]$	singular value of $\mathbf{A}$ (cf Appendix 2)
$\sigma_{\max}[\mathbf{A}]$, $\sigma_{\min}[\mathbf{A}]$	maximum, minimum singular value of $\mathbf{A}$ (cf Appendix 2)
$\|\mathbf{x}\|$	norm of vector $\mathbf{x}$
$\|\mathbf{A}\|_M$	norm of matrix $\mathbf{A}$ (for short also written as $\|\mathbf{A}\|$)
$\|\mathbf{G}\|_{\mathrm{Op}}$	norm of operator $\mathbf{G}$ (cf eqn (2.1-23))
$\lvert\mathbf{x}\rvert$, $\lvert\mathbf{A}\rvert$	vector/matrix comprising the moduli of the entries of vector $\mathbf{x}$ or matrix $\mathbf{A}$, respectively
$\mathrm{Re}\,[\lambda]$	real part of a complex number λ
$\mathbb{R}$	field of real numbers
$\mathbf{a} \leqq \mathbf{b}$	means $a_i \leqq b_i$ for $\mathbf{a} = (a_1, \ldots, a_n)'$, $\mathbf{b} = (b_1, \ldots, b_n)'$

System description

$\mathbf{G}$	system operator
$\mathbf{G}(p)$, $\hat{\hat{\mathbf{G}}}(p)$	transfer function matrix
$\mathbf{G}(t)$	impulse response matrix

$\mathcal{G}$	set of matrices $\mathbf{G}$
$\mathcal{D}$	Nyquist contour (Fig. 2.3)
$\hat{\mathbf{G}}$, $\delta\mathbf{G}$	approximate model, model error
$\mathbf{K}_s$	matrix of the static reinforcement
$\mathbf{G}_{ys}$, $\mathbf{G}_{zu}$, ...	matrices belonging to the approximate model of the plant
$\bar{\mathbf{G}}_{ys}$, $\bar{\mathbf{G}}_{zu}$, ...	matrices belonging to the closed-loop approximate model
$\delta(t)$, $\sigma(t)$	impulse function, step function

SUBJECT INDEX